Opioid Peptides in Substance Abuse

József I. Székely, M.D., D.Sc.

Chief
Division of Pharmacology
Institute for Drug Research
Budapest, Hungary

CRC Press
Boca Raton Ann Arbor London Tokyo

Library of Congress Cataloging-in-Publication Data

Opioid peptides in substance abuse/edited by József I. Székely.
p. cm. — (Physiology of drug abuse series)
Includes bibliographical references and index.
ISBN 0–8493–7937–7
1. Drug abuse—Pathophysiology. 2. Opioids. 3. Opioids—Receptors.
4. Endorphins. I. Székely, József I. (József Iván).
1937- . II. Series: CRC series in physiology of drug abuse.
[DNLM: WM 270 06158 1993]
RC564.059 1993
616.86'32--dc20 93-9526
CIP

Direct all inquiries to CRC Press, Inc., 2000 Corporate Blvd., N.W., Boca Raton, Florida 33431.

International Standard Book Number 0-8493-7937-7

Library of Congress Card Number 93-9526

Printed in the United States of America

1 2 3 4 5 6 7 8 9 0

Printed on acid-free paper

DEDICATION

To my wife
Mária Székely

CRC Series in
PHYSIOLOGY OF DRUG ABUSE
Edited by Ronald R. Watson

Published Titles:

Alcohol and Cancer

Alcohol and Neurobiology: Brain Development and Hormone Regulation

Alcohol and Neurobiology: Receptors, Membranes, and Channels

Drugs of Abuse and Neurobiology

Marijuana/Cannabinoids: Neurobiology and Neurophysiology

Neurobiology of Cocaine: Cellular and Molecular Mechanisms

Neurobiology of Opiates

Nutrition and Alcohol

Forthcoming Titles:

Alcohol, Drugs of Abuse, and Immune Functions

PREFACE

The discovery of opioid peptides, i.e., endogenous natural counterparts of morphine-like substances and their common receptors aroused great interest in the 1970s. Since then, many monographs and books have been published dealing with their biochemistry, function, distribution in the organism, etc. Nevertheless, to our knowledge, so far no monograph has specifically addressed one of the most pertinent issues: their role in substance abuse.

It was a great challenge for the author, a "life-long" pharmacologist with a 2-year tenure at the NIDA Addiction Research Center, Baltimore, MD, to provide a balanced survey of the topic. Efforts have been taken to include all relevant data in the literature pertaining to the subject. Primarily *in vivo* functional data have been surveyed. Closely related biochemical and morphological findings are shortly discussed. Drug addiction is dealt with as a "behavioral paradigm" in animals and a pathological entity in humans.

Dr. Székely is a member of the International Narcotics Research Club and three Hungarian scientific societies. He is interested in psychopharmacology, in general, but most of his publications relate to opioid peptides.

SERIES EDITOR

Ronald R. Watson, Ph.D., initiated and directs the Alcohol Research Center at the University of Arizona School of Medicine. The main theme of this NIAAA center grant is to understand the role of ethanol-induced immunosuppression on disease and disease resistance in animals. He has edited 14 books on drug and alcohol abuse. He has worked for several years on research for the U.S. Navy Alcohol and Substance Abuse Program.

Dr. Watson attended the University of Idaho, but graduated from Brigham Young University in Provo, Utah with a degree in Chemistry in 1966. He completed his Ph.D. in 1971 in Biochemistry from Michigan State University. His postdoctoral schooling was completed at the Harvard School of Public Health in Nutrition and Microbiology. He was an Assistant Professor of Immunology at the University of Mississippi Medical Center in Jackson from 1973 to 1974 and at the Indiana University Medical School from 1974 to 1978. He was an Associate Professor at Purdue University in the Department of Food and Nutrition from 1978 to 1982. In 1982, he joined the faculty at the University of Arizona in the Department of Family and Community Medicine and is a Research Professor.

Dr. Watson's major research interest involves effects of dietary materials on disease resistance, biochemical changes, and immune functions. This includes understanding the effects of alcohol, cocaine, and morphine on immune functions.

He has published 275 research papers and review chapters, and edited 25 books. He is a member of several national and international nutrition, immunology, and cancer societies.

THE AUTHOR

József I. Székely, M.D., D.Sc., is Chief, Division of Pharmacology at the Institute for Drug Research, Budapest, Hungary (HIDR).

Dr. Székely graduated from the Semmelweis University Medical School (SUMS), Budapest, Hungary in 1961 (summa cum laude). He subsequently obtained his M.S. degree in Psychology. From 1961 to 1969 he served as Assistant Professor in the Department of Physiology of SUMS. Since 1969 he has been employed on a permanent basis at the HIDR. From 1989 to 1991 he worked at the Addiction Research Center of the National Institute on Drug Abuse, Baltimore, MD, as a visiting scientist. At the HIDR he worked mostly in various psychopharmacological laboratories. Nevertheless, from 1982 to 1989 he was assigned to the Clinical Department of HIDR where he took part in the phase I or II clinical trials of some new drugs. He assumed his present position (Chief of Pharmacology) in 1992.

Dr. Székely obtained his D.Sc. degree from the Hungarian Academy of Sciences in 1986. Since he regularly gives lectures to medical students, the title of Titular Professor was awarded to him in 1990 at the SUMS.

Dr. Székely is (co)author of about 70 papers, the majority of them published in international journals. He is (co)author of 11 book chapters and holds more than a dozen pharmaceutical patents. He is first author of four monographs (*Opioid Peptides* I, II, III, and IV) written with Drs. A. Z. Rónai and K. Ramabadran.

ACKNOWLEDGMENT

This work was launched while tenured at the Addiction Research Center, National Institute on Drug Abuse, Baltimore, Maryland.

The author thanks Drs. Edythe London, David Newlin, Lino Covi and other colleagues at the Addiction Research Center for helpful discussions and encouragement.

Drs. Mary Pfeiffer, Judit Stverteczky, and Ms. Ildikó Süttó provided great help to procure reprints, journals, books, and other materials processed in the monograph. Ms. Julia Singer and Mr. János Horváth helped me in drawing the figures and preparing the manuscript by computer.

TABLE OF CONTENTS

1 Introduction: The Pharmacological Effects of μ- and κ-Agonists are Divergent, in Some Aspects Even Opposite

I. BACKGROUND

The postulation of multiplicity of opiate receptors coincided in time with the identification of the first endogenous opioids. Met- and Leu-enkephalin[1] and β-endorphin[2–4] were isolated in 1975 and 1976. A careful examination of the *in vitro* pharmacological properties of (nor)morphine and natural enkephalins in various opioid-sensitive isolated organs and receptor-binding assays and comparison of their sensitivity to naloxone quickly led to the conclusion that both the opiate receptors and their endogenous ligands are multiple.[5] Therefore, Lord et al.[5] postulated the existence of separate μ- and δ-receptors. On the other hand, practically in the same year, Martin and colleagues[6,7] differentiated at least three separate subtypes of opiate receptors (μ-, κ- and σ-ones) based primarily on the difference in the *in vivo* effects of classical opiates and certain benzomorphan derivatives in chronic spinal dogs. For a while it was difficult to decide whether or not the apparent divergences in their conclusions were due to the fact that in one of these laboratories natural enkephalins were compared to (nor)morphine *in vitro*,[1,5] whereas in the other one[6,7] the effects of various synthetic opiates were differentiated *in vivo*.

A further confounding factor is that only some classical morphine derivatives are relatively pure ligands of a single subtype of opiate receptors, i.e., that of the μ-ones,[6,8,9] whereas natural enkephalins have considerable affinity to both μ- and κ-receptors. On the other hand nowadays it is generally established (for review see Mussacchio)[10] that σ-receptors are not of opioid nature. Some years later Goldstein et al.[11] isolated the prototype of the third group of endogenous opioids, that is, dynorphin. A bit later it has been found to be a selective ligand of κ-receptors.[12] As for β-endorphin, it binds with similar affinity to μ- and δ-receptors, but its affinity to the κ-ones is rather weak.[9]

II. CURRENT RESEARCH

After 2 decades of opiate research it has now been well established that

- There are three separate families of endogenous opioids in the nervous system (of mammals): enkephalins, arising from proenkephalin;[13–15] β-endorphin (and its shorter fragments), originating from proopiomelanocortin;[16] and dynorphin, cleaved from prodynorphin.[17,18]

- On the other hand. opioid receptors have three main subtypes, μ-, δ-, and κ-ones as discussed above.

- The members of the former two families (enkephalins, β-endorphin) bind to μ- and δ-receptors, but not to κ-ones, whereas dynorphin is a rather selective ligand of κ-receptors.[9,12]

Many recent data prove that the κ-receptors consist of different subtypes[19–22] and some groups claim that neither the μ-receptors[23,24] nor the δ-ones[25] are uniform, i.e., they are also supposed to have subclasses. In the above cited papers the proposal on multiplicity of μ-, δ-, and κ-receptors is based mostly on receptor-binding experiments. Thus, the data provided in these papers still do not provide enough evidence for distinct physiological roles of various subtypes of μ-, δ-, and κ-receptors. (In rat vas deferens β-endorphin seems to have a further type of naloxone-sensitive selective binding site, termed ε.[26,27] Nevertheless, the existence of this fourth subtype of opiate receptors in the brain has not been convincingly demonstrated yet. At least the majority of authors, including the ones postulating the existence of μ-, δ-, and κ-receptors, do not pay particular attention to these peculiar ε-ones.)

In the meantime, several enkephalin analogs have been developed, binding selectively to μ- or δ-receptors. e.g., FK 33–824 (=[D-Ala^2,NMePhe^4,Met(O)^5]-enkephalin-ol)[28] binds selectively to the μ-receptors and DPDPE (=[D-Pen^2,D-Pen^5]-enkephalin)[29,30] is a selective δ-ligand. Nevertheless, there is still no convincing evidence for drastically different functional roles for μ- and δ-receptors in regard of their behavioral actions.[31]

Thus, the whole business with the multiplicity of opiate receptors and their ligands seems to boil down to the eventual difference in the μ-, δ-, and κ-receptor-mediated events.

For these reasons this chapter focuses on the eventual divergence in μ-, δ-, and κ-opiate receptor-mediated processes in the organism.

A. Biochemical Effects

μ-, δ-, and κ-agonists apparently elicit divergent biochemical effects. It is well established that opiates inhibit the adenylate cyclase in brain membranes[32] and the opiate receptors are negatively coupled to this enzyme in the NG 108–15 neuroblastoma × glioma hybrid cell line.[33–35] This biochemical action has been considered a characteristic effect of opiates for a long time in general. However, it has been found recently[36] that etorphine, morphine, and (D-Ala^2,NMePhe^4,Gly^5-ol)-enkephalin (DAGO), i.e., selective μ-agonists, but not the κ-preferring U-50488H[37] and U-69593[38] similarly inhibit the adenylate cyclase activity in rabbit cerebellar membrane preparation which contains mostly μ-receptors,[39] but not in the guinea pig cerebellum known to contain primarily κ-ones. Thus the inhibition by opiates of adenylate cyclase is mediated by μ-, but not κ-receptors. Since in the neuroblastoma × glioma cell line the opiate receptors are of δ-type,[40] both μ- and δ-receptors might be functional antagonists of the κ-ones in this regard.

There is a similar contrast in relation of another important "second messenger" system as well. In rat hippocampal slice preparation dynorphin$_{1-13}$, its D-Ala^2 analog, and U-50488H all stimulated the phosphoinositide turnover, whereas DAGO, morphine, and a δ-preferring enkephalin analog, DPDPE, were found inactive.[41]

Since two cardinal second messenger systems seem to be modulated in opposite directions by μ-,δ- and κ-receptors, respectively, it is not surprising that they induce divergent, eventually opposite effects in the living organism as well (vide infra) for some examples.

B. Electrophysiological Actions

There is no room here even for a short overview of the multiple and rather complex electrophysiological actions of selective μ-, δ-, and κ-opioids. Nevertheless, it is probably pertinent to mention here that they differ in their basic mechanisms of actions even when their effects are apparently similar. All the three main types of opioids shorten the calcium component of the action potentials of the mouse dorsal root (sensory) neurons in culture.[42–45] This effect is related to inhibition of Ca^{++} entry.[46] However, μ- and δ-agonists elicit this effect by activation of voltage- and/or Ca^{++}-dependent potassium channels, leading to hyperpolarization of the resting potential and more rapid depolarization during activation, whereas activation of κ-receptors decreases the voltage-sensitive Ca^{++} conductance (channels). The K^+ and Ca^{++} conductances also are differentially mediated in the locus coeruleus and myenteric plexus by μ-,δ- and κ-opioids, respectively.[47,48] According to the proposal of North,[48] different types of opiate receptors might coexist on the same neurons. Furthermore, according to a recent suggestion of Crain and Shen,[49] the same neurons might be both excited or inhibited by all the three different subtypes of opiate receptors, but the underlying mechanism is always the same: modulation of the K^+ and Ca^{++} conductance by the μ-,δ- and κ-opioids, respectively.

C. Release of Some Neurotransmitters

The release of some neurotransmitters also is dissimilarly modulated by the various subtypes of opiate receptors. Relatively many data refer to modulation of the release of dopamine and acetylcholine.

- As described by Mulder et al.,[50] in rat striatal slice preparation, $dynorphin_{1-13}$ potently inhibits dopamine release, whereas neither the δ-receptor-preferring (D-Ala^2,D-Leu^5)-enkephalin (DADLE) nor morphine, the "prototypic" μ-ligand, has any effect. On the other hand, the acetylcholine release from the same preparation was potently suppressed by δ-receptor-preferring agonists as DADLE, DPDPE, and μ-opioids as morphine and DAGO, but not by dynorphin.[50,51] In rat cortex preparation, just as in the striatum, primarily μ-agonists inhibit the release of acetylcholine.[52]

- In cat hypothalamic slice preparation the K^+-induced liberation of substance P was suppressed by both morphine and DADLE, but not by U-50488H.[53,54]

- In view of the well-known histamine-releasing action of classical opiates it is not surprising that they also suppress its K^+-evoked liberation from mouse cortical slices.[55] However, in rat brain slice preparation $dynorphin_{1-13}$, U-50488, and U-69593 suppressed the histamine release, whereas morphine, DAGO, and DPDPE were found completely inactive.[56] Thus regulation of histamine release seems to be modulated in the opposite direction by μ-,δ- and κ-agonists, respectively.

- In rat hippocampus, agonists of the μ- and δ-receptors inhibit serotonin release.[57] However, U-50488H has been found to enhance the K^+-induced serotonin release from mouse brain slice and spinal cord synaptosomal preparation.[58]

In some other *in vitro* studies no such drastic receptor-dependent differences were seen. In various areas of rabbit brain both μ- and κ-receptor agonists suppressed the acetylcholine release.[59] In guinea pig brain (in the cortex, cerebellum, and hippocampus) the potassium-induced release of norepinephrine is suppressed by DAGO, DADLE, and U-50488H as well,[60], i.e., no receptor selectivity can be seen. Met-enkephalinamide, β-endorphin, and $dynorphin_{1-17}$ all inhibit the stimulated dopamine release from the posterior pituitary, but the effect of dynorphin has been found the strongest.

Nevertheless, the eventual differences in the μ- and κ-receptor-induced modulation of neurotransmitter release seems to be more conspicuous *in vivo*.

Substance P — For example, examining the noxious stimulation-induced liberation of substance P in the superficial layers of dorsal horn in rabbits, morphine and Met-enkephalin depressed the release of this mediator of pain, but dynorphin had no effect[61] just as *in vitro* when using hypothalamic slice preparation.[53,54] Since substance P liberation (along the nociceptive pathways) is a manifestation of pain, these findings indicate that dynorphin might not function as an endogenous analgesic. (For further data see Chapters 2 and 3.)

Dopamine Release — Furthermore, as reviewed in Chapter 7, *in vivo* both μ- and δ-agonists enhance the dopamine turnover in the nigrostriatal (and mesolimbic) dopamine system, whereas κ-agonists are either inactive or induce inhibition. With regard to substance abuse, probably the most crucial issue is the modulation of dopamine release from the n. accumbens (and other mesolimbic terminal areas). As discussed in detail in Chapter 7, μ-agonists potently enhance the dopamine release, whereas U-50488, a selective synthetic agonist of κ-receptors,[37] suppresses the dopamine action.[62]

Corticosterone — Probably it is worth mentioning here that μ-, δ-, and κ-agonists all elevate the serum level of corticosterone, i.e., activate the hypothalamo-pituitary-adrenocortical axis, at least in rats.[63] Upon repeated administration, tolerance developed to this effect of morphine and U-50488H as well, but no cross tolerance was seen.[63] Morphine induces corticosterone release in doses well above the analgesic ED_{50}, whereas the similar action of U-50488H can already be detected in subanalgesic doses.[37] Thus μ- and κ-opiates apparently elicit this action via different pathways. (See the review of Illés[64] for further details on the modulation of transmitter release by multiple opioid receptors and their ligands.)

Summarily, these data indicate that μ- and/or δ-agonists on one side and κ-ones on the other side apparently induce qualitatively different molecular and biochemical effects; their interaction with ionic channels explaining their electrophysiologiocal actions also are different. In addition, their effects on the release of neurotransmitters are divergent, in some regards opposite. For these reasons it is probably not surprising that their effects in the living organisms also are different.

D. Autonomic Effects

The qualitative differences in the autonomic effects of various types of opioids also seem to be remarkable. Nevertheless, for lack of studies systematically comparing the autonomic effects of μ-, δ, and κ-opioids, only some examples can be referred to. Morphine and other classical opiates induce a strong antidiuretic effect,[65] which is attributed to liberation of antidiuretic hormone.[66,67] In contrast one of the most characteristic autonomic effects of κ-agonists is increased urination allegedly due to decrease in the blood level of antidiuretic hormone.[68]

In rats, μ-opioids as morphine induce hyperthermia or peculiar biphasic effect on body temperature.[69,70] Hyperthermia was seen in rats after s.c. injection of (D-Met^2,Pro^5)-enkephalinamide as well,[71] i.e., with an opioid peptide showing nearly equal affinity to μ- and δ-receptors.[40,72] However, dynorphin induces monophasic hypothermia.[73,74]

There is no room here to review the cardiovascular and respiratory effects of "prototypical" μ- and κ-agonists. Nevertheless, a reference is warranted to the fact that Martin et al.[6] postulated the existence of separate μ- and κ-receptors among others on the basis of qualitative differences in their cardiorespiratory actions.

E. Analgesic Effect

By definition all the opioids are analgesics. The classical opiates are highly selective ligands of μ-receptors.[9] The question whether or not δ-receptors are able to elicit analgesia by themselves is an unresolved issue even today. As discussed above, natural enkephalins bind to both μ- and δ-receptors.Examining the relative potencies of synthetic enkephalin analogs in guineapig ileum, i.e., in a preparation containing predominantly μ-receptors and in mouse vas deferens which only contains δ-ones, a surprising conclusion has arisen. Both Kosterlitz et al.[8] and our group[75] observed that for analgesia μ-receptors are more important than the δ-ones. Truly in the last decades numerous enkephalin analogs have been developed, some of them highly selective for the μ- or δ-receptors. Among them DPDPE is the most selective one for the δ-ones,[29,30] and it does have analgesic activity. Nevertheless, Murray and Cowan[31] reported that DPDPE elicited the phenomenon of "Straub-tail" (tail erection), which is regarded as a selective action of narcotic analgesics.[76,77] Furthermore, in this study[31] this effect of DPDPE could be prevented by β-funaltrexamine, i.e., selective alkylation of μ-receptors,[78] but not by ICI 174864, i.e., by a selective antagonist of δ-receptors.[79] Hence the possibility arises that even the "selective δ-agonists" might somehow activate μ-receptors. This apparent controversy can be resolved by accepting the proposition that at least a part of μ- and δ-receptors are "cross-linked".[80]

On the other hand, as summarized by Millan[81] and discussed in detail in Chapters 2 and 3, only μ-opioids seem to induce real analgesic action. In the case of κ-agonists it is more appropriate to speak of inhibition of certain polysynaptic (nociceptive) reflexes accompanied with dysphoria.

F. Behavioral Actions

The behavioral actions of μ- and κ-agonists are not simply different, but in most aspects opposite. The divergence in the behavioral-motivational effects of μ- and κ-agonists is subject to this monograph in general and is reviewed in detail in some subsequent chapters. For this reason only some aspects of these differences are listed here.

As discussed in Chapter 7, μ-agonists such as morphine induce in most laboratory species locomotor hyperactivity, stimulation similar to dopamine receptor stimulants (eventually after an initial phase of motor depression). In contrast, κ-agonists such as dynorphin, ketocyclazocine, and U-50488H cause an overall, lasting behavioral locomotor depression in mice,[37,82] in rats,[83] and in monkeys[84] as well. (For details also see Chapter 2.) As discussed in Chapter 7, the locomotor stimulation induced by μ-agonists can be antagonized by dopamine antagonists, i.e., it is due to facilitation of the (mesolimbic) dopaminergic system. However, the direction of interaction is just the opposite in the case of κ-agonists; dynorphin in mice antagonizes the locomotor stimulation elicited by apomorphine.[85]

It is well established that in most laboratory species morphine and other μ-agonists enhance the muscular tone. In contrast, κ-opioids induce ataxia, muscle weakness, and eventually flaccid paralysis in rodents (see Chapter 2) and monkeys[84] as well. (In some papers dynorphin is called "cataleptic". It is, however, wrong terminology. Namely by definition catalepsy means both immobility and increased muscular tone.)

As discussed in detail in Chapters 4 and 5, contrary to the euphoriant action of μ-opioids, κ-agonists are aversive, dysphoric in animals and humans as well. As discussed there, μ-opioids are rewarding, whereas κ-agonists are punishing; they induce avoidance.

Obviously for this reason, though physical dependence can be established in rats by continuous infusion of Met-enkephalin or β-endorphin into the brain,[86] κ-agonists such as dynorphin analogs[87] and U-50488H[37,84] do not induce physical dependence in rodents. μ- and κ-Agonists are easily discriminated from each other in drug-discrimination procedures.[84,88] Of course, U-50488H failed to generate self-administration behavior.[84]

Based on these findings it is not surprising that no cross tolerance can be established between the μ-and κ-agonists, whether their analgesic action[37] (see also Chapters 2 and 3), overall motor depressant,[89] or their ACTH and corticosterone mobilizing actions were examined.[90]

All these data confirm that just in various behavioral assays, reflecting their overall psychotropic actions and their motivational character, μ- and κ-agonists appear as functional antagonists. In this monograph the related data are reviewed and a hypothesis is advanced that posits that drug abuse is due to perturbation of the functional balance among the endogenous ligands of μ- and κ-receptors.

REFERENCES

1. **Hughes, J., Smith, T. W., Kosterlitz, H. W., Fothergill, L. A., Morgan, B. A., and Morris, H. R.,** Identification of two related pentapeptides from the brain with potent opiate agonist potency, *Nature*, 258, 577, 1975.
2. **Li, C. H. and Chung, D.,** Isolation and structure of an untriakontatpeptide with opiate activity from camel pituitary gland, *Proc. Natl. Acad. Sci. U.S.A.*, 73, 1145, 1976.
3. **Bradbury, A. F., Smyth, D. G., Snell, C. R., Birdsall, N. J. M., and Hulme, E. C.,** C-Fragment of lipotropin has a high affinity for brain opiate receptors, *Nature*, 260, 793, 1976.
4. **Chretien, M., Benjannet, S., Dragon, N., Seidah, N. G., and Lis, M.,** Isolation of peptides with opiate activity from sheep and human pituitaries, *Biochem. Biophys. Res. Commun.*, 72, 472, 1976.
5. **Lord, J. A. H., Waterfield, A. A., Hughes, J., and Kosterlitz, H. W.,** Endogenous opioid peptides: multiple agonists and receptors, *Nature*, 267, 495, 1977.
6. **Martin, W. R., Eades, C. G., Thompson, J. A., Huppler, R. E., and Gilbert, P. E.,** The effects of morphine- and naloxone-like drugs in the non-dependent and morphine-dependent chronic spinal dog, *J. Pharmacol. Exp. Ther.*, 197, 517, 1976.
7. **Gilbert, P. E. and Martin, W. R.,** The effects of morphine- and nalorphine-like drugs in the non-dependent, morphine-dependent and cyclazocine-dependent chronic spinal dog, *J. Pharmacol. Exp. Ther.*, 198, 66, 1976.
8. **Kosterlitz, H. W., Lord, J. A. H., Paterson, S. J., and Waterfield, A. A.,** Effects of changes in the structures of enkephalins and narcotic analgesic drugs on their interactions with μ- and δ-receptors, *Br. J. Pharmacol.*, 68, 333, 1980.
9. **Goldstein, A. and Naidu, A.,** Multiple opioid receptors: ligand specificity profiles and binding site signatures, *Mol. Pharmacol*, 36, 265, 1989.
10. **Mussacchio, J. M.,** The psychomimetic effects of opiates and the sigma receptor, *Neuropsychopharmacology*, 3, 191, 1990.
11. **Goldstein, A., Tachibana, S., Lowney, L. I., Hunkapiller, M., and Hood, L.,** Dynorphin-(1–13), an extraordinarily potent opioid peptide, *Proc. Natl. Acad. Sci. U.S.A.*, 76, 6666, 1979.
12. **Chavkin, C., James, I. F., and Goldstein, A.,** Dynorphin is a specific endogenous ligand of the κ-opioid receptor, *Science*, 215, 415, 1982.
13. **Comb, M., Seeburg, P. H., Adelman, J., Eiden, L., and Herbert, E.,** Primary structure of the human Met- and Leu-enkephalin precursor and its mRNA, *Nature*, 295, 663, 1982.
14. **Gubler, U., Seeburg, P. H., Hoffman, B. J., Gage, L. P., and Udenfriend, S.,** Molecular cloning establishes proenkephalin as precursor of enkephalin containing peptides, *Nature*, 295, 206, 1982.
15. **Noda, M., Furutani, Y., Takahshi, H., Toyosato, M., Hirose, T., Inayama, S., Nakanishi, S., and Numa, S.,** Cloning and sequence analysis of cDNA for bovine adrenal preproenkephalin, *Nature*, 295, 202, 1982.

16. **Nakanishi, S., Inoue, A., Kita, T., Nakamura, M., Chang, A. C. Y., Cohen, S. N., and Numa, S.,** Nucleotide sequence of cloned cDNA for bovine corticotropin-β-lipotropin precursor, *Nature*, 278, 423, 1979.
17. **Watson, S. J., Khatchaturian, H., Taylor, L., Fischli, W., Goldstein, A., and Akil, H.** Prodynorphin peptides are found in the same neurons throughout rat brain: immunocytochemical study, *Proc. Natl. Acad. Sci. U.S.A.*, 80, 891, 1983.
18. **Fallon, J. H. and Leslie, F. M.,** Distribution of dynorphin and enkephalin peptides in the rat brain, *J. Comp. Neurol.*, 249, 293, 1986.
19. **Attali, B., Gouarderes, C., Mazarguil, H., Audigier, Y., and Cros, J.,** Evidence of multiple 'kappa' binding sites by use of opioid peptides in the guinea pig lumbo-sacral spinal cord, *Neuropeptides*, 3, 53, 1982.
20. **Su, T.-P.,** Further demonstration of *kappa* opioid binding sites in the brain: evidence for heterogeneity, *J. Pharmacol. Exp. Ther.*, 232, 144, 1985.
21. **Iyengar, S., Kim, H. S., and Wood, P. L.,** Effects of kappa opiate agonists on neurochemical and neuroendocrine indices: evidence for kappa receptor subtypes, *Life Sci.*, 39, 637, 1986.
22. **Zukin, R. S., Eghbali, M., Olive, D., Unterwald, E. M., and Tempel, A.,** Characterization and visualization of rat and guinea pig brain κ opioid receptors: evidence for κ_1 and κ_2 opioid receptors, *Proc. Natl. Acad. Sci. U.S.A.*, 85, 4061, 1988.
23. **Pasternak, G. W.,** Multiple mu opiate receptors: biochemical and pharmacological evidence for multiplicity, *Biochem. Pharmacol.*, 35, 361, 1986.
24. **Pasternak, G. W. and Wood, P. J.,** Multiple mu opiate receptors, *Life Sci.*, 38, 1889, 1986.
25. **Rothman, R. B., Bowen, W. D., Bykov, V., Schumacher, U. K., Pert, C. B., Jacobson, A. E., Burke, T. R., and Rice, K. C.,** Preparation of rat brain membranes greatly enriched with either type-1-delta or type-2-delta opiate binding sites using site-directed alkylating agents: evidence for two-site allosteric model, *Neuropeptides*, 3, 201, 1984.
26. **Lemaire, S., Magnan, J., and Regoli, D.,** Rat vas deferens: specific bioassay for endogenous opioid peptides, *Br. J. Pharmacol.*, 64, 327, 1978.
27. **Schulz, R., Faase, E., Wüster, M., and Herz, A.,** Selective receptors for β-endorphin on the rat vas deferens, *Life Sci.*, 24, 843, 1979.
28. **Römer, D., Büscher, H. H., Hill, R. G., Pless, J., Bauer, W., Cardinaux, F., Closse, A., Hauser, D., and Huguenin, R.,** A synthetic enkephalin analogue with prolonged parenteral and oral analgesic activity, *Nature*, 268, 547, 1977.
29. **Mosberg, H. I., Hurst, R., Hruby, V. J., Gee, K., Yamamura, H. I., Galliga, J. J., and Burks, T. F.,** Bis-penicillamine enkephalins possess highly improved specificity towards δ-opioid receptors, *Proc. Natl. Acad. Sci. U.S.A.*, 80, 5871, 1983.
30. **Cotton, R., Kosterlitz, H. W., Patterson, S. J., Rance, M. J., and Traynor, J. R.,** The use of $[^{+}H]$-[D-Pen^2,D-Pen^5]enkephalin as a highly selective ligand for delta binding site, *Br. J.Pharmacol.*, 84, 927, 1985.
31. **Murray, C. W. and Cowan, A.,** [D-Pen^2,D-Pen^5]enkephalin, the standard delta opioid agonist induces morphine-like behavior in mice, *Psychopharmacology*, 102, 425, 1990.

32. **Collier, H. O. J. and Roy, A. C.,** Morphine-like drugs inhibit the stimulation by E prostaglandin of cyclic AMP formation by brain homogenates, *Nature*, 255, 159, 1974.
33. **Klee, W. A. and Nierenberg, M.,** A neuroblastoma X glioma hybrid cell line with morphine receptors, *Proc. Natl. Acad. Sci. U.S.A.*, 71, 3474, 1974.
34. **Sharma, S. K., Nierenberg, M., and Klee, W. A.,** Morphine receptors as regulators of adenylate cyclase activity, *Proc. Natl. Acad. Sci. U.S.A.*, 72, 590, 1975.
35. **Traber, J., Gullis, R., and Hamprecht, B. L.,** Influence of opiates on the levels of adenosine 3′-5′ cyclic monophosphate in neuroblastoma X glioma hybrid cells, *Life Sci.*, 17, 1863, 1975.
36. **Polastron, J., Boyer, M.-J., Quertermont, Y., Thouvenot, J.-P., Meunier, J.-C., and Jauzac, Ph.,** μ-Opioid receptors and not κ-opioid receptors are coupled to the adenylate cyclase in the cerebellum, *J. Neurochem.*, 54, 562, 1990.
37. **VonVoigtlander, P. F., Lahti, R. A., and Ludens, J. H.,** U-50,488: a selective and structurally novel non-mu (kappa) opioid agonist, *J. Pharmacol. Exp. Ther.*, 224, 7, 1983.
38. **Rothman, R. B., France, C. P., Bykov, V., De Costa, B. R., Jacobson, A. E., Woods, J. H., and Rice, K. C.,** Pharmacological activities of optically pure enantiomers of the κ-opioid agonist, U-504888, and its cis diastereomer: evidence for three κ receptor subtypes, *Eur. J. Pharmacol.*, 167, 345, 1989.
39. **Meunier, J. C., Kouakou, Y., Puget, A., and Moisand, C.,** Multiple opiate binding sites in the central nervous system of the rabbit. Large predominance of mu subtype in the cerebellum and characterization of a kappa subtype in the thalamus, *Mol. Pharmacol.*, 24, 23, 1983.
40. **Chang, K. J. and Cuatrecasas, P.,** Multiple opiate receptors: enkephalin and morphine bind with different specificity, *J. Biol. Chem.*, 254, 2610, 1979.
41. **Periyasamy, S. and Hoss, W.,** Kappa opioid receptors stimulate phosphoinositide turnover in rat brain, *Life Sci.*, 47, 219, 1990.
42. **Werz, M. A. and Macdonald, R. L.,** Opioid peptides selective for mu and delta receptors reduce calcium-dependent action potential duration by increasing membrane potassium conductance, *Neurosci. Lett.*, 42, 173, 1983a.
43. **Werz, M. A. and Macdonald, R. L.,** Opioid peptides with differential affinity for *mu* and *delta* receptors decrease sensory neuron calcium dependent action potentials, *J. Pharmacol. Exp. Ther.*, 227, 394, 1983b.
44. **Werz, M. A. and Macdonald, R. L.,** Dynorphin reduces calcium-dependent action potential duration by decreasing a voltage dependent conductance, *Neurosci. Lett.*, 46, 185, 1984.
45. **Shen, K.-F. and Crain, S. M.,** Dynorphin prolongs the action potential of mouse sensory ganglion neurons by decreasing potassium conductance whereas another specific kappa opioid does so by increasing a calcium conductance, *Neuropharmacology*, 29, 343, 1990.
46. **Werz, M. A., Grega, D. S., and Macdonald, R. L.,** Actions of *mu, delta* and *kappa* agonists and antagonists on mouse primary afferent neurons in culture, *J. Pharmacol. Exp. Ther.*, 243, 258, 1987.

47. **Cherubini, E. and North, R. A.,** μ And κ opioids inhibit transmitter release by different mechanisms, *Proc. Natl. Acad. Sci. U.S.A.*, 82, 1860, 1985.
48. **North, A. R.,** Opioid receptor types and membrane ion channels, *Trends Neurosci.*, 9, 114, 1986.
49. **Crain, S. M. and Shen, K.-F.,** Opioids can evoke direct receptor mediated excitatory effects on sensory neurons, *Trends Pharm. Sci.*, 11, 77, 1990.
50. **Mulder, A. H., Wardeh, G., Hogenboom, F., and Frankhuyzen, A. L.,** κ- And δ-opioid receptor agonists differentially inhibit striatal dopamine and acetylcholine release, *Nature*, 308, 278, 1984.
51. **Arenas, E., Alberch, J., Arroyos, R. S., and Marsal, J.,** Effects of opioids on acetylcholine release evoked by K^+ or glutamic acid from rat neostriatal slices, *Brain Res.*, 523, 51, 1990.
52. **Hagan, R. M. and Hughes, I. E.,** Opioid receptor subtypes involved in the control of transmitter release in the cortex of the brain of the rat, *Neuropharmacology*, 23, 491, 1984.
53. **Micevych, P. E., Yaksh, T. L., and Go, V. L. W.,** Opiate mediated inhibition of the release of cholecystokinin and substance P, but not neurotensin from cat hypothalamic slices, *Brain Res.*, 250, 283, 1982.
54. **Micevych, P. E., Yaksh, T. L., and Go, V. L. W.,** Studies on the opiate-receptor mediated inhibition of K^+-stimulated cholecystokinin and substance P from cat hypothalamus in vitro, *Brain Res.*, 290, 87, 1984.
55. **Itoh, Y., Oishi, R., Nishibori, M., and Saeki, K.,** Involvement of mu receptors in the opioid-induced increase in the turnover of mouse brain histamine, *J. Pharmacol. Exp. Ther.*, 244, 1021, 1988.
56. **Gulat-Marnay, C., Lafitte, A., Arrang, J.-M., and Schwartz, J.-C.,** Modulation of histamine release in the rat brain by κ-opioid receptors, *J. Neurochem.*, 55, 47, 1990.
57. **Passarelli, F. and Costa, F.,** *Mu* and *delta* opioid receptors inhibit serotonin release in rat hippocampus, *J. Pharmacol. Exp. Ther.*, 248, 299, 1989.
58. **Ho, B. Y. and Takemori, A. E.,** Release by U-50,488H of [$^+$H]serotonin from brain slices and spinal cord synaptosomes of U-50,488H tolerant and nontolerant mice, *J. Pharmacol. Exp. Ther.*, 254, 8, 1990.
59. **Jakisch, R., Geppert, M., Brenner, A. S., and Illés, P.,** Presynaptic opioid receptors modulating acetylcholine release in the hippocampus of the rabbit, *Naunyn-Schmiedeberg's Arch. Pharmacol.*, 332, 156, 1986.
60. **Werling, L. L., Brown, S. R., and Cox, B. M.,** Opioid receptor regulation of the release of norepinephrine in brain, *Neuropharmacology*, 26, 987, 1987.
61. **Hirota, N., Kuraishi, Y., Hino, Y., Sato, Y., Sato, M., and Takagi, H.,** Met-enkephalin and morphine but not dynorphin inhibit noxious stimuli-induced release of substance P from rabbit dorsal horn in situ, *Neuropharmacology,* 24, 567, 1985.
62. **Di Chiara, G. and Imperato, A.,** Opposite effects of mu and kappa agonists on dopamine release in the nucleus accumbens and in the dorsal caudate of freely moving rats, *J. Pharmacol. Exp. Ther.*, 244, 1067, 1988.
63. **Iyengar, S., Kim, H. S., and Wood, P. L.,** *Kappa* opiate agonists modulate the hypothalamic-pituitary-adrenocortical axis in the rat, *J. Pharmacol. Exp. Ther.*, 238, 429, 1986.

64. **Illés, P.,** Modulation of transmitter and hormone release by multiple opioid receptors, *Rev. Physiol. Biochem. Pharmacol.*, 12, 139, 1989.
65. **De Bodo, R. C.,** The antidiuretic action of morphine and its mechanism, *J. Pharmacol. Exp. Ther.*, 82, 74, 1944.
66. **Weitzman, R. E., Fisher, D. A., Minick, S., Ling, N., and Guillemin, R.,** β-Endorphin stimulates secretion of arginin-vasopressin *in vivo, Endocrinology*, 101, 1643, 1977.
67. **Bisset, G. W., Chowdrey, H. S., and Feldberg, W.,** Release of vasopressin by enkephalin, *Br. J. Pharmacol.*, 62, 370, 1978.
68. **Leander, J. D.,** A *kappa* opioid effect: increased urination in the rat, *J. Pharmacol. Exp. Ther.*, 224, 89, 1983.
69. **Gunne, L. M.,** The temperature response in rats during acute and chronic morphine administration: a study of morphine tolerance, *Arch. Int. Pharmacodyn.*, 129, 416, 1960.
70. **Tulunay, F. C.,** The effects of morphine and various agonist type analgesics on body temperature in rats, *Life Sci.*, 27, 511, 1980.
71. **Szikszay, M., Benedek, G., and Székely, J. I.,** Thermoregulatory effect of D-Met^2-Pro^5-enkephalinamide, *Neuropeptides*, 3, 465, 1983.
72. **Székely, J. I., Rónai, A. Z., Dunai-Kovács, Z., Miglécz, E., Berzétei, I., Bajusz, S., and Gráf, L.,** (D-Met^2,Pro^5)-enkephalinamide: a potent morphine-like analgesic, *Eur. J. Pharmacol.*, 43, 293, 1977.
73. **Cavacchini, E., Candeletti, S., Spampinato, S., and Ferri, S.,** Hypothermia elicited by prodynorphin-related peptides: opioid and non-opioid actions, *Neuropeptides*, 14, 45, 1989.
74. **Cavacchini, E., Spampinato, S., and Ferri, S.,** G_1 proteins and calcium in dynorphin-induced hypothermia and behaviour, *Eur. J. Pharmacol.*, 180, 357, 1990.
75. **Rónai, A. Z., Berzétei, I., Székely, J. I., Miglécz, E., Kurgyis, J., and Bajusz, S.,** Enkephalin-like character and analgesia, *Eur. J. Pharmacol.*, 69, 263, 1981.
76. **Straub, W.,** Eine empfindliche biologische Reaktion auf Morphin, *Dtsch. Med. Wochenschr.*, 37, 1462, 1911.
77. **Bilbey, D. L. J., Salem, H., and Grossman, M. H.,** The anatomical basis of the Straub phenomenon, *Br. J. Pharmacol.*, 15, 540, 1960.
78. **Takemori, A. E., Larson, D. L., and Porthogese, P. S.,** The irreversible narcotic antagonistic and reversible agonistic properties of fumarate methyl ester derivative of naltrexone,*Eur. J. Pharmacol.*, 70, 445, 1981.
79. **Cotton, R., Giles, M. G., Miller, L., Shaw, J. S., and Timms, D.,** ICI 174864: a highly selective antagonist for the opioid δ-receptor, *Eur. J. Pharmacol.*, 97, 331, 1984.
80. **Howard, A. D., Sarne, Y., Gioannini, T. L., Hiller, J. M., and Simon, E. J.,** Identification of distinct binding site subunits of mu and delta opioid receptors, *Biochemistry*, 25, 357, 1986.
81. **Millan, M. J.,** κ-Opioid receptors and analgesia, *Trends Pharm. Sci.* 11, 70, 1990.
82. **Ukai, M., Yamada, S., and Kameyama, T.,** Naloxone reverses the inhibitory effect of dynorphin A on motor activity in the mouse, *Pharmacol. Biochem. Behav.*, 20, 815, 1984.

83. **Iwamoto, E. T.,** Locomotor activity and antinociception after putative *mu, kappa* and *sigma* opioid receptor agonists in the rat: influence of dopaminergic agonists and antagonists, *J. Pharmacol. Exp. Ther.*, 217, 451, 1981.
84. **Tang, A. H. and Collins, R. J.,** Behavioral effects of novel kappa opioid analgesic, U-50488H, in rats and rhesus monkeys, *Neuropharmacology*, 85, 309, 1985.
85. **Ukai, M., Toyoshi, T., and Kameyama, T.,** Dynorphin A(1–13) modulates apomorphine-induced behaviors using multidimensional behavioral analyses in the mouse, *Brain res.*, 499, 299, 1989.
86. **Wei, E. and Loh, H.,** Physical dependence on opiate-like peptides, *Science*, 193, 1262, 1976.
87. **Nakazawa, T., Kaneko, T., Yoshino, H., Tachibana, S., Goto, M., Taki, T., and Yamatsu, K.,** Physical dependence liability of dynorphin A analogs in rodents, *Eur. J. Pharmacol.*, 201,185, 1991.
88. **Negus, S. S., Picker, M. J., and Dykstra, L. A.,** Kappa antagonist properties of buprenorphine in non-tolerant and morphine-tolerant rats, *Psychopharmacology*, 98, 141, 1989.
89. **Herman, B. H. and Goldstein, A.,** Cataleptic effects of dynorphin-(1–13) in rats made tolerant to *mu* opiate receptor agonist, *Neuropeptides*, 2, 13, 1981.
90. **Ignar, D. M. and Kuhn, C. M.,** Effects of specific *mu* and *kappa* opiate tolerance and abstinence on hypothalamo-pituitary-adrenal axis secretion in the rat, *J. Pharmacol. Exp. Ther.*, 255, 1287, 1990.

2 Aspecific Character of the Analgesia Induced by κ-Agonists

I. INTRODUCTION

A considerable body of evidence (see Section II) indicates that under experimental conditions κ-agonists might not induce true analgesia unless they are administered directly into the spinal cord and, instead of heat, other modalities of nociceptive stimuli are used for noxious stimulation. Such a surprising conclusion seems to apply primarily to dynorphin and to a lesser degree to various synthetic κ-agonists. However, it is well established that classical opiates induce analgesia whatever assay is used and they are active whether injected supraspinally or spinally. (Namely, by definition, just those biological assays are considered reliable measures of analgesic effect which are sensitive to classical (μ)-opiates.) Therefore, the findings that dynorphin(s) only induces analgesia if given spinally and nociceptive stimuli other than heat are used as noxious stimulus suggest (Section II) that they cannot be real analgesics, i.e., endogenous analgesics of the organism. At the beginning of the "dynorphin research", in view of the extremely strong *in vitro* efficacy of natural dynorphin(s), which is actually the most potent endogenous opioid in some opiate-sensitive isolated organ preparations,[1] it was difficult to believe that it is practically inactive in various analgesic assays. Nevertheless, there are further lines of evidence indicating that dynorphins cannot be truly selective endogenous opioid agonists mediating analgesia. As reviewed in this chapter there is considerable ambiguity about the reversibility by opiate antagonists of the alleged analgesic action of dynorphin (Section II); its antinociceptive effect is accompanied with motor paralysis (Section III), and its electrophysiological actions are qualitatively different from those of the classical opiates (Section IV). The related data are summarized along these lines in this chapter.

II. EFFECTS OF κ-AGONISTS IN CONVENTIONAL ANALGESIA ASSAYS

Truly in the first year following the isolation of dynorphin(s), two groups reported that they failed to find any analgesic effect with dynorphin,[2,3] but a third one claimed to demonstrate it.[4] The subsequent studies also generated apparently contradictory results.

However, a careful review of the related publications reveals that there is no real contradiction among the related studies if the site of application and the test used are also taken into consideration. As shown in Table 1, in tail flick and hot plate tests, $dynorphin_{1-17}$, $dynorphin_{1-13}$, and their shorter fragments fail to induce significant analgesic activity either in mice or rats,[2,3,5–9] if administering them i.c.v. However, dynorphin does induce antinociception if the assay is based on nonthermal modalities of pain. Thus, dynorphin is capable of eliciting antinociception if mechanical stimulation of the tail is used for noxious stimulation[4,8–10] or electrical stimuli are applied.[11] A further crucial factor is the site of application. Namely, antinociception can be evoked even in the tail flick test if, instead of i.c.v. administration, dynorphin is injected into the n. reticularis (para)gigantocellularis, i.e., into a medullary center[10] or intrathecally.[12–16] Dynorphin has not been examined for analgesia upon peripheral administration due to its well-known metabolic instability.[4,14,17]

Regarding the dependence of the outcome of these studies on the assay used, it must be added that though many procedures have been described to test drugs for eventual analgesic effect, the most extensively used ones, i.e., the tail flick,[18] hot plate,[19] and tail immersion[20] tests, are based on thermal stimulation. Obviously, inhibition of excessive heat-induced pain reactions in rodents correlates better with the clinical efficacy of classical opiates than other procedures which use either mechanical stimulation (e.g., tail pinch[21] or paw pressure[22] tests), chemical irritation, i.e., writhing test,[23,24] formalin test,[25] or electrical stimulation of the tail[26] or tooth, and eventually the whole body,[27] as noxious stimuli. The high predictive value of inhibition of thermally elicited nociceptive reflexes (with regard to clinical analgesic efficacy) is actually not so evident. Submammalian species are poikilothermic and obviously not so sensitive to heat stimuli. Because of the lack of adequate analgesia assays, there is a distinct paucity of data relating to the eventual efficacy of analgesic drugs in poikilotherm animals. In this respect, it is worth recalling that, in frogs, levorphanol inhibits the acetic acid-induced wiping reflex only in surprisingly high doses,[28] and according to other authors the opiate receptors in this species are of the κ-type[29] or they belong to a subclass different from the ones characterized in various mammalian species.[30] As discussed in Chapter 3, under certain experimental conditions morphine decreases the pain threshold instead of eliciting analgesia. Thus, the eventuality arises that, spe-

TABLE 1
Aspecific Character of Analgesia Induced by Dynorphin

Species	Assay	Drug[a]	Analgesia	Reversal by opiate antagonists	Ref.
Mouse and rat	Tail flick	Dynorphin$_{1-13}$ (20 and 25 μg i.c.v.)	No	—	2
Rat	Tail clip-induced vocalization	Dynorphin$_{1-13}$ (100 nmol i.c.v.)	Yes	No	4
Rat	Vocalization threshold to electric stimulation	Dynorphin$_{1-13}$ (60 nmol i.c.v.)	No	—	3
Rat	Tail flick	Dynorphin$_{1-13}$ (40 μg i.c.v.)	No	—	6
Rat	Tail flick	Dynorphin$_{1-13}$ (20 μg i.c.v.)	No	—	5
Rat	Tail flick	Dynorphin$_{1-13}$ (40 μg i.c.v.)	Yes	Partial	43
Mouse	Tail flick	Dynorphin$_{1-13}$ (80 nmol i.c.v.)	No	—	7
Rat	Flinch/jump to electric stimulation	Dynorphin$_{1-13}$ (20 nmol i.c.v.)	Yes	No	11
Rat	Paw pressure Hot plate	Dynorphin$_{1-17}$ (10–80 μg i.c.v.)	Yes No	No —	8
	Tail pinch	Dynorphin$_{1-13}$ (1–10 nmol into NRGC-NRPG[b])	Yes	Not examined	10
Rat	Tail flick	Dynorphin$_{1-13}$ (1–10 intrathecally)	Yes	Yes	12
Rat	Tail flick	Dynorphin$_{1-17}$, dynorphin B (2.5–20.0 nmol intrathecally)	Yes in high doses	Partially	13

TABLE 1 (continued)
Aspecific Character of Analgesia Induced by Dynorphin

Species	Assay	Drug[a]	Analgesia	Reversal by opiate antagonists	Ref.
Mouse	Tail flick	$Dynorphin_{1-17}$ (15–50 nmol i.c.v.)	Weak	Yes but only with high doses	9
	Tail pinch	(4–15 nmol i.c.v.)	Yes		
	Writhing	(1–8 nmol i.c.v.)	Yes		
Rat	Tail flick and vocalization threshold to electric stimulation	$Dynorphin_{1-17}$ (25 nmol intrathecally)	Yes Yes	Yes, only in high intraspinal doses	14
Rat	Tail flick	$Dynorphin_{1-17}$ (6 nmol intrathecally)	Yes	No	15
Rat	Tail flick Hot plate Writhing	$Dynorphin_{1-17}$ (30 nmol intrathecally)	No	—	16

[a] Subscript indicates the longest fragment examined (shorter fragments showed much weaker effect in all studies).
[b] NRGC and NRPG = nucleus reticularis giganto- or paragigantocellularis.

cifically, mammals are the most sensitive to heat, which might be somehow related to their thermoregulatory capacity and the preponderance of μ-receptors along the pain pathways.

Nevertheless, in this regard more informative are the data on the relative effectiveness of dynorphin and some highly selective agonists of κ-receptors in various analgesic assays. Tyers[20] first observed that two κ-receptor-preferring analgesics, ethylketazocine and M-2034, were effective primarily in assays based on noxious stimulation of mechanical or chemical nature (inflamed paw pressure and writhing tests) and much less potent in tail flick or hot plate assays. On the other hand, the ED_{50} values of morphine and some of its classical derivatives have been found nearly identical in the thermal and nonthermal analgesia assays. For this reason he postulated that heat-induced pain reactions are mediated by μ-receptors, whereas κ-receptors mediate nociceptive reactions induced by stimuli of other modalities.[20] As summarized in Table 2, the subsequent studies basically confirmed this postulation, i.e., in the related works, listed there, dynorphin and some synthetic κ-agonists have been found more potent in tail pinch, paw pressure, or writhing assays than in tail flick and hot plate, i.e., heat-dependent tests. A notable exception seems to be the efficacy of dynorphin in a test based on eliciting vocalization by electrical stimulation of the tail.[14] The finding is surprising since this test reflects the affective/emotional aspects of the pain.[26] However, this effect could only be elicited by higher doses of dynorphin, which also causes motor paralysis,[14] and in another work[31] synthetic κ-agonists were completely inactive in this test.

Some other observations point to the same conclusion, namely, whatever assay was used the effect of κ-ligands proved to be much weaker than that of the μ-agonists.

Using the rat tooth pulp (electrical) stimulation procedure, U-50488H and tifluadom (given s.c.) generated very shallow dose-response curves similar to partial agonists of μ-receptors (e.g., pentazocine), but contrary to morphine and its agonists congeners, with which very steep dose-effect curves were obtained.[32] Similarly, in a very careful study, Millan et al.[33] demonstrated that whether they were applied i.c.v. or intrathecally, the dose-response curves of κ-agonists (U-50488H, tifluadom, U-69593) were very shallow in the heat-elicited tail flick and mechanically induced paw pressure tests as compared to those of some prototypical μ-agonists including morphine and (D-Ala^2,NMePhe^4,Gly^5)-enkephalin-ol (DAMGO).

Furthermore, these studies show that the maximal possible analgesic effect, seen with classical morphine congeners, cannot be reached with the κ-ligands and partial agonists/antagonists of the μ-receptors.[32,33] In a subsequent work Millan[31] showed that the effectiveness of s.c. injected morphine and fentanyl in the tail flick test did not change upon varying the intensity of heat stimulation or the strength of mechanical stimulation eliciting tail withdrawal. The analgesic efficacy of above mentioned synthetic κ-agonists also was independent of the stimulus strength if the reaction was elicited by pressure, but their"analgesic" potency decreased upon intensifying the heat stimulation. Other authors emphasize that the effect of U-50488 (given i.c.v. or i.t.) is very short lasting in the hot plate test.[34]

Surprisingly, there is one condition in which the efficacy of κ-agonists is beyond doubt. Already, Tyers[20] noticed that in the inflamed foot pressure test of a dozen μ- and κ-ligands, ethylketazocine and buprenorphine are far the strongest analgesics (eliciting inflammation of the paw by local injection of yeast), but they are much weaker than morphine and codeine in the tail flick and hot plate tests. In another study,[35] ethylketocyclazocine was more effective in suppressing the behavioral signs of pain induced by intraplantar injection of formalin (i.e., against pain due to formalin-induced

TABLE 2
Relative Efficacies of Kappa Agonists in Different Analgesic Assays

Species	Drug	Route of administration	Order of potency in various analgesic assays	Ref.
Mouse and	EKZ	s.c.	$IPP_r > W_m > HP_m > TI_r$	20
rat	MR-2034	s.c.	$IPP_r \cong W_m > HP_m > TI_r$	
Rat	Dynorphin	i.c.v.	PP > HP	8
Mouse	Dynorphin	i.t.	TF > HP	45
	U-50488H	i.p.	TF > HP	
	U-50488H	i.c.v.	TF ≅ HP	
	U-50488H	i.t.	TF > HP	
	EKCZ	i.p.	TF ≅ HP	
		i.c.v.	TF ≅ HP	
		i.t.	TF > HP	
Mouse	U-50488H	s.c.	W ≅ TF ≅ HP	44
Rat	U-50488H	s.c.	W > TI > HP > TF	
Rat	U-50488H	i.t.	W > TF ≅ HP	46
	EKCZ	i.t.	HP > W > TF	
Mouse	Dynorphin	i.c.v.	W > TP > TF	9
Rat	Dynorphin	i.t.	ESV ≅ TF	14
	Dynorphin	i.t.	W > HP > ≅ TF	16
	U-50488H	i.t.	W > HP	34
	U-50488H	i.c.v.	HP > W	
	EKCZ	s.c.	FT > TI	35
	Tifluadom	i.t.	TP ≅ TF	33
	Tifluadom	i.c.v.	TP ≅ TF	
	U-50488H	i.t.	TP ≅ TF	
	U-50488H	i.c.v.	TP ≅ TF	
	U-69593	i.t.	TP ≅ TF	
	U-69593	i.c.v.	TP ≅ TF	
	U-50488H	s.c.	TP > TF > ESV	31
	U-69593	s.c.	TP > TF > ESC	
	U-50488H	intraplantar	IPP > PP	37
	Tifluadom		IPP > PP	
	PD-117302	s.c.	PFT > TI	36

Note: ESV = vocalization induced by electric stimulation of the tail; FT = formalin test (behavioral scoring after injecting formalin into the paw); HP = hot plate test; (I)PP = (inflamed) paw pressure test; PFT = paw formalin test; TF = tail flick test; TI = tail immersion into hot water; TP = tail pressure (or pinch) test; W = writhing induced by (chemical) irritation of the peritoneum; EKZ = ethylketazocine; EKCZ = ethylketocyclazocine; m = mouse; r = rat.

local inflammation) than in the tail immersion test. In line with these findings, PD 117302, a highly selective κ-agonist, has recently been reported completely inactive in the tail immersion test (where morphine is very active), but, using the above mentioned formalin test, its efficacy exceeded that of morphine.[36] Finally, Stein et al.[37] reported that the enhanced sensitivity of the paw to pressure, due to inflammation induced by local application of Freund's adjuvant, could also be reversed by local (intracutaneous) application of U-50488H. Truly, in this study (D-Ala^2,NMePhe^4,Gly^5-ol)-enkephalin (DAGO), a μ-selective enkephalin analog, and (D-Pen^2,D-Pen^5)-enkephalin (DPDPE), a δ-specific enkephalin analog, also showed considerable efficacy. These data indicate that under certain experimental condition κ-agonists are quite effective in mitigation of nociceptive reaction elicited by inflammation.

As shown in Table 1, dynorphin is only active if given intrathecally. Actually, already Wood et al.[38] proposed that κ-agonists act at the spinal level, since ethylketazocine was much more effective upon intrathecal than more rostral application. As mentioned above,[37] analgesia can also be elicited by local injection of opioids into the inflamed tissue, and, using this way of administration, analgesia can be elicited by κ-, μ-, and δ-agonists as well. Furthermore, Abbott[35] found that ethylketocyclazocine was only partly effective in the formalin test if applied i.c.v., but complete analgesia could be attained by systemic (s.c.) administration. Thus, the possibility arises that the κ-agonists act as caudally as the peripheral nervous system. Some electrophysiological data reviewed later (Section IV) are in agreement with such a proposition.

Truly, synthetic ligands of κ-receptors are not completely inactive upon i.c.v. application. However, comparing the corresponding ED_{50} values U-50488H, tifluadom, and U-69593 have been found much more active upon spinal than supraspinal application, and using the latter route of administration the slopes of their dose-response curves were much shallower.[33,39] The weak but indisputable efficacy of supraspinally applied synthetic κ-agonists as opposed to the complete ineffectiveness of dynorphin (given i.c.v.) seems to be a contradiction. However, it is not necessarily so. It can be due not only to the well-established multiplicity of κ-receptors,[40–42] but the phenomenon might also be explained by supposing that even the most selective κ-agonists have a slight affinity to the μ-receptors.

The lack of cross tolerance between μ- and κ-agonists is a further proof that they act in different ways. If mouse vas deferens preparation is made of mice previously chronically treated with fentanyl or (D-Ala^2,D-Leu^5)-enkephalin (DADLE) the potency of dynorphin remains unaltered.[3] Thus *in vitro* there is apparently no cross tolerance between the μ-, δ-, and κ-receptors. *In vivo* in rats, partial cross tolerance was demonstrated between i.c.v.-administered morphine and dynorphin in one study,[11] but not in another one upon intrathecal application.[43] Since dynorphin *in vivo* is quickly metabolized,[4,17] it cannot be discounted that the positive finding was due to its shorter metabolites (i.e., extended Leu-enkephalin containing) sequences which also have some analgesic activity mediated by μ- and δ-receptors. Therefore, more convincing are the studies where synthetic ligands were used. In pioneering work of Von Voigtlander et al.,[44] an absolute lack of any cross-tolerance between morphine and U-50488H was reported. Since then the first finding on the lack of cross tolerance between μ- and κ-agonists in analgesia (and other) assays has been repeatedly confirmed.

As summarized in Table 1, the data on the eventual modulation by naloxone or naltrexone of dynorphin-induced analgesia are ambiguous. Most groups could not reverse the alleged analgesic action of dynorphin by these antagonists:[4,8,11,15] however, others found partial[13,43] or "complete" reversal, but generally only in high doses.[9,12,14]

The contradiction can partly be explained by the relative insensitivity of κ-receptors to classical nonselective antagonists as naloxone (vide infra).[44] Thus it may be that in studies reporting negatively, the dose of naloxone was too low. In addition, another factor might also explain the controversy. As discussed below (Section III), the issue that dynorphin also causes motor paralysis, incapacitation, and the eventual reversibility of these motor disturbances by opiate antagonists is really obscure. Thus, it is difficult to find out whether the reversal by opiate antagonists of dynorphin "analgesia" is a specific effect or if it is simply due to abolition of motor incapacitation leading to "false analgesia". However, the synthetic ligands of κ-receptors do not cause motor impairment.

As soon as U-50488H was developed, the weaker antagonism of its analgesic effect by naloxone also became evident.[44,45] In the latter study the pA_2 value of naloxone against morphine and U-50488H were compared. Numerically, upon s.c. administration, 36 times more naloxone was needed to reverse the analgesic effect of U-50488H in mouse tail flick assay ($pA_2 = 5.28$) than needed against morphine ($pA_2 = 6.84$).[44] Injecting the agonists into the spinal cord and naloxone systemically, similar data have been obtained. The pA_2 values of naloxone against κ-ligands ranged from 5.9 to 6.6, whereas against μ-agonists these values were greater than 7.[46] On the other hand, norbinaltorphimine, a preferential κ-receptor antagonist, reversed more easily the analgesic effect of U-50488H than that of morphine.[31,33] Of course, the differential sensitivity to opiate antagonists, per se, does not indicate that the κ-agonists induced "analgesia" is not genuine, but it indicates that the underlying mechanisms are different.

Specifically, in the case of dynorphin the relative insensitivity to opiate antagonists might be also related to the data on the "analgesic efficacy" of its Des-Tyr^1 analogs in analgesia assays.[5,14] This finding prompted Walker et al.[5] to propose already in 1982 that dynorphin, beside κ-ones, does interact with nonopioid receptors as well. An important point since the mechanism of dynorphin-induced motor incapacitation and the lack of similar effect of highly selective synthetic ligands of κ-receptors can only be interpreted by supposing such an interaction (see the next Section).

III. MOTOR IMPAIRMENT INDUCED BY κ-AGONISTS

One of the most puzzling features of "dynorphin analgesia" is that it cannot be separated from motor paralysis or incapacitation. Practically in all analgesia assays, pain is measured by the latency of a special movement (e.g., tail withdrawal) or its frequency (e.g., writhing), i.e., a motor reaction is used to indicate that the pain threshold has been reached. Therefore, even today it is impossible to find out whether or not the alleged antinociceptive effect of dynorphin, if recorded, was not due to motor impairment. A further unsolved issue is the lack of any similar impairment with the synthetic ligands of κ-receptors. κ-Agonists, contrary to morphine, induce monophasic behavioral depression and sedation,[44] but no motor disturbances.

Actually, the motor impairment elicited by dynorphin was already mentioned in the first related papers. Herman et al.[4] and Walker et al.[2] described as a collateral finding the strange behavior of rats consisting of "barrel rolling", rigid hindlimb extension, catalepsy, profound immobility, and in the case of unilateral central injection ipsilateral flexion and contralateral extension of the limbs, head turning to the contralateral side, etc. In the more detailed studies the syndrome is characterized as flaccid hindlimb paralysis in extended position, loss of muscle tone, disappearance of placing, stepping even the writhing reflexes.[12,14,15,47] This syndrome can only be elicited by dynorphins, whereas U-50488H failed to induce such symptoms even in the heroic dose of 300 nmol i.c.v.[47] or 100 nmol intrathecally[16] in rats. In a careful study, Stevens and Yaksh[16] demonstrated that in various analgesic assays (writhing, hot plate, tail flick) dynorphin_{1-17} given intrathecally to rats was effective only in doses which also induced motor incapacitation. Thus, virtually no analgesia can be induced by dynorphin, but only motor impairment, which results in "false" positivity in these tests. Dynorphin given intrathecally causes an initial facilitation, then a permanent block of tail flick reaction.[48] In parallely conducted electrophysiological experiments, a simultaneous activation, then permanent (lasting for weeks) abolition of C-fiber reflex response was seen.[48] Lasting inhibition of any reflex response indicates cell death and the subsequent experiments partly revealed the mechanism of neuronal damage seen after dynorphin treatment (vide infra).

In one early study,[11] naloxone reversed the motor effects of dynorphin without antagonizing the apparent analgesic action, and in some others works[9,12,14] both seemed to be eliminated if the dose used was high enough.

However, in the more detailed (mostly subsequent) studies the paralytic effect of dynorphin could not be reversed by naloxone, even in doses up to 32 mgkg s.c.[2,4,15,16,47]

A further interesting property of hindlimb paralysis is that it can also be elicited by des-Tyr[1] fragments of dynorphin, which of course have no opioid activity.[5] In another frequently quoted study,[14] the analgesic/paralytic effect of $dynorphin_{1-17}$ lasted much longer than the time needed (10 min) for spontaneous degradation of $dynorphin_{1-17}$ to $dynorphin_{2-17}$. Thus the question arises whether the paralytic effect of dynorphin (not seen with synthetic κ-agonists) is mediated by κ-opiate receptors or not. In one study,[49] intrathecal pretreatment with nor-binaltorphimine, a selective irreversible ligand of κ-receptors,[50] failed to reverse the motor paralysis and spinal cord injury seen after dynorphin-A (also given intrathecally). However, in another work,[51] nor-binaltorphimine did reverse the motor impairment caused by $dynorphin_{1-17}$. Truly, in this study of Faden,[51] contrary to other ones made in the same laboratory,[47,52] $dynorphin_{2-17}$, a non-opioid fragment, induced no motor impairment. Thus the dynorphin-induced motor paralysis might be mediated partly by κ-opioid and partly by nonopioid mechanisms. Some recent studies indicate that the dynorphin-induced motor paralysis is mediated by some elements of the NMDA-receptor complex. *In vitro* in a receptor binding assay, $dynorphin_{1-13}$ has been found to partially displace labeled glutamate from its NMDA-sensitive binding site.[53] Thus it is not surprising that in one study both 2-amino-5-phosphonovalerate and γ-D-glutamylglycine, i.e., antagonists of the NMDA-sensitive excitatory amino acid receptors attenuated the motor paralytic and "analgesic" effect of $dynorphin_{1-13}$.[54] Actually, $dynorphin_{1-13 \text{ or } 1-17 \text{ or } 2-17}$-elicited motor paralysis could be reversed by both competitive and noncompetitive antagonists of the NMDA receptors such as 4-(3-phosphonopropyl)-2-piperazine-carboxylic acid and dextrorphan, MK-801, respectively.[52,55] Even strychnine, i.e., an antagonist of glycine receptors, reversed the dynorphin-induced paralysis and increase in tail flick latency.[56] These data indicate that certain kinds of paralysis seen after spinal trauma might be mediated by excessive release of endogenous dynorphin, which in turn activates the NMDA receptors, resulting in cell death. Though these interesting observations do not belong to the subject of this monograph they also prove that dynorphin can hardly function as an endogenous analgesic.

IV. ELECTROPHYSIOLOGICAL STUDIES

The related electrophysiological observations also confirm that activation of κ-receptors might initiate processes different from the ones seen after the classical opiates. Furthermore, according to the electrographic experiments the effect of dynorphin is even more complex as indicated by the behavioral investigations. According to some studies, as summarized in Table 3, both μ- and κ-agonists inhibit the evoked activity of convergent (multireceptive) neurons, and this inhibition can be reversed by naloxone.[57–60] However, in other studies[61–63] intrathecally administered dynorphin and synthetic κ-analogs as U-50488H and U-69593 induced variable response; the firing of individual units induced by nociceptive stimulation was either facilitated or inhibited (vide infra).

Furthermore, in apparent contradiction to the behavioral experiments, μ- and κ-agonists were found equipotent in suppression of noxious heat- and pinch-induced discharges if the stimuli of different modalities were matched for intensity, i.e., if the thermal and mechanical stimuli induced the same degree of firing.[64] Nevertheless, the data on the eventual selectivity of κ-agonists are far from unequivocal. Examining the stimulation-induced discharges of lumbar motoneurones, the effect of μ-agonists was selective for noxious stimuli, whereas κ-agonists also inhibited the responses induced by tactile stimulation.[59] Actually, in the first related study[65] the motoneurons of spinalized rats were more sensitive to the inhibitory effect of systemic U-50488H than the dorsal horn units. Thus, the effect of κ-agonists is not really selective.

On the other hand, if inflammation was induced by local injection of carragenan into the knee joint of cats, the spontaneous discharges of corresponding ipsilateral units were more easily inhibited by U-50488H and ethylketocyclazocine than by morphine and a μ-selective enkephalin analog, DAGO.[57] In agreement with this finding, if dorsal horn neurons were activated by injecting formalin s.c. into their receptive field in the plantar region, the resulting firing could effectively be inhibited by dynorphin and U-50488H, applied directly into the site of formalin injection, whereas morphine and DSTBUTLET had no inhibitory action.[60] The inhibitory effect of U-50488H was found naloxone reversible, but not accompanied by any anti-inflammatory action.[60] Consequently, if the painful stimulus is related to inflammation, just the κ-agonists seem to be the most effective in electrophysiological experiments just as observed in behavioral assays (Section II).

Thus, the electrophysiological data are in good agreement with the behavioral ones: κ-agonists are ''selective'' in suppressing antinociceptive reflexes initiated by stimulation of inflamed tissue (see Table 1). However, the issue is further complicated by observations that spinal cord application of dynorphin or U-50488H might induce not only inhibition, but also facilitation. Knox and Dickenson,[61] followed by others,[62,63] reported that intrathecal application of these substances caused both inhibition and facilitation of C-fiber-evoked responses and that the units responding with excitation prevailed.

Thus, administration of these κ-agonists results in a hyperalgesic effect, i.e., lowering of the threshold needed for activation of superficial dorsal horn nociceptive neurons by thermal or mechanical stimuli, en-

TABLE 3
Comparison of the Effects of Kappa and Mu Agonists on Nociceptive Reflexes in Electrophysiological Studies

Species	Drug, route of administration	Experimental paradigm	Observation	Ref.
Rat	U-50488H i.v.	Effect on thermally or mechanically induced unit activity in the spinal cord (dorsal horn units and motoneurons)	Preferential inhibition of the motoneurons	65
Rat	Dynorphin U-50488H directly on the surface of the spinal cord	Cutaneous C-fiber-evoked responses in the dorsal horn	Naloxone-insensitive facilitation, or inhibition, or biphasic modulation	61
Cat	U-50488H, EKC;[a]	Resting activity of type-IV afferent units arising from inflamed knee joint	Strong inhibition of spontaneous discharges	
	Morphine DAGO[b] intraarterially		Much weaker inhibition	
Cat	Dynorphin U-50488H, DAGO, DADLE[c] phoretically into the:	Multireceptive sensory responses recorded from the spino-cervical tract		58
	Laminae III-V:		Only the κ-ligands inhibit the nociceptive responses	
	Substantia gelatinosa:		Only DAGO and DADLE are effective	
Rat (spinal)	U-50488H, tifluadom, fentanyl i.v.	Stimulus-evoked discharges from the lumbar motoneurons evoked by noxious heat and mechanical stimuli	The κ- and μ- agonists are equally effective against heat- and pinch-induced discharges if the stimuli are matched for intensity	59
Rat (spinal)	U-50488H fentanyl, morphine i.v.	Stimulus-evoked discharges from the lumbar motoneurons evoked by noxious heat and non-noxious tactile stimuli	μ-agonists selectively inhibit the noxious heat-induced responses, but the effect of U-50488H is less selective	64

TABLE 3 (continued)
Comparison of the Effects of Kappa and Mu Agonists on Nociceptive Reflexes in Electrophysiological Studies

Species	Drug, route of administration	Experimental paradigm	Observation	Ref.
Rat (spinal)	U-50488H, tifluadom, morphine fentanyl i.v.	Stimulus-evoked discharges from the lumbar motoneurons evoked by noxious heat and non-noxious mechanical stimuli	The order of potency of the κ- and μ-agonists is the same against stimuli of different modalities but the effect of κ-ligands is more resistant to reversal by naloxone	68
Rat	Dynorphin U-50488H DPDPE[d] DAGO phoretically into Lamina I:	Evoked responses of multireceptive cells	DAGO and DPDPE, but not the κ-agonists inhibit the nociceptive responses;	66
	Laminae III-V:		Only the κ-agonists inhibit the responses	
Rat	U-50488H on the surface of the spinal cord	Evoked activity of the convergent neurons in the dorsal horn	Suppression of responses induced by both thermal and mechanical stimuli	72
Rat	U-50488H, morphine, fentanyl i.v.	Stimulus-evoked discharges from the lumbar motoneurons by noxious stimuli in intact animals and after spinalization	Spinalization does not alter the efficacy of morphine, but reduces the effects of fentanyl and U-50488H	67
Rat	U-50488H morphine DSTBULET[e] s.c.	Spontaneous activity of convergent units in the dorsal horn; the effect of formaline and the drugs injected s.c. into the paw in the receptive field of cells	Formalin injection activates the cells; this effect can be reversed by U-50488H, but not by the other two drugs	60

TABLE 3 (continued)
Comparison of the Effects of Kappa and Mu Agonists on Nociceptive Reflexes in Electrophysiological Studies

Species	Drug, route of administration	Experimental paradigm	Observation	Ref.
Rat	Dynorphin, U-50488H into the spinal cord	Multireceptive cells in the superficial area of dorsal horn	Expansion of the receptive area of the skin from where responses can be elicited; lowering of the threshold to mechanical and thermal stimuli; the effect is not reversible by naloxone	62
Rat (pups and adults)	U-50488H, U-69593 i.v. and intrathecally	Modulation of the C-fiber-evoked discharges of the dorsal horn neurons	Inhibition, prevented by intrathecally given norbinaltorphimine in 21-d-old pups, but not abolished by the same pretreatment in adult animals	63

[a] EKC: ethylketocyclazocine.
[b] DAGO: (D-Ala^2,N-MePhe^4,Gly^5)-enkephalin-ol.
[c] DADLE: (D-Ala^2,D-Leu^5)-enkephalin.
[d] DPDPE: (D-Pen^2,D-Pen^5)-enkephalin.
[e] DSTBULET: Tyr-D-Ser-(Otbu)-Gly-Phe-Leu-Thr.

hanced firing to stimulation, and an expansion of the receptive field of these cells.[61–63] This effect is naloxone irreversible.[62] Thus κ-agonists can cause both inhibition and facilitation of the convergent neurons and act primarily in the spinal cord or at the periphery.

Furthermore, the site of action of κ-agonists is surely different from those of the μ-ones, even within the spinal cord. U-50488H and dynorphin inhibit the multireceptive units if iontophoresed into the deeper part of the dorsal horn, i.e., into laminae III to V, whereas μ-agonists are effective if applied into the substantia gelatinosa.[58,66] Spinalization did not alter the inhibition by morphine and fentanyl of stimulus-evoked discharges from lumbar motoneurons, but the similar effect of U-50488H was attenuated.[67] Thus, the anatomical substrate of μ- and κ-agonist-induced inhibition must be different, and obviously the descending pathways also contribute to the effects of the latter ones.

Otherwise, just as in the case of behavioral experiments, more naloxone is needed to reverse the inhibitory effect of U-50488H and tifluadom than that of morphine and fentanyl,[68] indicating the involvement of different subtypes of opiate receptor inhibition and, as seen above,[62] the excitatory action of κ-ligands is naloxone insensitive. Probably it is worth recalling here that both dynorphin and U-50488H have been reported to induce naloxone-insensitive changes in the activity of hippocampal neurons, too.[5,69]

A further confounding factor is the eventual role of κ-receptors in their electrophysiological actions. According to a recent study,[63] intrathecal pretreatment with norbinaltorphimine prevents the eventual inhibitory effects of U-50488H and U-69593 in rat pups, but not in adult animals. Rat pups are known to have higher κ-receptor density in the spinal cord than adult ones.[70,71] Thus, the possibility arises that κ-agonists interact selectively with their own receptors in neonatal rats, but not in adults ones, which have been used for the electrophysiological studies. In other words, a part of the findings might be related to activation of non-κ-receptors upon administration of ''selective'' κ-agonists or at least a subtype of the multiple κ-receptors which is not sensitive to binaltorphimine. Thus, it is pretty difficult to interpret the heterogeneous data.

Summarily, these data indicate that the alleged analgesic effect of dynorphins and κ-agonists in general can hardly be specific. Namely, as discussed above,

- Their efficacy is ambiguous just in the most selective animal models of analgesia.
- The dynorphin-induced antinociception is difficult to separate from motor incapacitation and paralysis.
- Some of their effects are independent of opiate receptors.
- In some types of electrophysiological experiments they actually facilitate the nociceptive reactions.

The findings about their hyperalgesic effect are very surprising and potentially important physiologically. Therefore, the related data are reviewed in a separate chapter (Chapter 3).

REFERENCES

1. **Goldstein, A., Tachibana, S., Lowney, L. I., Hunkapiller, M., and Hood, L.**, Dynorphin-(1–13) an extraordinarily potent opioid peptide, *Proc. Natl. Acad. Sci.*, 76, 6666, 1979.
2. **Walker, J. M., Katz, R. J., and Akil, H.**, Behavioral effects of dynorphin$_{1-13}$ in the mouse and rat: initial observations, *Peptides*, 1, 341, 1980.
3. **Wüster, M., Schulz, R., and Herz, A.**, Opiate activity and receptor selectivity of dynorphin$_{1-13}$ and related peptides, *Neurosci. Lett.*, 20, 79, 1980.
4. **Herman, B. H., Leslie, F., and Goldstein, A.**, Behavioral effects and *in vivo* degradation of intracerebroventricularly administered dynorphin-(1–13) and D-Ala^2-dynorphin-(1–11) in rats, *Life Sci.*, 27, 883, 1980.
5. **Walker, J. M., Moises, H. C., Coy, D. H., Baldrighi, G., and Akil, H.**, Non-opiate effects of dynorphin and des-Tyr-dynorphin, *Science*, 218, 1136, 1982.
6. **Friedman, E. J., Jen, M.-F., Chang, J. K., Lee, N. M., and Loh, H. H.**, Dynorphin: possible modulatory peptide on morphine and β-endorphin analgesia in mouse, *Eur. J. Pharmacol.*, 69, 357, 1981.
7. **Höllt, V., Tulunay, F. C., Woo, S. K., Loh, H. H., and Herz, A.**, Opioid peptides derived from pro-enkephalin A but not from proenkephalin B are substantial analgesics after administration into brain of mice, *Eur. J. Pharmacol.*, 85, 355, 1982.
8. **Hayes, A. G., Skingle, M., and Tyers, M. B.**, Antinociceptive profile of dynorphin in the rat, *Life Sci.* 33 (Suppl. I), 657, 1983.
9. **Nakazawa, T., Ikeda, M., Kaneko, T., and Yamatsu, K.**, Analgesic effects of dynorphin-A and morphine in mice, *Peptides*, 6, 75, 1985.
10. **Kaneko, T., Nakazawa, T., Ikeda, M., Yamatsu, K., Iwama, T., Wada, T., Satoh, M., and Takagi, H.**, Sites of analgesic action of dynorphin, *Life Sci.*, 33 (Suppl. I), 661, 1983.
11. **Petrie, E. C., Tiffany, S. T., Baker, T. B., and Dahl, J. L.**, Dynorphin (1–13): analgesia, hypothermia, cross-tolerance with morphine and β-endorphin, *Peptides*, 3, 41, 1982.
12. **Przewlocki, R., Shearman, G. T., and Herz, A.**, Mixed opioid/non-opioid effects of dynorphin and dynorphin related peptides after their intrathecal injection in rats, *Neuropeptides*, 3, 233, 1983.
13. **Han, J.-S., Xie, G.-X., and Goldstein, A.**, Analgesia induced by intrathecal injection of dynorphin B in the rat, *Life Sci.*, 34, 1573, 1984.
14. **Spampinato, S., and Candeletti, S.**, Characterization of dynorphin A-induced antinociception at spinal level, *Eur. J. Pharmacol.*, 110, 21, 1985.
15. **Herman, B. A. and Goldstein, A.**, Antinociception and paralysis induced by intrathecal dynorphin A, *J. Pharmacol. Exp. Ther.*, 232, 27, 1985.
16. **Stevens, C. W. and Yaksh, T. L.**, Dynorphin A and related peptides administered intrathecally in the rat: a search for putative *kappa* receptor activity, *J. Pharmacol. Exp. Ther.*, 238, 833, 1986.
17. **Leslie, F. M. and Goldstein, A.**, Degradation of dynorphin-(1–13) by membrane-bound rat brain enzymes, *Neuropeptides*, 2, 185, 1982.

18. **D'Amour, F. E. and Smith, D. L.,** A method for determining loss of pain sensation, *J. Pharmacol. Exp. Ther.*, 72, 74, 1941.
19. **Eddy, N. B., Fuhrmeister Touchberry, C., and Lieberman, J. E.,** Synthetic analgesics. I. Methadone isomers and derivatives, *J. Pharmacol. Exp. Ther.*, 98, 121, 1950.
20. **Tyers, M. B.,** Classification of opiate receptors that mediate antinociception in animals, *Br. J. Pharmacol.*, 69, 503, 1980.
21. **Haffner, F.,** Experimentelle Prüfung schmerzstillender Mittel, *Dtsch. Med. Wochenschr.*, 55, 731, 1929.
22. **Winter, C. A. and Flataker, L.,** Nociceptive thresholds as affected by parenteral administration of irritants and various antinociceptive drugs, *J. Pharmacol. Exp. Ther.*, 148, 373, 1965.
23. **Siegmund, E., Cadmus, R., and Lu, G.,** A method for evaluating both non-narcotic and narcotic analgesics, *Proc. Soc. Exp. Biol. Med.*, 95, 729, 1957.
24. **Witkin, L. B., Heubner, C. F., Galdi, F., O'Keefe, E., Spitaletta, P., and Plummer, A. J.,** Pharmacology of 2-amino-indane hydrochloride (SU-8629): a potent non-narcotic analgesic, *J. Pharmacol. Exp. Ther.*, 133, 400, 1961.
25. **Dobuisson, D. and Dennis, S. G.,** The formalin test: a quantitative study of the analgesic effects of morphine, meperidine, and brain stem stimulation in rats and cats, *Pain*, 4, 161, 1977.
26. **Paalzow, G. and Paalzow, L.,** Morphine-induced inhibition of different pain responses in relation to the regional turnover of rat brain noradrenaline and dopamine, *Psychopharmacology*, 45, 9, 1975.
27. **Evans, W. O.,** A new technique for the investigation of some analgesic drugs on a reflexive behavior in the rat, *Psychopharmacology*, 2, 318, 1961.
28. **Stevens, C. W. and Pezalla, P. D.,** Naloxone blocks the analgesic action of levorphanol but not of dextrorphan in the leopard frog, *Brain Res.*, 301, 171, 1984.
29. **Simon, J., Szücs, M., Benyhe, S., Borsodi, A., Zeman, P., and Wollcman, M.,** Solubilization and characterization of opioid binding sites from frog *(Rana esculenta)* brain, *J. Neurochem.*, 43, 957, 1984.
30. **Mollereau, C., Pascaud, A., Baillat, G., Mazarguil, H., Puget, A., and Meunier, J.-C.,** Evidence for a new type of opioid binding site in the brain of the frog *Rana ridibunda, Eur. J. Pharmacol.*, 150, 75, 1988.
31. **Millan, M. J.,** *Kappa*-opioid receptor mediated antinociception in the rat. I. Comparative actions of *mu*- and *kappa*-opioids against noxious thermal, pressure and electrical stimuli, *J. Pharmacol. Exp. Ther.*, 251, 334, 1989
32. **Steinfels, G. F. and Cook, L.,** Antinociceptive profiles of *mu* and *kappa* opioids in a rat tooth pulp stimulation procedure, *J. Pharmacol. Exp. Ther.*, 236, 111, 1986.
33. **Millan, M. J., Clonkowski, A., Lipkowski, A., and Herz, A.,** *Kappa*-opioid receptor-mediated antinociception in the rat. II. Supraspinal in addition to spinal sites of action, *J. Pharmacol. Exp. Ther.*, 251, 342, 1989.
34. **Porreca, F., Mosberg, H. I., Hurst, R., Hruby, V. J., and Burks, T. F.,** Roles of *mu, delta* and *kappa* opioid receptors in spinal and supraspinal mediation of gastrointestinal transit effects and hot-plate analgesia in the mouse, *J. Pharmacol. Exp. Ther.*, 230, 341, 1984.

35. **Abbott, F. V.,** Peripheral and central antinociceptive actions of ethylketocyclazocine in the formalin test, *Eur. J. Pharmacol.*, 152, 93, 1988.
36. **Wheeler-Aceto, H. and Cowan, A.,** Standardization of the rat paw formalin test for evaluation of analgesics, *Psychopharmacology*, 104, 35, 1991.
37. **Stein, C., Millan, M. J., Shippenberg, T. S., Peter, K., and Herz, A.,** Peripheral opioid receptors mediating antinociception in inflammation. Evidence for involvement of *mu, delta* and *kappa* receptors, *J. Pharmacol. Exp. Ther.*, 248, 1269, 1989.
38. **Wood, P. L., Rackham, A., and Richard, J.,** Spinal analgesia: comparison of the mu agonist morphine and the kappa agonist ethylketocyclazocine, *Life Sci.*, 28, 2119, 1981.
39. **Piercey, M. F., Lahti, R. A., Schroeder, L. A., Einspahr, F. J., and Barsuhn, C.,** U-50488H, a pure kappa receptor agonist with spinal analgesic loci in the mouse, *Life Sci.*, 31, 1197, 1982.
40. **Attali, A., Gouarderes, V., Mazarguil, H., and Cros, J.,** Evidence for multiple kappa binding sites by use of opioid peptides in the guinea pig lumbosacral spinal cord, *Neuropeptides*, 3, 53, 1982.
41. **Su, T.-P.,** Further demonstration of kappa binding sites in the brain: evidence for heterogeneity, *J. Pharmacol. Exp. Ther.*, 232, 144, 1985.
42. **Zukin, R. S., Eghbali, M., Olive, D., Unterwald, E. M., and Tempel, A.,** Characterization and visualization of rat and guinea pig brain κ opioid receptors: evidence for κ_1 and κ_2 opioid receptors, *Proc. Natl. Acad. Sci. U.S.A.*, 85, 4065, 1988.
43. **Han, J.-S. and Xie, C.-W.,** Dynorphin: potent analgesic effect in spinal cord of the rat, *Life Sci.*, 31, 1781, 1982.
44. **Von Voigtlander, P. F., Lahti, R. A., and Ludens, J. H.,** U-50,488H: a selective and structurally novel non-mu (kappa) opioid agonist, *J. Pharmacol. Exp. Ther.*, 224, 7, 1983.
45. **Piercey, M. F., Varner, K., and Schroeder, L. A.,** Analgesic activity of spinally administered dynorphin and ethylketocyclazocine, *Eur. J. Pharmacol*, 80, 283, 1982.
46. **Schmauss, C. and Yaksh, T. L.,** *In vivo* studies on spinal opiate receptor systems mediating antinociception. II. Pharmacological profiles suggesting differential association of *mu*, *delta*, and *kappa* receptors with visceral, chemical and cutaneous thermal stimuli in the rat, *J. Pharmacol. Exp. Ther.*, 228, 1, 1984.
47. **Faden, A. I. and Jacobs, T. P.,** Dynorphin-related peptides cause motor dysfunction in the rat through non-opiate action. *Br. J. Pharmacol.*, 81, 271, 1984.
48. **Caudle, R. M. and Isaac, L.,** Influence of dynorphin(1–13) on spinal reflexes in the rat, *J. Pharmacol. Exp. Ther.*, 246, 508, 1988.
49. **Long, J. B., Rigamonti, D. D., De Costa, B., Rice, K. C., and Martinez-Arizala, A.,** Dynorphin A-induced rat hindlimb paralysis and spinal cord injury are not altered by the κ-opioid antagonist nor-binaltorphimine, *Brain Res.*, 497, 155, 1989.
50. **Takemori, A. E., Ho, B. Y., Naeseth, J. S., and Portoghese, P. S.,** Norbinaltorphimine, a highly selective kappa-opioid antagonist in analgesic and receptor binding assays, *J. Pharmacol. Exp. Ther.*, 246, 255, 1988.

51. **Faden, A. I.,** Opioid and nonopioid mechanisms may contribute to dynorphin's pathophysiological actions in spinal cord injury, *Ann. Neurol.*, 27, 67, 1990.
52. **Bakshi, R. and Faden, A. I.,** Competitive and non-competitive NMDA antagonists limit dynorphin-A induced rat hindlimb paralysis, *Brain Res.*, 507, 1, 1990.
53. **Massardier, D. and Hunt, P. F.,** A direct non-opiate interaction of dynorphin-(1–13) with the N-methyl-D-aspartate (NMDA) receptor, *Eur. J. Pharmacol.*, 170, 125, 1989.
54. **Caudel, R. M. and Isaac, L.,** A novel interaction between dynorphin (1–13) and an N-methyl-D-aspartate site, *Brain Res.*, 443, 329, 1988.
55. **Isaac, L., Van Zandt O'Malley, T., Ristic, H., and Stewart, P.,** MK-801 blocks dynorphin A(1–13)-induced loss of the tail flick reflex in the rat, *Brain Res.*, 531, 83, 1990.
56. **Stewart, P. and Isaac, L.,** A strychnine-sensitive site is involved in dynorphin-induced paralysis and loss of the tail-flick reflex, *Brain Res.*, 543, 322, 1991.
57. **Russel, N. J. W., Schaible, H.-G., and Schmidt, R. F.,** Opiates inhibit the discharges of fine afferent units from inflamed knee joint of the cat, *Neurosci. Lett.*, 76, 107, 1987.
58. **Fleetwood-Walker, S. M., Hope, P. J., Mitchell, R., El-Yassir, N., and Molony, V.,** The influence of opioid receptor subtypes on the processing of nociceptive inputs in the spinal dorsal horn of the cat, *Brain Res.*, 451, 213, 1988.
59. **Parsons, C. G. and Headley, P. M.,** On the selectivity of intravenous μ- and κ-opioids between nociceptive and non-nociceptive reflexes in the spinalized rat, *Br. J. Pharmacol.*, 98, 544, 1989.
60. **Haley, J., Ketchum, S., and Dickenson, A.,** Peripheral κ-opioid modulation of the formalin response: an electrophysiological study in the rat, *Eur. J. Pharmacol.*, 191, 437, 1990.
61. **Knox, R. J. and Dickenson, A. H.,** Effects of selective and non-selective κ-opioid receptor agonists on cutaneous C-fibre-evoked responses of rat dorsal horn neurones, *Brain Res.*, 415, 21, 1987.
62. **Hylden, J. L. K., Nahin, R. L., Traub, R. J., and Dubner, R.,** Effects of spinal kappa-opioid receptor agonists on the responsiveness of nociceptive superficial dorsal horn neurons, *Pain*, 44, 187, 1991.
63. **Sullivan, A. F. and Dickenson, A. H.,** Electrophysiologic studies on the spinal antinociceptive action of kappa opioid agonists in the adult and 21-day-old rat, *J. Pharmacol. Exp. Ther.*, 256, 1119, 1991.
64. **Parsons, C. G. and Headley, P. M.,** Spinal antinociceptive actions of μ- and κ-opioids: the importance of stimulus intensity in determining 'selectivity' between reflexes to different modalities of noxious stimulation, *Br. J. Pharmacol.*, 98, 523, 1989.
65. **Headley, P. M., Parsons, C. G., and West, D. C.,** Comparison of mu, kappa and sigma preferring agonists for effect on spinal nociceptive and other responses in rats, *Neuropeptides*, 5, 249, 1984.

66. **Hope, P. J., Fleetwood-Walker, S. M., and Mitchell, R.,** Distinct antinociceptive actions mediated by different opioid receptors in the region of lamina I and laminae III-V of the dorsal horn of the rat, *Br. J. Pharmacol.*, 101, 477, 1990.
67. **Herrero, J. F. and Headley, P. M.,** The effects of sham and full spinalization on the systemic potency of μ- and κ-opioids on spinal nociceptive reflexes in rats, *Br. J. Pharmacol.*, 104, 166, 1991.
68. **Parsons, C. G., West, D. C., and Headley, M.,** Spinal antinociceptive actions and naloxone reversibility of intravenous μ- and κ-opioids in spinalized rats: potency mismatch with values reported for spinal administration, *Br. J. Pharmacol.*, 98, 533, 1989.
69. **Alzheimer, C. and Bruggencate, G. T.,** Nonopioid actions of κ- opioid receptor agonists, U-50488H and U-69593, on electrophysiological properties of hippocampal CA3 neurons *in vitro*, *J. Pharmacol. Exp. Ther.*, 255, 900, 1990.
70. **Allerton, C. A., Smith, J. A. M., Boden, P. R., Hunter, J. C., Hill, R. G., and Hughes, J.,** Correlation of ontogeny with function of [^{3}H]U69593 labelled κ opioid binding sites in the rat spinal cord, *Brain Res.*, 502, 149, 1989.
71. **James, I. F., Bedttaney, J., Perkins, M. N., Ketchum, S. B., and Dray, A.,** Opioid receptor ligands in the neonatal rat spinal cord: binding and in vitro depression of the nociceptive responses, *Br. J. Pharmacol.*, 99, 503, 1990.
72. **Hernández, A., Paelie, C., Pérez, H., Pelissier, T., and Soto-Moyano, R.,** κ-Opioid receptor-mediated depression of activity evoked in convergent dorsal horn cells by thermal and non-thermal noxious stimulation, *Eur. J. Pharmacol.*, 186, 323, 1990.

3 The Antianalgesic Effect of κ-Agonists

I. ANTIANALGESIC EFFECT OF κ-AGONISTS IN ACUTE EXPERIMENTS

The ability of dynorphin to antagonize the analgesic effect of morphine was described by one group already in the second year of "dynorphin research".[1,2] This surprising finding was subsequently confirmed and extended in several other laboratories. As summarized in Table 1, the related data are almost unequivocal but not numerous. Obviously, it was not easy to accept that dynorphin, the *in vitro* highly active endogenous opioid, behaves as an opiate antagonist in analgesia assays. Furthermore, as reviewed below, the data regarding the mechanism of this action are highly controversial.

As shown in Table 1, in drug-naive, i.e., morphine-nontolerant animals, i.c.v.-injected dynorphin inhibits the analgesic action of morphine whether the latter also is given i.c.v.[1,3] or systemically (s.c.).[2] Intrathecally administered dynorphin also inhibits the antinociceptive action of morphine given either i.c.v.[4] or s.c.[5] When both were injected intrathecally, synergism was recorded in one study,[3] but antagonism in other laboratories.[6,7] No explanation can be given for this contradiction, but the majority of data confirm that opiates and dynorphin antagonize each other's effects even if both are injected intraspinally. For example, in a related electrophysiological study the inhibition by local iontophoresis of morphine of evoked responses of nociceptive neurons in the dorsal horn of spinal cord was reversed also by phoretically applied dynorphin or U-50488H.[8] Intraspinally administered dynorphin antiserum enhances the analgesic effect of i.c.v.-injected U-50488H, morphine, and (D-Ala^2,NMePhe^4,Gly^5-ol)-enkephalin (DAGO).[9]

These findings have been extended by other studies showing that activation of κ-receptors by synthetic agonists also results in a decrease in pain threshold, i.e., antianalgesic effect. Injecting them i.p., synthetic κ-agonists as U-50488H and bremazocine, but not tifluadom, antagonizes the analgesic effect of morphine.[10] Intravenous infusion of ethylketazocine induces hyperalgesia in decerebrated dogs, which is reversible by tran-

TABLE 1
Effect of κ-Agonists on Opiate Analgesia

Animals examined	Experimental paradigm	Observed effect	Ref.
Mouse	Modulation by i.c.v.-injected dynorphin$_{1-13}$ of analgesia (tail flick) induced by		1
	Morphine (i.c.v.)	Inhibition	
	β-Endorphin (i.c.v.)		
	In low doses	Potentiation	
	In high doses	Inhibition	
	FK-33824[a] (i.c.v.)	No effect	
Drug-naive mouse	Modulation by i.c.v.-injected dynorphin$_{1-13}$ of analgesia (tail flick) induced by		2
	morphine (s.c.)	Inhibition	
	β-endorphin (i.c.v.)	Inhibition	
	FK-33824 (s.c.)	No effect	
	DADLE[b] (i.c.v.)	No effect	
Morphine tolerant mouse	As above		
	Morphine (s.c.)	Potentiation	
	β-Endorphin	Potentiation	
Rat	Effect of des-Tyr-dynorphin (given i.c.v.) on morphine (s.c.)-induced analgesia	Inhibition	15
	Effect of dynorphin$_{1-13}$ on morphine-induced analgesia administering them		3
	intrathecally	Synergism	
	i.c.v.	Antagonism	
Drug naive rat	Interaction of intrathecal dynorphin and morphine in tail flick test	Antagonism	6
Morphine-tolerant rat	As above	Synergism	
Rat	Modulation of the inhibition by DAGO[c] of C- and Aβ-fiber-evoked responses of nociceptive neurons in the spinal cord (dorsal horn) upon local (phoretic) administration of		8
	Dynorphin$_{1-13}$	Reversal	
	U-50488H	Reversal	
Mouse	Effect of intrathecal dynorphin$_{1-17}$ on analgesia induced by morphine also injected intrathecally	Antagonism	7
	Modulation by intrathecally applied dynorphin$_{1-17}$ of analgesia induced by i.c.v.		4
	Morphine	Antagonism	
	β-Endorphin	Antagonism	
	DAGO	Antagonism	
	DPDPE[d]	Antagonism	
	U-50488H	Antagonism	

TABLE 1 (continued)
Effect of κ-Agonists on Opiate Analgesia

Animals examined	Experimental paradigm	Observed effect	Ref.
	Effect of intrathecally applied dynorphin antiserum on analgesia induced by i.c.v. application of		9
	U-50488H	Potentiation	
	DAGO	Potentiation	
	Morphine	Moderate enhancement	
	β-Endorphin	No effect	
	DPDPE	No effect	
Mouse	Effect of intrathecal dynorphin$_{1-17,1-13,1-8}$ on the analgesic effect of morphine given s.c.	Inhibition: rightward shift of the dose-response curve	5
Decerebrated dog	Effect of ethylketazocine (i.v.) on pain threshold	Hyperalgesia reversible by spinal cord transection	11
Dog	Effect of ethylketazocine infused into the fourth cerebral ventricle on pain threshold	Hyperalgesia	12
Drug-naive rat	Modulation of morphine (i.p.) analgesia by i.p.		10
	Tifluadom	No effect	10
	U-50488H	Antagonism	
	Bremazocine	Antagonism	
Morphine tolerant rat	As above		
	Tifluadom	No effect	
	U-50488H	Potentiation	
	Bremazocine	Potentiation	
Rat	Effect of medical thalamic injection of U-50488H on pain threshold	Hyperalgesia	13

[a] (D-Ala^2,NMePhe^4,Met(O)^5)-enkephalin-ol.
[b] (D-Ala^2,D-Leu^5)-enkephalin.
[c] (D-Ala^2,NMePhe^4, Gly^5)-enkephalin-ol.
[d] (D-Pen^2,D-Pen^5)-enkephalin.

section of the spinal cord.[11] Hyperalgesia can be elicited by infusing ethylketazocine into the fourth ventricle[12] or U-50488H into the medial thalamus.[13] Summarily, these data indicate the existence of dynorphin/κ-receptor-mediated antianalgesic mechanisms in the brain and even in the spinal cord. Nevertheless, the intricate mechanism of this antianalgesic effect is far from clarified yet.

A. The Effects of Short Fragments

Shorter fragments of dynorphin, e.g., dynorphin$_{1-9}$, were found inactive in one study,[1] but dynorphin$_{1-8}$ in another one[5] proved to be as active as dynorphin$_{1-17}$ in antagonizing the effect of morphine. Since shorter fragments of dynorphin, e.g., Leu-enkephalin, are not selective for κ-receptors,[14] the involvement of other types of receptors cannot be excluded. Actually, the effect might be independent of opiate receptors in general, since in one study,[15] which has not been repeated yet, des-Tyr-dynorphin, i.e., a nonopiate fragment, was also claimed to attenuate the analgesia induced by systemic morphine. On the other hand, injecting them intrathecally, both naloxone, i.e., nonselective opiate antagonist, and nor-binaltorphimine, a κ-selective opiate antagonist, reversed the morphine-antagonist effect of dynorphin$_{1-17}$ in mice.[7] Similarly, reversal of analgesia was seen with naloxone and nor-binaltorphimine (but not by the δ-selective naltindrole) if morphine was injected systemically.[5] Furthermore, the hyperalgesia induced by i.v. ethylketazocine in decerebrated dogs was found naloxone reversible.[11] Thus, primarily κ-receptors might mediate the antianalgesic effect of both dynorphin and its synthetic analogs.

In one related study,[7] smaller doses of naloxone or nor-binaltorphimine were needed to reverse the antianalgesic effect of dynorphin than to reverse the analgesic effect of U-50488H (injecting all these drugs intrathecally). Thus the possibility arises that the antianalgesic effect of dynorphin might be mediated by a subpopulation of κ-receptors, at least in the spinal cord. Of course, such a proposition needs further experimental data. Anyhow, the data on the multiplicity of κ-opiate receptors[16–19] and the data in Chapter 2 on the analgesic effect of intrathecally or systemically injected κ-agonists indirectly support such a proposition.

B. κ-Receptors and Location

A related issue is the anatomical location of κ-receptors responsible for the purported antianalgesic action. The data summarized in Table 1 indicate that κ-receptors might induce an antianalgesic effect at both spinal and multiple supraspinal levels. However, as discussed in Chapter 2, κ-agonists also have analgesic effect if given spinally, but not supraspinally. Martin's group[20] postulated the existence of a medullary hyperalgesic center, since hyperalgesia can be elicited by injecting ethylketazocine into the fourth ventricle[12] or the similar effect of its i.v. application in decerebrated dogs is reversible by spinal transection.[11] Actually, Fujimoto et al.[9] suppose the existence of an "antianalgesic descending system" which releases dynorphin in the spinal cord. In a recent study, Hutchison et al.[21] demonstrated that there is basal release of dynorphin in the lamina I region of spinal cord in the cat which is enhanced upon electrical stimulation of unmyelinated afferent fibers, i.e., upon noxious stimuli, and abolished by spinal cord transection. Putting these data together, the activation of the κ-receptors supraspinally seems to activate a pathway which releases dynorphin in the spinal cord and induces hyperalgesia. On the other hand, a subpopulation of κ-receptors, intrinsic to the spinal cord, apparently elicit "analgesia" or at least inhibition of nociceptive polysynaptic reflexes activated by nonthermal nociceptive stimuli. However, as speculative as these propositions are,[9,20] the data reviewed in Chapter 2 and in this one are consistent with such a proposal.

C. Further Considerations

A further puzzling finding is that dynorphin only inhibits unequivocally the antinociception induced by morphine, but the data regarding its effect on the analgesic actions of β-endorphin, μ-, and δ-selective enkephalin analogs as FK-33824, DAGO, and (D-Ala^2,D-Leu^5)-enkephalin (DADLE), respectively, are controversial.[1,2,4,9]

Therefore, it has been proposed that some opiates such as morphine, and also the μ-preferring DAGO activate the descending antianalgesic pathways, whereas other analgesics such as β-endorphin or (D-Pen^2,D-Pen^5)-enkephalin (DPDPE) do not.[9] These puzzling data are in part explained and in part further complicated by the observations that, at least partially different descending pathways, mediate the analgesic effects of morphine and β-endorphin. Mapping the sensitive areas in the brain stem, analgesia can be elicited from different structures by morphine and β-endorphin, respectively.[22,23] β-Endorphin and DAGO are the most effective if microinjected into the caudal medulla (raphe pallidus nucleus, raphe obscurus nucleus, and adjacent areas), whereas morphine acts preferentially in the rostral ventromedial medulla (raphe magnus nucleus, gigantocellular reticular nucleus, and neighboring reticular areas).[22,23] The analgesia elicited by brain stem microinjection of β-endorphin is accompanied by Met-enkephalin release from the spinal cord.[22,24] Thus, it can activate descending enkephalinergic neurons.[25,26] On the other hand, the antinociception elicited by injecting morphine into the periaqueductal gray (a main relay station of descending pathways) or gigantocellular reticular nucleus (the area most sensitive to it) can be reversed by intrathecally injected α_2-adrenergic and serotonin antagonists as yohimbine and methysergide,[23,27,28] and it does not induce Met-enkephalin release from the spinal cord.[22,24] Tseng and co-workers attribute this difference to activation by β-endorphin of the putative ϵ-receptors.[29,30] However, the existence of ϵ-receptors in the brain has never been substantiated. Therefore, the differences in their modes of action can only be explained by differential involvement of various subpopulations of opiate receptors. Morphine is a rather selective ligand of μ-receptors,[14,31] whereas β-endorphin shows nearly identical affinity to μ- and δ-receptors.[14,32] Hence the difference might only be explained by supposing that morphine acts practically only via the μ-receptors, whereas both μ- and δ-receptors are involved in the analgesic effect of i.c.v.-applied β-endorphin and the enkephalin analogs listed in Table 1.

The situation is further complicated by the data on the intricate, but unclarified interrelation among corticotropin-releasing factor (CRF), κ-receptors, and stress reactions. Of the various subtypes of opiate agonists, κ-agonists seem to be the most potent inducers of stress reactions.[33–35] For example, U-50488H has been found to elevate the serum corticosterone level in lower doses than needed to elicit any behavioral action.[35] The first hormone in the neurohumoral chain of events, called stress reaction, CRF has been found to induce analgesia upon systemic,[36,37] but not after i.c.v. application.[38] According to a recent study,[39] CRF induces analgesia if administered intrathecally, and this effect can be reversed by nor-binaltorphimine. Thus the antianalgesic effect of κ-agonists might be masked by the analgesic effect of CRF, liberated due to their dysphoric action.

It is probably relevant to mention here that systemically administered morphine induces hyperalgesia in some strains of domestic fowl.[40] Since under special experimental conditions morphine also interacts with κ-receptors,[41,42] the phenomenon might be mediated by them. However, the hyperalgesic effect of morphine in fowls can be reversed both by β-funaltrexamine and nor-binaltorphimine, i.e., via μ- and κ-receptor blockade as well.[43]

Finally, it is worth mentioning here that the existence of "peripheral analgesic mechanisms" has also been demonstrated. Local administration of opiates reverses the PGE- and bradykinin-induced hyperalgesia of the rat paw,[44,45] antagonizes the pain induced by chemical stimulation of the peritoneum of mice,[46] and induces elevation of the pain threshold in the rat paw.[47,48] In these models of peripheral analgesia there is no antagonism between κ-receptors and other types of opiate receptors; κ-, δ-, and μ-agonists all induce real naloxone-reversible analgesia and their effectiveness is by and large proportional with their efficacy upon systemic administration.[45,46,48] Thus, the antianalgesic effect of κ-agonists can only be detected if administering them into the central nervous system, but not in the periphery. This is a surprising finding in view of the activation of dynorphinergic system in some models of peripheral inflammation (vide infra).

II. SYNERGISM BETWEEN κ- AND μ-AGONISTS IN OPIATE-TOLERANT ANIMALS

As summarized in Table 1, both dynorphin[2,6] and synthetic κ-agonists[10] enhance the analgesic effect of morphine and β-endorphin in morphine-tolerant animals. A broad array of data confirm these findings, indicating that in morphine-tolerant subjects the antagonism between κ- and μ-agonists is replaced by synergism. Furthermore, in sufentanyl-tolerant rats the "catalepsy" induced by $dynorphin_{1-13}$ (given i.c.v.) was also enhanced.[49]

In line with the above mentioned initial findings in morphine-dependent monkeys, $dynorphin_{1-13}$, injected i.v., has been found capable of suppressing the withdrawal signs, i.e., it substituted for morphine.[50] In a yet unconfirmed study, $dynorphin_{1-13}$ suppressed the withdrawal in heroin addicts as well.[51] In rats, $dynorphin_{1-13}$ inhibits the withdrawal if injected i.v. or spinally, but it is inactive upon i.c.v. application.[52]

The data relating to the interaction of κ- and μ-agonists in tolerance development are highly controversial. Repeated treatments with U-50488H, resulting in tolerance to its analgesic and hypothermic actions, does not modify the morphine-induced analgesia, hyperthermia, and catalepsy in rats.[53] Furthermore, in some studies,[54,55] coadministration of U-50488H failed to modify the development of tolerance to the analgesic and hyperthermic effects of morphine. Nevertheless, in other laboratories, U-50488H blocked the development of tolerance to the analgesic effect of morphine in rats if they were coadministered.[56] Examining various routes of administration, i.p. coadministration of U-50488H suppressed the tolerance development to s.c. or i.t., morphine without modifying the tolerance development to i.c.v. morphine.[57] Furthermore, the suppressant effect of U-50488H can be abolished by nor-binaltorphimine given i.p. or i.t. but not by i.c.v. application.[57] Thus the site of "synergistic" interaction upon chronic application of κ- and μ-agonists might be the spinal cord.

Nevertheless, there are more peripheral sites of synergism as well. *In vitro,* in morphine-tolerant guineapig ileum, $dynorphin_{1-13}$ is capable of restoring the sensitivity of the preparation, though in acute experiments these agents do not modify the effect of each other.[58]

The situation is further complicated by data on antagonism between dynorphin and morphine (in morphine-tolerant animals) in experimental conditions where synergism is seen in acute experiments. Dynorphin enhances the morphine-elicited respiratory depression and hypothermia in drug-naive animals, but antagonism was seen in both aspects in tolerant subjects.[59]

For the time being, no explanation can be given for the findings that κ- and μ-receptor-mediated mechanisms are mostly antagonistic in drug-naive animals, but synergistic in opiate-tolerant subjects. It is plausible to ask whether or not these changes are due to alterations at the receptor level. Bhargava et al.[53] demonstrated that chronic administration of U-50488H resulted in downregulation of κ-receptors in various brain regions, but striatum and the B_{max} of δ-receptors decreased in the spinal cord. In addition, it is well known[60] that in the state of opiate tolerance and withdrawal there is a general hypersensitivity of the dopamine system.

Since dopamine is involved in both the analgesic and behavioral effects of classical opiates (for review see Székely)[61], among others, altered interaction among the dopamine, κ-, and μ-opiate systems might account for the above mentioned seemingly paradoxical findings.

Some behavioral experiments indirectly indicate eventual changes in receptor sensitivity. In one study,[62] selective blockade of the μ-receptors by β-funaltrexamine resulted in enhancement of the analgesic effect of U-50488H in mice. In another study the antianalgesic effect of intrathecally administered dynorphin$_{1-17}$ was abolished in mice pretreated with a single high dose of morphine.[4] The dose of morphine used in this study was high enough to induce acute tolerance[63] and a decrease of the pA_2 of naloxone against morphine, i.e., the enhanced sensitivity of μ-receptors to opiate antagonists.[64] Furthermore, administering naloxone in a single high dose a couple of hours or days after morphine injection (giving the latter also in a single high dose) develops protracted tolerance to both morphine and i.c.v.-, but not i.t.-administered U-50488H.[65] Thus eventual changes in the density and/or sensitivity of μ- or κ-receptors might account for the above reviewed changes in interactions.

In addition, alterations in the relative contributions of various brain regions to the analgesic actions of μ- and/or κ-agonists might also explain some of these effects. Namely, it is well established that in drug-naive animals simultaneous spinal and supraspinal microinjections of morphine results in more intense analgesia then the sum of the analgesic effects attainable by single injections.[66] This kind of synergism, called "multiplicative interaction" between spinal and supraspinal sites, is lost in morphine-tolerant animals.[67] In a more recent study using more selective ligands, fentanyl for the μ- and DADLE for the δ-selective ones, the additive interaction seen in control animals turned to antagonistic interaction for fentanyl and multiplicative for DADLE in morphine-tolerant mice.[68] Thus the eventual changes in receptor density and sensitivity might in turn change the interaction among the various brain areas mediating the analgesic, and eventually antianalgesic effects of drugs under study. Of course, the exact delineation of the changes turning the antagonism in drug-naive animals to synergism in morphine-tolerant ones requires further laborious studies.

III. HYPERALGESIA AND ENHANCED DYNORPHINERGIC TONE IN EXPERIMENTALLY INDUCED (POLY)ARTHRITIS

It is well established that the dorsal horn of spinal cord contains dynorphin in relatively high concentration. The density is the highest in laminae I–II and also considerable in laminae V–VI.[69–73] The former laminae constitute the main input area of peripheral afferents carrying nociceptive information. As demonstrated by Hutchison et al.[21] in cats, electrical stimulation of unmyelinated primary afferent fibers results in enhanced release of dynorphin in the region of lamina I, and to a lesser degree in laminae V–VI, whereas the stimulation of myelinated efferents (including A_{δ} ones) has no effect. The basal release (but not the noxious stimulation-evoked one) is abolished by spinalization.[21] These experiments provide direct evidence for the existence of an ascending dynorphinergic system and also prove the role of dynorphin in pain perception in the spinal cord. Of course, the presence of dynorphin along the main pain pathways and its release upon painful stimuli does not necessarily prove that it is an algogenic agent. Hypothetically, it might be an endogenous agent, which, similarly to enkephalins,[74] inhibits the release of substance P and other known mediators of pain. However, a nearly unequivocal line of evidence based on the same experimental model proves that it is not the case.

Probably the best model of dynorphin-mediated hyperalgesia is just a kind of experimentally induced polyarthritis in rats[75] which has been used for a long time in pharmacological examination of analgesic and anti-inflammatory drugs. Using this model, complete Freund adjuvant (CFA) (suspension of killed *Mycobacterium butyricum* in paraffin oil) is injected either unilaterally into the hindpaw (intracutaneously) or the phlogistic agent is inoculated at the base of the tail. It is well known that in rats, sensitized by CFA, chronic inflammation (monoarthritis) develops first at the site of injection (during the first week), which spreads to the contralateral side (second week), then affects the whole body (third week), peaks about 18 to 21 d after inoculation, then the symptoms spontaneously recede in 2 to 4 weeks. The functional changes accompanying the process were difficult to interpret before realizing that dynorphin is an algogenic agent. Actually, as summarized below, the whole process might be regarded a simple ''spinal'' model of drug addiction.

The inflammation is accompanied with its classical manifestations (hyperemia; swelling), and hyperalgesia also is observable. There might be nothing peculiar about hyperalgesia of inflamed tissue, since many known (and unknown) algogenic agents participate in the genesis of inflammation. Nevertheless, as described carefully in detail by Millan and co-workers,[76,77] this type of pain is unusual. This kind of hyperalgesia means primarily a decrease in the threshold of paw pressure, whereas the sensitivity to heat stimuli is reduced. The altered sensitivity, i.e., enhanced responsiveness to pressure and decreased sensitivity to other types of noxious stimuli was also demonstrated in electrophysiological experiments examining nociceptive units in dorsal horn[79] and ventrobasal thalamic neurons of rats. Examining the inflamed paw, the analgesic effect of μ- and δ-agonists as morphine, DAGO, and DADLE is enhanced, whereas that of U-50488H is decreased, but no changes in sensitivity can be detected on the untreated side.[76,80] In arthritic animals neither naloxone, nor ICI-

154,159, a selective antagonist of δ-receptors, modified significantly the pressure threshold, but MR-2266, a κ-preferring agonist, further reduced the pain threshold.[76,80] Surprisingly, if naloxone was administered continuously by osmotic minipump, the swelling of the inflamed paw was reduced and the decrease in pain threshold seen in control (untreated arthritic) rats in acute experiments tapered off.[81] In addition, animals receiving continuous naloxone infusion did not show the usual behavioral manifestation of hyperalgesia as vocalization, aggression, immobility, hyperventilation, etc.[81]

In this model, besides the nonspecific signs of pathological state (reduced body weight, altered diurnal rhythmicity of food intake, and core temperature), some symptoms, such as increase in adrenal weight, higher β-endorphin level in the anterior pituitary and the plasma, indicate activation of the pituitary-adrenal system, i.e., stress.[76,77,82] Nearly all these observations can be explained by supposing that

- The hyperalgesia (and the inflammation itself) is aggravated by excessive release of endogenous substances enhancing the pain sensitivity, especially to nonthermal noxious stimuli.
- These endogenous substances are expected to be stressful and aversive.
- The functional preponderance of these algogenic mechanisms render the organism more responsive to classical μ-opiates, such as morphine.

Obviously, excessive dynorphin release might explain hyperalgesia, especially to nonthermal stimuli as discussed in the previous sections of this Chapter and Chapter 2. The beneficial effect of chronic naloxone can be attributed to antagonism of the dynorphin system which prevails over the endorphin/enkephalin-mediated ones. Furthermore, the stressful and aversive character of κ-agonists is discussed in detail in Chapter 4. In addition, a functional hyperreactivity of analgesic mechanisms under the control of μ-receptors can be expected at the time of dynorphin overproduction.

Using radioimmunoassay during the first week of inflammation, a rise in the level of immunoreactive dynorphin was observed, but this increase was restricted to the dorsal right quadrant of the spinal cord ipsilateral to the inflamed paw.[77,82] According to immunocytochemical studies, the increase can be seen primarily in laminae I–II.[83–85] A higher dynorphin level might reflect either enhanced production or decreased utilization. However, *in situ* hybridization and immunocytochemistry identified an increase in transcription of preprodynorphin mRNA in laminae I–II, V–VI, and dorsolateral to the central canal in the lumbar segment receiving innervation from the inflamed paw.[78,84,86,87] Thus, the higher dynorphin level indicates functional hyperactivity and not decreased release/utilization. Both these immunocytochemical and cDNA hybridization studies indicate that the dynorphin content is multiplied in laminae I–II. However, it is not clarified yet whether the number of neurons expressing dynorphin is also enhanced or only dynorphin synthesis is increased above the detection level in many neurons.[85] According to cDNA hybridization studies, the transcription of dynorphin is enhanced 24 h after inoculation of CFA and it is maximal (about six- to eightfold elevation) on days 2 to 5.[82] The dynorphin-containing neurons in lamina I–II are of Golgi type II, i.e., they have rather short axons and are confined to the same segment.[84] Nevertheless, the dynorphin content is increased in both local circuit and projection neurons.[85] The involvement of laminae V–VI also indicates that enhanced nociceptive input might activate not only intraspinal reflexes, but the message is also conveyed rostrally.[87] Actually, dynorphin-containing cells in lamina I project to the nucleus of the solitary tract and connections have been detected to the lateral reticular nucleus and mesencephalon as well.[88] Furthermore, a rise of immunoreactive dynorphin was observed in polyarthritic rats, even at the thalamic level. The mechanism of enhanced expression of dynorphin in polyarthritic rats has not been clarified yet. Nevertheless, there are certain analogies which seem to be helpful. Activation of small-diameter cutaneous sensory afferents by noxious heat or chemical stimuli or electrical stimulation results in rapid appearance of c-fos-proteinlike immunoreactivity in the superficial layers of the dorsal horn (layers I, II, V, and X).[89] It has also been suggested that proto-oncogenes c-fos and c-myc participate in the process leading to prolonged enhancement of genetic expression.[90,91] The increase in mRNA level of c-fos mRNA has been detected in this model, too, and it was followed in several hours by enhancement of dynorphin content.[92] Actually, in many dorsal horn neurons c-fos protein, dynorphin, and enkephalins are colocalized.[87] Thus, probably the permanent painful stimulation, induced by CFA, results in genetic alteration, leading to enhanced expression of dynorphin. Actually, Barthó et al.[93] demonstrated recently that desensitization of capsaicine-sensitive neurons by a large dose of capsaicine before injection of CFA abolished the enhanced sensitivity to morphine-induced analgesia of inflamed paw (without modifying the other symptoms of inflammation). Since capsaicine affects selectively the chemosensitive C-fiber neurones,[94] these might initiate the process resulting in overproduction of dynorphin.

It is probably logical to mention here that similar elevation of the spinal cord dynorphin content was detected in another model of chronic pain. The dynorphin level was enhanced in laminae I–II and V–III in the corresponding segments of spinal cord of rats upon inducing painful neuropathy by constriction by ligature of the sciatic nerve.[95]

On the other hand, the level of Met- and Leu-enkephalin started to increase in the same region of spinal cord at 5 weeks, i.e., at the time of spontaneous recovery.[77] In hybridization studies the enhanced transcription of dynorphin also preceded the enhanced transcription of enkephalins.[78] The very fact that elevated dynorphin level coincides in time with the generalization of inflammation, whereas Met- and Leu-enkephalins are released in the phase of recovery, indicates that the former ones are involved in the pathology, whereas the latter ones might contribute to the spontaneous reversal of the process.

A careful examination of the receptor density in the afflicted regions of the spinal cord revealed a moderate decrease in the B_{max} of κ-receptors,[77] but the density of other subclasses of opiate receptors (μ- and δ-ones) apparently remained the same.[77,78] Obviously, enhanced release of dynorphin might result in secondary downregulation of the κ-receptors.

The weight loss and altered diurnal rhythmicity in food intake are analogous to changes seen in rats exposed to recurrent footshock[86] and also seen in humans in chronic pain conditions. Thus the weight loss might be due to liberation of CRF, which also is a powerful anoretic agent[96] and supposed to mediate the stress-induced anorexia in humans.[97] Interestingly, in arthritic animals the analgesic effect of clonidine, an α_2-adrenergic receptor agonist, was also enhanced.[80] Thus it was proposed that during inflammation the opiate-activated descending pain-inhibitory pathway is more active since it is well established that noradrenaline release mediates the analgesia brought about by the antinociceptive effect of this pathway.[28]

Some data indicate that this form of chronic pain leads to changes of motivational behavior as well. Providing them a free choice between water and drug solutions, arthritic rats were found to consume more suprofen and fentanyl than healthy controls.[98,99] Since both suprofen and fentanyl have analgesic action, their enhanced spontaneous intake might be simply a conditioned phenomenon. However, in another study[100] the self-administration of morphine in previously trained rats significantly decreased if adjuvant arthritis was induced; it then returned to the initial level as the pathology waned. Furthermore, using the conditioned place preference for measuring the reinforcing property of the drug under study (for details see Chapters 4 and 5) in arthritic rats the rewarding potency of morphine was not modified, but the aversive effect of a κ-agonist, U-69593, was lost.[101] It is tempting to speculate that the latter findings are related to the above mentioned enhanced sensitivity to morphine and downregulation of κ-receptors upon excessive dynorphin release. Of course, it is only a conjecture.

However, summarily these data show that chronic arthritis is the first animal model where pain, distress, enhanced dynorphin production, and spontaneous intake of an addictive substance[98] seem to coincide and the motivational character of μ- and κ-agonists are also altered.[100,101] It is an important finding, since, as discussed in the subsequent chapters of this monograph, analogous alterations can be detected upon drug addiction in the brain as well.

REFERENCES

1. **Friedman, H. J., Jen, M.-F., Chang, J.-K., Lee, N. M., and Loh, H. H.,** Dynorphin: a possible modulatory peptide on morphine or β-endorphin analgesia in mice, *Eur. J. Pharmacol.*, 69, 357, 1981.
2. **Tulunay, F. C., Jen, M.-F., Chang, J.-K., Loh, H. H., and Lee, N. M.,** Possible regulatory role of dynorphin on morphine and β-endorphin-induced analgesia, *J. Pharmacol. Exp. Ther.*, 219, 296, 1981.
3. **Ren, M. F., Lu, C. H., and Han, J. S.,** Dynorphin-A-(1–13) antagonizes morphine analgesia in the brain and potentiates morphine analgesia in the spinal cord, *Peptides*, 6, 1015, 1985.
4. **Fujimoto, J. M. and Holmes, B.,** Systemic single dose morphine pretreatment desensitizes mice to the spinal antianalgesic action of dynorphin A (1–17), *J. Pharmacol. Exp. Ther.*, 254, 1, 1990.
5. **Song, Z. H. and Takemori, A. E.,** Involvement of spinal kappa opioid receptors in the antagonistic effect of dynorphin on morphine antinociception, *Life Sci.*, 48, 1447, 1991.
6. **Schmauss, C. and Herz, A.,** Intrathecally administered dynorphin-(1–17) modulates morphine-induced antinociception differently in morphine-naive and morphine-tolerant rats, *Eur. J. Pharmacol.*, 135, 429, 1987.
7. **Fujimoto, J. M. and Arts, K. S.,** Clonidine, administered intracerebroventricularly in mice, produces an anti-analgesic effect which may be mediated spinally by dynorphin A (1–17), *Neuropharmacology*, 29, 351, 1990.
8. **Dickenson, A. H. and Knox, R. J.,** Antagonism of μ-opioid receptor-mediated inhibition of nociceptive neurones by U50488H and $dynorphin_{1-13}$ in the rat dorsal horn, *Neurosci. Let.*, 75, 229, 1987.
9. **Fujimoto, J. M., Arts, K. S., Rady, J. J., and Tseng, L. F.,** Spinal dynorphin A (1–17): possible mediator of antianalgesic action, *Neuropharmacology*, 29, 609, 1990.
10. **Ramaro, P., Jablonski, H. I., Jr., Rehder, K. R., and Bhargava, H. N.,** Effect of κ-opioid receptor agonists on morphine analgesia in morphine-naive and morphine-tolerant rats, *Eur. J. Pharmacol.*, 156, 239, 1988.
11. **Wu, K. M., Martin, W. R., Kamerling, S. G., and Wettstein, J. G.,** Possible medullary κ-hyperalgesic mechanism. I. A new potential role for endogenous opioid peptides in pain perception, *Life Sci.*, 33, 1831, 1983.
12. **Kamerling, S. G., Martin, W. R., Wu, K. M., and Wettstein, J. G.,** Medullary κ hyperalgesic mechanisms. II. The effects of ethylketocyclazocine administered into the fourth cerebral ventricle of the conscious dog, *Life Sci.*, 33, 1839, 1983.
13. **Carr, K. D. and Bak, T. H.,** Medial thalamic injection of opioid agonists: μ-agonist increases while κ-agonist decreases stimulus threshold for pain and reward, *Brain Res.*, 441, 173, 1988.
14. **Lord, J. A. H., Waterfield, A. A., Hughes, J., and Kosterlitz, H. W.,** Endogenous peptides: multiple agonists and receptors, *Nature*, 267, 495, 1977.

15. **Walker, J. M., Tucker, D. E., Coy, D. H., Walker, B. B., and Akil, H.,** Des-tyrosine-dynorphin antagonizes morphine analgesia, *Eur. J. Pharmacol.*, 85, 121, 1982.
16. **Attali, B., Gouarderes, V., Mazarguil, H., and Cros, J.,** Evidence for multiple kappa binding sites by use of opioid peptides in the guinea pig lumbosacral spinal cord, *Neuropeptides,* 3, 53, 1982.
17. **Su, T.-P.,** Further demonstration of kappa binding sites in the brain: evidence for heterogeneity, *J. Pharmacol. Exp. Ther.*, 232, 144, 1985.
18. **Zukin, R. S., Eghbali, M., Olive, D., Unterwald, E. M., and Tempel, A.,** Characterization and visualization of rat and guinea pig brain κ opioid receptors: evidence for κ_1 and κ_2 opioid receptors, *Proc. Natl. Acad. Sci. U.S.A.*, 85, 4061, 1988.
19. **Clark, J. A., Liu, L., Price, M., Hersh, B., Edelson, M., and Pasternak, G. W.,** Kappa opiate receptor multiplicity: evidence for two U50,488-sensitive κ_1 subtypes and a novel κ_3 subtype, *J. Pharmacol. Exp. Ther.*, 251, 461, 1989.
20. **Martin, W. R., Kumar, S., and Sloan, J. W.,** Opioid and nicotinic medullary hyperalgesic influences in the decerebrated rat, *Pharmacol. Biochem. Behav.*, 29, 725, 1988.
21. **Hutchison, W. D., Morton, C. R., and Terenius, L.,** Dynorphin A: in vivo release in the spinal cord of the cat, *Brain Res.*, 532, 299, 1990.
22. **Tseng, L. F., Tang, R., Stackman, R., Camara, A., and Fujimoto, J. M.,** Brainstem sites differentially sensitive to β-endorphin and morphine for analgesia and release of Met-enkephalin in anesthetized rats, *J. Pharmacol. Exp. Ther.*, 253, 930, 1990.
23. **Tseng, L. F. and Collins, K. A.,** Different mechanisms mediating tail-flick inhibition induced by β-endorphin, DAMGO and morphine from ROb and GiA in anesthetized rats, *J. Pharmacol. Exp. Ther.*, 257, 530, 1991.
24. **Tseng, L. F., Higgins, M. J., Hong, J. S., Hudson, P. M., and Fujimoto, J. M.,** Release of immunoreactive Met-enkephalin from the spinal cord by intraventricular β-endorphin but not by morphine in anesthetized rats, *Brain Res.*, 343, 60, 1985.
25. **Hökfelt, T., Terenius, L., Kuypers, H. G. J. M., and Dan, O.,** Evidence for enkephalin immunoreactive neurons in the medulla oblongata projection to the spinal cord, *Neurosci. Lett.*, 14, 55, 1979.
26. **Beitz, A. J., Clements, J. R., Ecklund, L. J., and Mulletett, M. M.,** The nuclei of brainstem enkephalin and cholecystokinin projections to the spinal trigeminal nucleus of the rat, *Neuroscience*, 20, 409, 1987.
27. **Yaksh, T. L.,** Direct evidence that spinal serotonin and noradrenaline terminals mediate the spinal antinociceptive effects of morphine in PAG. *Brain Res.*, 160, 180, 1979.
28. **Camerata, P. and Yaksh, T. L.,** Characterization of the spinal adrenergic receptors mediating the spinal effects produced by microinjection of morphine into the periaqueductal grey, *Brain Res.*, 336, 133, 1985.
29. **Lemaire, S., Magnan, J., and Regoli, D.,** Rat vas deferens: a specific bioassay for endogenous opioid peptides, *Br. J. Pharmacol.*, 64, 327, 1978.
30. **Schulz, R., Faase, E., Wüster, M., and Herz, A.,** Selective receptors for β-endorphin on the rat vas deferens, *Life Sci.*, 24, 843, 1979.

31. **Martin, W. R., Eades, C. G., Thompson, J. A., Huppler, R. E., and Gilbert, P. E.,** The effects of morphine- and nalorphine-like drugs in the nondependent and opiate-dependent chronic spinal dog, *J. Pharmacol. Exp. Ther.*, 197, 517, 1976.
32. **Rónai, A. R., Székely, J. I., Bajusz, S., and Berzétei, I.,** The in vitro pharmacology of D-Met^2,Pro^5-enkephalinamide, *J. Pharm. Pharmacol.*, 33, 534, 1981.
33. **Lahti, R. A. and Collins, R. J.,** Opiate effects on plasma corticosteroids: relationship to dysphoria and reinforcement, *Pharmacol. Biochem. Behav.*, 17, 107, 1982.
34. **Fuller, R. W. and Leander, J. D.,** Elevation of serum corticosterone in rats by bremazocine, a κ-opioid agonist, *J. Pharm. Pharmacol.*, 36, 345, 1984.
35. **Iyengar, S., Kim, H. S., and Wood, P. L.,** Kappa opiate agonists modulate the hypothalamo-pituitary-adrenocortical axis in the rat, *J. Pharmacol. Exp. Ther.*, 238, 429, 1986.
36. **Wei, E. T., Kiang, J. G., Buchan, P., and Smith, T. W.,** Corticotropin-releasing factor inhibits neurogenic plasma extravasation in the rat paw, *J. Pharmacol. Exp. Ther.*, 238, 783, 1986.
37. **Hargreaves, K., Costello, A., and Dubner, R.,** Corticotropin releasing factor (CRF) produces analgesic and antiinflammatory effects by multiple mechanisms of action, *Soc. Neurosci. Abstr.*, 15, 471, 1989.
38. **Sherman, J. E. and Kalin, N. H.,** ICV-CRF potently affects behavior without altering antinociceptive responding, *Life Sci.*, 39, 433, 1986.
39. **Song, Z. H. and Takemori, A. E.,** Involvement of spinal kappa opioid receptors in the antinociception produced by intrathecally administered corticotropin-releasing factor in mice, *J. Pharmacol. Exp. Ther.*, 254, 363, 1990.
40. **Hughes, R. A.,** Strain-dependent morphine-induced analgesic and hyperalgesic effects on thermal nociception in domestic fowl (Gallus gallus), *Behav. Neurosci.*, 104, 619, 1990.
41. **Takemori, A. E. and Portoghese, P. S.,** Evidence for the interaction of morphine with kappa and delta opioid receptors to induce analgesia in β-funaltrexamine-treated mice, *J. Pharmacol. Exp. Ther.*, 243, 91, 1987.
42. **Goldstein, A. and Naidu, A.,** Multiple opioid receptors: ligand selectivity profiles and binding site signatures, *Mol. Pharmacol.*, 36, 265, 1989.
43. **Sufka, K. J., Hughes, R. A., and Giordano, J.,** Effect of selective opiate antagonists on morphine-induced hyperalgesia in domestic fowl, *Pharmacol. Biochem. Behav.*, 38, 49, 1991.
44. **Ferreira, S. H., Molina, N., and Vettore, O.,** Prostaglandin hyperalgesia V: a peripheral analgesic receptor for opiates, *Prostaglandins*, 23, 53, 1982.
45. **Taiwo, Y. O. and Levine, J. D.,** K- and δ-opioids block sympathetically dependent hyperalgesia, *J. Neurosci.*, 11, 928, 1991.
46. **Bentley, G. A., Newton, S. H., and Starr, J.,** Evidence for an action of morphine and enkephalins on sensory nerve endings in the mouse peritoneum, *Br. J. Pharmacol.*, 73, 325, 1981.
47. **Stein, C., Millan, M. J., Yassouridis, A., and Herz, A.,**Antinociceptive effects of μ- and κ-agonists in inflammation are enhanced by a peripheral opioid-receptor-specific mechanism, *Eur. J. Pharmacol.*, 155, 255, 1988.

48. **Stein, C., Millan, M. J., Shippenberg, T. S., Peter, K., and Herz, A.,** Peripheral opioid receptors mediating antinociception in inflammation. Evidence for involvement of mu, delta and kappa receptors, *J. Pharmacol. Exp. Ther.*, 248, 1269, 1989.
49. **Herman, B. H. and Goldstein, A.,** Cataleptic effects of dynorphin-(1–13) in rats made tolerant to mu-opioid agonists, *Neuropeptides*, 2, 13, 1981.
50. **Aceto, M. D., Dewey, W. L., Chang, J.-K., and Lee, N. M.,** Dynorphin-(1–13): effects in non-tolerant and morphine-dependent rhesus monkeys, *Eur. J. Pharmacol.*, 83, 139, 1982.
51. **Wen, H. L. and Ho, W. K. K.,** Suppression of withdrawal symptoms by dynorphin in heroin addicts, *Eur. J. Pharmacol.*, 82, 183, 1982.
52. **Green, P. G. and Lee, N. M.,** Dynorphin A-(1–13) attenuates withdrawal in morphine-dependent rats: effect of route of administration, *Eur. J. Pharmacol.*, 145, 267, 1988.
53. **Bhargava, H. N., Ramaro, P., and Gulati, A.,** Effects of morphine in rats tested chronically with U-50,488H, a kappa opioid receptor agonist, *Eur. J. Pharmacol*, 162, 257, 1989.
54. **Fukugawa, Y., Katz, R. J., and Suzuki, T.,** Effects of a selective κ-opioid agonist, U-50,488H, on morphine dependence in rats, *Eur. J. Pharmacol.*, 170, 47, 1989.
55. **Bhargawa, H. N., Matwyshyn, G., and Ramaro, P.,** The effect of U-50,488H, a κ-opiate receptor agonist on tolerance to the analgesic and hyperthermic effects of morphine in the rat, *Gen. Pharmacol.*, 22, 429, 1991.
56. **Yamamoto, T., Ohno, M., and Ueki, S.,** A selective κ-opioid agonist, U-50,488H, blocks the development of tolerance to morphine analgesia in rats, *Eur. J. Pharmacol.*, 156, 173, 1988.
57. **Takahashi, M., Senda, T., and Kaneto, H.,** Role of spinal κ opioid receptor in the blockade of development of antinociceptive tolerance to morphine, *Eur. J. Pharmacol.*, 200, 293, 1991.
58. **Rezvani, A. and Way, E. L.,** Dynorphin-(1–13) restores the potency of morphine on the tolerant guinea pig ileum, *Eur. J. Pharmacol.*, 102, 475, 1984.
59. **Lee, N. M. and Smith, A. P.,** Possible regulatory function of dynorphin and its clinical implications, *Trends Neurosci.*, 5, 108, 1984.
60. **Bhargava, H. N. and Gulati, A.,** Modification of brain and spinal cord dopamine D_1 receptors labelled with [^{3}H]SCH 23390 after morphine withdrawal from tolerant and physically dependent rats, *J. Pharmacol. Exp. Ther.*, 252, 901, 1990.
61. **Székely, J. I.,** The most characteristic in vivo effects of opiates, in *Opioid Peptides,* Vol. I. Research Methods, Székely, J. I. and Rónai, A. Z., Eds., CRC Press, Boca Raton, FL, 1982, 45–63.
62. **Walker, M. J. K., Lê, A. D., Poulos, C. X., and Cappell, H.,** Chronic selective blockade of μ opioid receptors produces analgesia and augmentation of the effects of κ-agonists, *Brain Res.*, 538, 181, 1991.
63. **Saelens, J. K., Granat, F. R., and Sawyer, W. K.,** The mouse jumping test — a simple screening method to estimate the physical dependence capacity of analgesics, *Arch. Int. Pharmacodyn.*, 190, 213, 1971.

64. **Tulunay, F. C. and Takemori, A. E.,** The increased efficacy of narcotic antagonists induced by various narcotic analgesics, *J. Pharmacol. Exp. Ther.*, 190, 395, 1974.
65. **Sofuoglu, M. and Takemori, A. E.,** Naloxone produces protracted tolerance in mice treated with a single dose of morphine, *J. Pharmacol. Exp. Ther.*, 254, 836, 1990.
66. **Yeung, J. C. and Rudy, T. A.,** Multiplicative interaction between narcotic agonists expressed at spinal and supraspinal sites of antinociceptive action as revealed by concurrent intrathecal and intracerebroventricular injections of morphine, *J. Pharmacol. Exp. Ther.*, 215, 663, 1980.
67. **Roerig, R. S., O'Brien, S. M., Fujimoto, J. M., and Wilcox, G. L.,** Tolerance to morphine analgesia: decreased multiplicative interaction between spinal and supraspinal sites, *Brain Res.*, 308, 360, 1984.
68. **Roerig, S. C., Hoffman, R. G., Takemori, A. E., Wilcox, G. L., and Fujimoto, J. M.,** Isobolographic analysis of analgesic interactions between intrathecally and intracerebroventricularly administered fentanyl, morphine and D-Ala^2-D-Leu^5-enkephalin in morphine-tolerant and non-tolerant mice, *J. Pharmacol. Exp. Ther.*, 257, 1091, 1991.
69. **Botticelli, L. J., Cox, B. M., and Goldstein, A.,** Immunoreactive dynorphin in mammalian spinal cord and dorsal root ganglia, *Proc. Natl. Acad. Sci. U.S.A.*, 78, 7783, 1981.
70. **Vincent, S. R., Hökfelt, T., Christensen, I., and Terenius, L.,** Dynorphin-immunoreactive neurons in the central nervous system, *Neurosci. Lett.*, 33, 185, 1982.
71. **Cho, H. J. and Basbaum, A. I.,** Ultrastructural analysis of dynorphin-B-immunoreactive cells and terminals in the superficial dorsal horn of the deafferented spinal cord of the rat, *J. Comp. Neurol.*, 281, 193, 1989.
72. **Przewlocki, R., Gramsch, C., Pasi, A., and Herz, A.,** Characterization of immunoreactive dynorphin, α-neo-endorphin, Met-enkephalin and substance P in human spinal cord, *Brain Res.*, 280, 95, 1983.
73. **Miller, K. E. and Seybold, V. S.,** Comparison of met-enkephalin, dynorphin A-, and neurotensin-immunoreactive neurons in the cat and rat spinal cords. I. Lumbar cord, *J. Comp. Neurol.*, 255, 293, 1987.
74. **Jessel, T. M. and Iversen, L. L.,** Opiate analgesics inhibit substance P release from rat trigeminal nucleus, *Nature*, 268, 549, 1977.
75. **Newbold, B. B.,** Chemotherapy of arthritis induced in rats by mycobacterial adjuvant, *Br. J. Pharmacol.*, 21, 127, 1963.
76. **Millan, M. J., Czlonkowski, A., Pilcher, C. W. T., Almeida, O. F. X., Millan, M. H., Colpaert, F. C., and Herz, A.,** A model of pain in the rat: functional correlates of alterations in activity of opioid system, *J. Neurosci.*, 7, 77, 1987.
77. **Millan, M. J., Czlonkowski, A., Morris, B., Stein, C., Arendt, R., Huber, A., Höllt, V., and Herz, A.,** Inflammation of the hind limb as a model of unilateral, localized pain: influence on multiple opioid systems in the spinal cord of the rat, *Pain*, 35, 299, 1988.
78. **Iadarola, M. J., Brady, L. S., Draisci, G., and Dubner, R.,** Enhancement of dynorphin gene expression in spinal cord following experimental inflammation: stimulus specificity, behavioral parameters and opioid receptor binding, *Pain*, 33, 313, 1988.

79. **Menétrey, D. and Besson, J. M.,** Electrophysiological characteristics of dorsal horn cells in rats with cutaneous inflammation resulting from chronic arthritis, *Pain*, 13, 343, 1982.
80. **Hylden, J. L. K., Thomas, D. A., Iadarola, M. J., Nahin, R. L., and Dubner, R.,** Spinal analgesic effects are enhanced in a model of unilateral inflammation/hyperalgesia: possible involvement of noradrenergic mechanisms, *Eur. J. Pharmacol.*, 194, 135, 1991.
81. **Millan, M. J. and Colpaert, F. C.,** The influence of sustained opioid receptor blockade in a model of long-term, localized inflammatory pain in rats, *Neurosci. Lett.*, 113, 50, 1990.
82. **Iadarola, M. J., Douglass, J., Civelli, O., and Naranjo, J. R.,** Differential activation of spinal cord dynorphin and enkephalin neurons during hyperalgesia: evidence using cDNA hybridization, *Brain Res.*, 455, 205, 1988.
83. **Takahashi, O., Traub, R. J., and Ruda, M. A.,** Demonstration of calcitonin-gene related immunoreactive axons contacting dynorphin A(1–8) immunoreactive spinal neurons in a rat model of peripheral inflammation and hyperalgesia, *Brain Res.*, 475, 168, 1988.
84. **Weihe, E., Millan, M. J., Höllt, V., Nohr, D., and Herz, A.,** Induction of the gene encoding pro-dynorphin by experimentally induced arthritis enhances staining for dynorphin in the spinal cord of rats, *Neuroscience*, 31, 77, 1989.
85. **Nahin, R. L., Hylden, J. L. K., Iadarola, M. J., and Dubner, R.,** Peripheral inflammation is associated with increased dynorphin immunoreactivity in both projection and local circuit neurons in the superficial dorsal horn of the rat spinal lumbar cord, *Neurosci. Lett.*, 96, 247, 1989.
86. **Höllt, V., Haarmann, I., Millan, M. J., and Herz, A.,** Prodynorphin gene expression is enhanced in the spinal cord of chronic arthritic rats, *Neurosci. Lett.*, 73, 90, 1987.
87. **Ruda, M. A., Iadarola, M. J., Cohen, L. V., and Young, W. S., III,** In situ hybridization cytochemistry and immunocytochemistry reveal an increase in spinal dynorphin biosynthesis in a rat model of peripheral inflammation and hyperalgesia, *Proc. Natl. Acad. Sci. U.S.A.*, 85, 622, 1988.
88. **Leah, J., Menétrey, D., and de Pommery, J.,** Neuropeptides in long ascending spinal tract cells in the rat: evidence for parallel processing of ascending information, *Neuroscience*, 24, 195, 1988.
89. **Hunt, S. P., Pini, A., and Evan, G.,** Induction of c-fos-like protein in spinal cord neurons following sensory stimulation, *Nature*, 328, 632, 1987.
90. **Berridge, M.,** Second messenger dualism in neuromodulation and memory, *Nature*, 323, 294, 1986.
91. **Goelet, P., Castellucci, V. F., Schacher, S., and Kandel,E. C.,** The long and the short of the long-term memory — a molecular framework, *Nature*, 322, 419, 1986.
92. **Fitzgerald, M.,** c-Fos and the changing face of pain, *Trends Neurosci.*, 13, 439, 1990.
93. **Barthó, L., Stein, C., and Herz, A.,** Involvement of capsaicine-sensitive neurones in hyperalgesia and enhanced opioid antinociception in inflammation, *Naunyn-Schmiedeberg's Arch. Pharmacol.*, 342, 666, 1990.

94. **Jancsó, G., Király, E., and Jancsó-Gábor, A.,** Pharmacologically induced selective degeneration of chemosensitive primary sensory neurones, *Nature*, 270, 741, 1977.
95. **Kajander, K. C., Sahara, Y., Iadarola, M. J., and Bennett, G. J.,** Dynorphin increases in the dorsal spinal cord in rats with a painful peripheral neuropathy, *Peptides*, 11, 719, 1990.
96. **Morley, J. E., Bartness, T. J., Gosnell, B. A., and Levine, A. S.,** Peptidergic regulation of feeding, *Int. Rev. Neurobiol.*, 27, 207, 1985.
97. **Donohoe, T. P.,** Stress-induced anorexia: implications for anorexia nervosa, *Life Sci.*, 34, 203, 1984.
98. **Colpaert, F. C., De Witte, Ph., Maroli, A. N., Awouters, F., Niemegeers, C. J. E., and Janssen, P. A. J.,** Self-administration of the analgesic suprofen in arthritic rats: evidence of *mycobacterium butyricum* induced arthritis as an experimental model of chronic pain, *Life Sci.*, 27, 921, 1980.
99. **Colpaert, F. C., Meert, Th., De Witte, Ph., and Schmitt, P.,** Further evidence validating adjuvant arthritis as an experimental model of chronic pain in the rat, *Life Sci.*, 31, 67, 1982.
100. **Lyness, W. H., Smith, F. L., Heavner, J. E., Iacono, C. U., and Garvin, R. D.,** Morphine self-administration in the rat during adjuvant-induced arthritis, *Life Sci.*, 45, 2217, 1989.
101. **Shippenberg, T. S., Stein, C., Huber, A., Millan, M. J., and Herz, A.,** Motivational effects of opioids in an animal model of prolonged inflammatory pain: alteration in the effects of κ- but not of μ-receptor agonists, *Pain*, 35, 179, 1988.

4 μ-Agonist Induced Euphoria as Opposed to Dysphoria Elicited by κ-Agonists in Humans and Experimental Animals

I. μ-AGONISTS INDUCE EUPHORIA, WHEREAS κ-AGONISTS DYSPHORIA IN HUMANS

The peculiarly "pleasant" subjective state induced by classical morphine congeners is *per definitionem* euphoria. Actually, it is difficult to trace back in the history of human culture when and who identified and termed the subjective feelings induced by opium extracts as euphoria. Of course, their endogenous counterparts, endorphins and enkephalins, were also expected to induce euphoria, and the inability of natural enkephalins and their synthetic derivatives to elicit similar subjective feelings surprised many researchers. Apparently, even more difficult was to accept the fact that synthetic agonists of κ-receptors produce a subjective state which is more or less the opposite of that seen after morphine. The authors were apparently biased by the findings that dynorphins are in some assays the most potent endogenous ligands of opiate receptors.[1,2] However, a careful search of the related literature reveals that data of critical importance were published long before drawing the evident conclusion (Table 1).

The first pioneering study was made by Haertzen[3] in 1970, i.e., before the discovery of opiate receptors and their endogenous ligands. In his study[3] the subjective effects of two "narcotic agonists" cyclazocine and nalorphine were examined in opiate addicts using special questionnaires designed to detect drug-induced subjective effects. As stated, "...Effects of tiredness, anxiety, dysphoria, absence of euphoria, headaches, inefficiency of cognitive, social and physical functioning mark some common features of cyclazocine and nalorphine and withdrawal...".[3] It was revealed later that cyclazocine and nalorphine are not solely partial agonist/antag-

TABLE 1
Subjective Effects Induced by Some Prototypic Ligands of μ- δ- and κ-Receptors in Humans

Drug examined	Receptor involved	Subjective effect	Ref.
Classical morphine like opiates	μ-Agonists	Euphoria	See the relevant literature
Nalorphine, cyclazocine	κ-Agonist and μ-agonist-antagonist	Dysphoria, various symptoms also seen in opiate withdrawal	3
(Keto) cyclazocine	κ-Agonist and μ-agonist-antagonist	Dysphoric state which is clearly discriminated from morphine	9
MR 2033	κ-Agonist	Dysphoria, anxiety, bodily complaints, severe discomfort, psychotomimesis	10
FK 33-824, BW942C, and other enkephalin analogs	μ-and δ-agonist and κ-agonist-antagonist	Dysphoria, a cohort of unusal autonomic reactions	See Table 2 for details
Naloxone, naltrexone	Nonselective opiate receptor antagonists	No distinct subjective effects per se in drug-free subjects	7–8

onists of μ-receptors, but they also act as agonists of the κ-ones.[4,5] Since the study was conducted in opiate (heroin) addicts, the symptoms were apparently due to antagonism of μ-opiate receptors, i.e., withdrawal. However, it is well established that both these substances induce dysphoric symptoms in general, i.e., in nonaddicted subjects as well.[6] Furthermore, in healthy volunteers pure opiate antagonists as naloxone and naltrexone induce very weak subjective symptoms, if at all.[7,8] Thus this study was the first which described the subjective effects of κ-agonists, but it equated them with the withdrawal of addictive μ-agonists. More than a decade later Kumor et al.[9] compared the subjective and autonomic effects of morphine with those of cyclazocine and ketocyclazocine using a battery of psychometric and routine clinical measures to clarify whether or not these drugs can be subjectively discriminated. Their conclusion was, "...The evidence shows that ketocyclazocine causes a dysphoric state and the drug was clearly discriminated from morphine. On the basis of these results we conclude that distinct differences in the activities of ketocyclazocine and morphine are observed in man. Taken at face value our data suggest that kappa activity is dysphoric and probably hallucinatory..''.[9] In the same year, Pfeiffer et al.[10] examined the subjective and autonomic effect of a new κ-selective agonist, MR 2033, in healthy volunteers. The drug induced bodily complaints such as weakness, sweating, vertigo, dizziness, an increase in anxiety, and severe discomfort. These symptoms were blocked by prior treatment with naloxone. Therefore, the authors concluded, ''...Thus κ-receptors seem to mediate psychotomimetic effects. In view of the euphorigenic properties of μ-agonists, our results imply the existence of opposed opioid systems affecting emotional and perceptual experiences...''.[10] Thus the reviewer only wonders why these seminal papers, published in most prestigious medical journals, have not had greater impact.

A related issue is the apparent lack of euphorigenic potency of β-endorphin and all the enkephalin analogs examined so far in humans (applying them either i.v., s.c., or i.m.). Due to their well-known metabolic instability[11] and weak capability to pass through the blood-brain barrier,[12–14] instead of β-endorphin and natural enkephalins, primarily the synthetic analogs of the latter ones have been examined. von Graffenried et al.[15] first reported that FK 33–824 ([D-Ala^2,NMePhe^4,Met(O)^5]-enkephalin-ol), probably the most potent μ-selective enkephalin analog, did not elicit any morphine-like subjective and autonomic effect in humans. Instead of euphoria, an array of peculiar subjective symptoms/autonomic signs were recorded, such as feeling of heaviness (paresthesia) in the limbs, redness of the face, injection of the conjunctivae, epigastrial oppression, or dryness of the mouth.[15] Similar symptoms were found in another study[16] and the drug was found unpleasant and dysphoric in clinical settings as well when trying to use it in the treatment of some mental disorders.[17–19] The same symptoms were detected by us while examining the effects of another enkephalin analog, (D-Met^2,Pro^5)-enkephalinamide, in healthy volunteers.[20,21] In addition to the above listed symptoms, we detected initial polyuria followed by oliguria, a phenomenon which we could not explain at that time.[20] Similar ''side effects'' were detected with another structurally closely related enkephalin analog, derived from the previous one, BW942C ([D-Met(O)^2,pNO$_2$Phe^4,Pro^5-enkephalinamide]). In addition, it was found dysphoric by opiate abusers.[22] Finally, metkephamide ([D-Ala^2,NMePhe^4,Met^5)enkephalinamide]) also was found to elicit these strange dysphoric autonomic signs in patients receiving the compound as an analgesic.[23] On the other hand, the well-known characteristic autonomic effects of classical morphine derivatives as miosis, respiratory depression, nausea, etc., were not seen with these enkephalin analogs.

The non morphine like subjective effects are difficult to interpret since, of these enkephalin analogs, FK-33824, (D-Met^2,Pro^5)-enkephalinamide, and metkephamide did induce an elevation of the pain threshold, the former ones in healthy volunteers[21,24] and the latter one in patients needing postoperative analgesia,[23,25] as expected of "true narcotic analgesics". As for their pure "psychic" effects, in many of the studies reviewed in Table 2, the authors reported drowsiness and recorded either a decrease or an increase in psychic tension, sometimes emotional detachment, in other cases just excitation.

A putative interpretation of these surprising findings was already put forward in the first related study of von Graffenried et al.,[15] who proposed that "...An ultimate explanation for the non-morphine-like type of effects seen might be the presence of enkephalin receptors in other tissues than those now known, with a specificity differing from that of morphine...". Recently, Vaupel et al.[35] showed that one of these enkephalin analogs, BW942C, induced polyuria both in humans and rats and the dose-response curves were of inverted U-shape. In this elegant pharmacological study they proved that BW942C behaves as a partial κ-agonist, whereupon in small doses the agonist and in high doses the antagonist effect prevails.[35] The authors also demonstrated a relatively weak, but detectable binding of BW942C to κ-receptors.[35] It is an important finding, since natural enkephalins only bind to δ- and μ-receptors[36] and opioid pentapeptides in general are supposed to be too short for binding to κ-receptors. Thus, the dysphoric effects of enkephalin analogs are probably mediated by κ-receptors. Hence, the initial polyuria followed by oliguria seen with (D-Met^2,Pro^5)-enkephalinamide[20] is easily interpreted by supposing an initial κ-receptor stimulation followed by μ-agonist and/or κ-antagonist effect. In general, the divergence of autonomic symptoms seen after classical opiates and the previously mentioned enkephalin analogs are obviously due to the fact that these opioid peptides activate not only μ-, but also κ- and δ-receptors.

There is an interesting corollary of the above mentioned human observations. The peculiar autonomic symptoms elicited by FK 33–824[15] and (D-Met^2,Pro^5)-enkephalin[20] appeared 3 to 5 min after injection and peaked about 10 to 15 min after systemic administration, then tapered off within 30 to 45 min. However, their analgesic and some other central effects (moderate improvement of attention and short-term memory) developed later.[21] Thus, the theoretical possibility arises that the dysphoric effect of κ-agonists are elicited via peripheral receptors and they are somehow related to the autonomic reactions elicited by them. Since there are many κ-receptors in the autonomic nervous system and many vegetative functions are regulated by them,[37] the above reviewed dysphoric effects might be central subjective epiphenomena of peripheral events.

II. SOME GENERAL CONSIDERATIONS

The data reviewed in the previous section collectively indicate that

- Activation of κ-receptors is aversive, dysphoric.

TABLE 2
Autonomic and Subjective Effects of Systemically Administered Enkephalin Analogs and β-Endorphin in Humans

Autonomic Symptoms Recorded by the Observer

Conjunctival injection
Facial pallor
Occasionally: erythemas and flare reactions
Moderate increase in heart rate
Orthostatic hypotension at higher dose levels (the dose-limiting adverse reaction)
Initial polyuria followed by oliguria
Borborygmus

Autonomic Reactions Experienced Subjectively

Feeling of heaviness in the limbs (paresthesia)
Dry mouth
Feeling of warmth mostly in the face
Feeling of nasal stuffiness
Tingling sensation in the eyes

No Consistent Changes in

Respiratory rate
Pupillary diameter
Full blood count
Conventional liver function tests
Serum electrolytes

Behavioral Symptoms Recorded by the Observer

Decreased vigilance
Slightly slurred speech
Deceleration of movements

Subjectively Experienced Psychic Effects

Slight drowsiness, fatigue, tiredness
Dizziness
Mostly decrease in psychic tension, but occasionally anxiety over the subjectively experienced unpleasant feelings
Dysphoria

The Substances Examined (i.v. or s.c. or i.m.)

β-Endorphin;
FK 33–824 = (D-Ala_2,$NMePhe^4$,Met^5(O)-ol)-enkephalin;

TABLE 2 (continued)
Autonomic and Subjective Effects of Systemically Administered Enkephalin Analogs and β-Endorphin in Humans

The Substances Examined (i.v. or s.c. or i.m.) (continued)

GYKI-14238 = (D-Met^2,Pro^5)-enkephalinamide;
BW942C = (D-Met(O)^2,pNOPhe^4,Pro^5)-enkephalinamide;
Metkephamide = (D-Ala$_2$,NMePhe^4,Met^5)-enkephalinamide

See References 15 through 35.

- κ-Agonists can be considered functional antagonists of μ-ones since they induce emotional changes opposite in direction to those elicited by μ-agonists.

Several further lines of evidence also indicate that μ- and κ-receptors activate distinct mechanisms, though all the phenomena elicited by them are reversible by opiate antagonists.

A. Locomotor Activity

As discussed in Chapter 7, μ-agonists induce locomotor hyperactivity, especially if injected into the rewarding brain areas. Thus in animal experiments they have an excitant character. Truly, in humans classical opiates are considered drugs of sedative character. However, just in humans their sedative effect might be due to abolition of excruciating pain and suffering (under clinical conditions) or to their anxiolytic action. There is a line of important observations which cannot be reconciled with an overall sedative effect, attributed to classical opiates. Namely, as described by Kay and co-workers,[40–42] morphine, methadone, and heroin, i.e., pure μ-agonists induce insomnia in nondependent opiate addicts (although the name *morphine* is derived from Morpheus, the Roman deity of dreams). The opiate-induced insomnia manifests itself among others in a decrease in REM sleep and an increase in sleep latency. Morphine has been found to induce similar sleep disturbances in rats,[43], rabbits,[44] and cats.[45] Thus, in spite of the general belief, there is a kind of underlying excitant action of classical opiates in humans, too.

Furthermore, actually all the other main classes of abused substances induce locomotor facilitation if injected into the rewarding brain areas in appropriate doses (Chapter 10).

However, κ-agonists such as dynorphin (Chapter 7) and U-50,488H induce dose-dependent behavioral depression in rats[38] and monkeys,[39] as well. It is well established that nonabused psychopharmaca such as the dopamine receptor-blocking neuroleptics, most antidepressants, most antiepileptic agents, etc., also induce locomotor depression in animals and a variety of unpleasant side effects in the clinical practice. Thus the failure of κ-agonists to induce locomotor stimulation seems to be somehow related to their aversive character.

B. Nonsubstitutability

It has been well established that κ-agonists do not substitute for morphine and its congeners in opiate-tolerant and -dependent animals.[4,5,38,46]

C. Nondependence

Dysphoric drugs are expected to induce no physical dependence. Really, U-50488H, probably the most selective κ-agonist,[38] has been found incapable of eliciting morphinelike dependence in rodents upon peripheral administration[38,47] or infusing it into the Sylvian aqueduct.[48] Injecting naloxone to rats receiving previously chronic s.c. infusion of μ-selective morphine or (D-Ala^2,NMePhe^4,Gly-ol)-enkephalin (DAGO), the withdrawal syndrome was serious, but only moderate symptoms were seen in subjects pretreated with δ-selective (D-Pen^2,D-Pen^5)-enkephalin and practically nothing after dynorphin-A and U-50488H.[48] However, one group did describe a range of symptoms elicitable in monkeys by chronic administration and withdrawal of U-50488H (or precipitation by opiate antagonists).[39] Nevertheless, besides general restlessness, the withdrawal symptoms seen after U-50488H (excessive scratching, grooming, yawning, unusual tongue movements) were qualitatively different from those seen after morphine.[39] Therefore, it is tempting to speculate that these symptoms were due to spinal and lower brain stem mechanisms as opposed to some characteristic symptoms seen during opiate withdrawal, e.g., jumping in mice[49] or vocalization in monkeys,[39] which are brought about by higher brain mechanisms.

III. REWARDING EFFECT OF μ-AGONISTS AS OPPOSED TO THE AVERSIVE EFFECTS OF κ-AGONISTS IN SELECTIVE ANIMAL MODELS

A. Self-Administration Procedure (SA)

SA is probably one of the oldest and most reliable way to examine whether a drug under study has rewarding potency or not. Weeks[50] demonstrated first that morphine-addicted rats could be trained to press a lever by rewarding the conditional responses with further i.v. injection(s) of morphine. The frequency of responses emitted increased upon changing the reinforcement ratio from continuous reinforcement (1:1) to a higher one (i.e., 5:1 or 10:1), i.e., the animals kept the total amount of morphine injected daily (hourly) relatively stable.[50] A significant decrease in the amount of morphine voluntarily ingested was only seen when elevating the reinforcement ratio above 75.[51] Responding could also be maintained by codeine and meperidine, i.e., other classical opiates known to induce euphoria. These studies were extended to monkeys as well.[52,53] Examining a series of classical opiates, i.e., μ-agonists, close correlation was found between their relative potencies in maintaining SA and to suppress the morphine abstinence syndrome in rhesus monkeys.[54] Comparison of stereoisomers (e.g., dextromethorphan vs. levomethorphan) revealed that only the pharmacologically active levo-isomers were capable of maintaining SA.[54] Thus the ability to generate or maintain SA is a reliable qualitative and quantitative indicator of the narcotic agonist potency of drugs.

Animals (mostly rats) can be trained to self-administer morphine (and other opiates) via other routes as well. Animals also self-administer morphine and other opiates intraperitoneally,[55] intragastrically,[56] in aerosol form,[57] and even intracerebro ventricularly.[58] Of course, the simplest way of self-treatment is the oral route when the subjects have free choice between morphine-containing and drug-free drinking water.[59–61] In the latter paradigm, drug-naive rats initially avoid the drug (e.g., morphine) solution because of its bitter taste. Subsequently, after some sessions of ''forced drinking'' when the drug solution is the only source of fluid, the bitter taste becomes a secondary reinforcer, but it is not strong enough to maintain SA for a longer period if the solution offered is bitter, but does not contain rewarding drug.[60]

In drug-naive animals no SA is induced if opiate antagonists are offered for reinforcement.[53] On the other hand, abrupt withdrawal of opiates in trained animals, i.e., lack of reinforcement, results in an initial increase, then gradual decline of response as seen during extinction of various forms of conditioned behavior.[50,62] The same pattern can be detected upon injecting opiate antagonists instead of or in combination with morphine.[50,53,54] Naltrexone shifts the dose-response curve of morphine to the right, but it does not modify the (maximal) rate of responding.[63] On the other hand, the intake of morphine decreases if etonitazene, another potent opiate, is added to the drinking water.[51] Collectively, the latter observations indicate that addicted animals emit the conditional responses in order to maintain an optimal level of stimulation of opiate receptors.

The potency of opiate antagonists has been found enhanced in animals in morphine-dependent rats.[64] This finding is consistent with the earlier ones[65,66] on enhancement of antagonist potency of naloxone after agonists treatment. Unfortunately, no data are available as to whether or not classical opiates also enhance the aversive effect of κ-agonists (vide infra).

On the other hand, κ-agonists such as nalorphine, cyclazocine, ketazocine, ethylketocyclazocine, and other benzomorphans, which are aversive as discussed above and share a discriminative stimulus effect with ethylketazocine (vide infra), failed to maintain SA in some studies conducted in monkeys.[54,67,68] Since these κ-agonists are aversive, it is logical to propose that stimulation of κ-receptors does not maintain SA due to the accompanying dysphoria. However, these κ-agonists also have additional μ-antagonist properties[4,5,8] which might also account for inhibition of SA. Thus, these data do not confirm unequivocally that κ-agonists are unable to maintain SA. Furthermore, Khazan's group claimed that in rats even κ-agonists were self-administered. They found that ketocyclazocine and ethylketocyclazocine,[69] in addition to dynorphin$_{1-13}$ and its D-Ala^2-analog,[70] are self-administered in morphine-dependent rats. Since both studies[69,70] were conducted in morphine-dependent rats, the possibility arises that either the κ-agonists are less aversive in rodents than in monkeys and humans, or the previously established morphine dependence alters the sensitivity of the organism. Probably the data on the altered interaction of μ- and κ-agonists in opiate-tolerant subjects might explain these discrepancies. Namely, as discussed in Chapter 3, in animals chronically pretreated with morphine, synergism replaces the antagonism detectable in drug-naive subjects. The controversy was finally solved by Tang and Collins,[47] who demonstrated that ethylketocyclazocine is self-administered in drug-naive rats also, but the same subjects did not self-administer U-50488H. Obviously, the self-administration of ethylketocyclazocine is due to its weak μ-agonist property, whereas the really selective κ-agonists, U-50488H, has no reinforcing property in this paradigm.

Finally, supplementing the not too numerous data on the SA of non-μ ligands, Belluzzi and Stein[71] demonstrated i.c.v. SA of natural enkephalins. Unfortunately, the experiments described in this frequently quoted paper[71] have not been confirmed in many independent laboratories and the authors' published findings indicate that Met- and Leu-enkephalins are not very potent reinforcers. Since they bind not only to μ-, but also to δ-receptors (and are equally effective in μ-receptor-containing guineapig ileum and δ-receptor-carrying mouse vas deferens),[36,72,73] their repeated examination seems to be warranted, especially in view of the distinctly dysphoric effects of synthetic enkephalin analogs in humans (vide supra).

On the other hand β-endorphin, an agent with significant affinity to both μ- and δ-receptors, but not to the κ-ones,[36,73] is self-administered into the cerebral ventricle by rats.[74] Furthermore, some opioid peptides have been found to be easily self-administered into areas which are part of the rewarding circuitry of the brain (see Chapter 6). Thus rats readily self-administer (D-Ala^2)-Met-enkephalinamide into the ventral tegmental area[75] and the lateral hypothalamus,[76] and Met-enkephalin into the n. accumbens[77] (Table 3).

B. Conditioned Place Preference (CPP)

CPP is another reliable tool to test the drugs under study for eventual rewarding and aversive properties. Using this procedure, rats are confined

TABLE 3
Self-Administration (SA) of Some Prototypic Ligands of μ- δ- and κ-Receptors

Drug examined[a]	Receptor involved	Species used	Induction of SA	Ref.
Morphine	μ-Agonist	Rat	Yes	50
Morphine (orally)	μ-Agonist	Rat	Yes	60
Morphine (i.c.v.)	μ-Agonist	Rat	Yes	58
Morphine	μ-Agonist	Monkey	Yes	52
Codeine, meperidine, methadone, heroin, levorphanol, azidomorphine, etorphine, fentanyl	μ-Agonist	Monkey	Yes, with potency in proportion with their agonist efficacy	53–54
Ethylketazocine, cyclazocine, nalorphine, oxilorphan	μ-Agonist-antagonist	Monkey	No	53–54 68
Ethylketo-cyclazocine, ketocyclazocine	μ-Agonist-antagonist	Rat	Yes	47–69
U-50488H	κ-Agonist	Rat	No	47
Dynorphin$_{1-13}$, (D-Ala^2)-dynorphin$_{1-13}$	κ-Agonist	Morphine-dependent rat	Yes	70
Dynorphin, U-50488H into: C3 area of hippocampus	κ-Agonist	Rat	Yes	78
Met-and Leu-enkephalin i.c.v.	μ-and δ-agonist	Rat	Yes	71
β-endorphin i.c.v.	μ-, δ-, and ε(?)-agonist	Rat	Yes	74
Met-enkephalin into: n. accumbens	μ-and δ-agonist	Rat	Yes	77
DALA[b] into: VTA[c]	μ-and δ-agonist	Rat	Yes	75
DALA into: LH[d]	μ-and δ-agonist	Rat	Yes	76

[a] The route of SA is i.v. unless indicated otherwise.
[b] DALA = (D-Ala^2)-Met-enkephalin.
[c] VTA = ventral tegmental area.
[d] LH = lateral hypothalamus.

to distinctly different environments (boxes) after the injections of drugs to be compared. After establishing contextual conditioning between the drug-induced subjective state and environmental stimuli, free choice is given to the subjects. Offering the rats free choice, they select the environment (one of the two boxes) paired with the rewarding drug.[79–82] A conceptually related technique is the taste preference-aversion procedure, whereupon an indifferent or bitter, eventually sweet taste (e.g., saccharine-flavored drinking water) is paired with one of the drugs and plain water with the other one. Morphine, other classical μ-agonists, and some potent enkephalin analogs have been found to induce place preference (CPP) whether they were administered peripherally[79,81–84] or into brain areas known as parts of the rewarding circuitry.[75,85]

On the other hand, opiate antagonists such as naloxone induce conditioned state aversion, i.e., the animals select the environment paired with saline over the one associated with the narcotic antagonists.[81,86] The conditioned place aversion induced by opiate antagonists indicates that functional elimination of endogenous opioids is dysphoric, i.e., they might maintain a ‘‘physiological level of euphoria’’. Bilateral radiofrequency lesion of the medio-basal hypothalamus, i.e., n. arcuatus, where the cell bodies of β-endorphin-containing neurons are located, resulted in significant reduction of the aversive potency of naloxone.[87] Thus β-endorphin might be at least one of the endogenous opioids responsible for a ‘‘basal’’ degree of euphoria. Of the selective antagonists of various subtypes of opiate receptors, CTOP, a preferential antagonist of μ-receptors, induced place aversion similarly to naloxone, whereas no conditioning could be produced with nor-binaltorphimine and ICI 174864, i.e., selective antagonists of the κ- and δ-receptors, respectively.[88] Thus, whichever is the endogenous opioid maintaining a basal level of euphoria (well-being) (enkephalins, β-endorphin, or something else), the effect seems to be mediated by μ-receptors.

Probably it is pertinent to mention here that beside the eventual (μ)-agonist or antagonist character of the drug under study, the location of receptors presumably affected also is an important factor. Bechara and van der Kooy[89] demonstrated in an ingenious study that methylnaltrexone (which does not pass the blood-brain barrier) and morphine in very small i.p. doses (which are not high enough to stimulate the opiate receptors in the brain) induce place aversion. In the same study,[89] subdiaphragmatic vagotomy abolished their aversive effects. The unpleasant side effects of opiates of gastrointestinal origin (primarily nausea) are well known clinically. Thus, their centrally mediated euphoric action might be counterbalanced by their uncomfortable adverse effects of peripheral origin. The above discussed aversive character of enkephalin analogs examined so far might also be due to preponderant stimulation of peripheral opiate receptors, since they hardly pass the blood-brain barrier. Furthermore, the question arises whether or not the individual differences in propensity to drug addiction can be explained by differences in sensitivity to the peripherally induced reactions of opiates.

Probably it is pertinent to mention here that significant strain-related differences were detected in rats in the CPP paradigm.[90] During breeding the rats were selected bidirectionally: establishing separate lines of subjects displaying high and low rates of lateral hypothalamic self-stimulation (another model of euphoria; vide infra). Animals less prone to operant responding for "rewarding" hypothalamic electrical self-stimulation showed less preference to the environment paired with morphine and heroin and-vice-versa.[90] Thus common genetic factors might underlay the individual inclination to these apparently different kinds of heuristic behavior.

As for other types of opioids, (D-Ala^2)-Met-enkephalin (given i.c.v.) induced place preference in rats trained previously with i.p. morphine.[91] I.c.v.-administered β-endorphin also induced CPP, reversible by naloxone.[92] Furthermore, (D-Ala^2,NMePhe^4,Gly^5)-enkephalin-ol (DAGO), a μ-selective and (D-Pen^2,D-Pen^5)-enkephalin (DPDPE), a δ-preferring enkephalin analog, both induced CPP.[93] However, pretreatment with CTOP, a μ-selective antagonist, only eliminated the reinforcing effect of the former one, whereas ICI 174864, a δ-selective antagonist, blocked the rewarding action of the latter one. Collectively these data indicate that both μ- and δ-receptors are able to mediate CPP, probably independently of each other.

No data are available on the intravenous SA of β-endorphin, obviously due to its inability to pass through the blood-brain barrier. Nevertheless, it is worth noting that in one study[94] the β-EP content decreased in various brain regions belonging to the mesolimbic dopamine system (n. accumbens, septum, hippocampus) upon i.v. SA of heroin and cocaine.

On the other hand, κ-agonists such as U-50488H, tifluadom, and bremazocine have been shown to induce conditioned place and/or taste aversion.[84,88] So did E-2078, a synthetic dynorphin derivative.[88] In apparent contradiction with the latter data, Iwamoto[95] induced CPP (instead of aversion) by i.c.v. injection of dynorphin$_{1-17}$. The pharmacological specificity of the effect was proven by the ability of naloxone (given in high doses) to reverse the CPP induced by dynorphin and the inefficacy of Leu-enkephalin, its eventually "active metabolite", and the lack of effect of dynorphin$_{2-17}$ in the same study.[95] Subsequently, the same author induced CPP by injecting dynorphin into the dorsal hippocampus.[96] Furthermore, it is probably pertinent to mention here that even U-50488H is self-administered into the C3 region of the hippocampus.[78] In the same laboratory[97] nor-binaltorphimine failed to antagonize the SA of dynorphin into the hippocampus, similarly to ICI 174864, but β-funaltrexamine did antagonize the rewarding effect of dynorphin. Therefore, Stevens and Stein[97] proposed that specifically in the C3 area of the hippocampus μ-receptors activated by dynorphin mediate the rewarding effect. Since dynorphins are considered selective ligands of κ-receptors,[2] the latter data[78,95–97] are not easy to interpret. For the time being probably the only possible explanation is to suppose that the aversive effect of κ-agonists is peripherally mediated, whereas that of the μ-ones is generated centrally as reviewed above. To decide this important issue, the motivational effects of κ-agonists should be compared in the same experimental setup upon central and peripheral administration, respectively.

The situation is apparently further complicated by some recent findings that morphine-induced CPP can also be abolished by ICS 205–903 and MDL 722–222,[98] i.e., selective antagonists of the serotonin$_3$ receptors and devazepide,[99] a specific CCK$_A$ receptor antagonist. Both cholecystokinin and the serotonin$_3$ receptors are known to induce very strong motivational actions. Thus, without questioning the reliability of these findings, the possibility arises that κ-opiates might lose their rewarding potency if their effect is obscured by another pharmacological intervention suppressing their euphorigenic action (Table 4).

C. Intracranial Self-Stimulation (ICSS)

The data obtained in this probably oldest experimental paradigm for measurement of eventual rewarding potency of drugs are in line with those recorded in the other two assays. Nevertheless, the initial findings, due to some pitfalls involved in the technique, are apparently contradictory. As demonstrated first by Olds and Milner[100] and Delgado et al.,[101] rats can be trained to press an operant manipulandum (lever) to obtain electric stimulation of certain areas in the brain (ICSS). (These areas are various parts of the rewarding circuitry, as discussed in Chapter 6, but in most cases the electrodes are implanted into the lateral hypothalamus to stimulate the medial forebrain bundle.) The effects of morphine on ICSS were examined first by Olds and Travis,[102] recorded suppression, whereas Adams et al.[103] reported facilitation after an initial period of suppression. As clarified later, an initial phase of depression is followed by facilitation and the dose-response curve is an inverted U-shape, just as in relation to spontaneous locomotor activity (for review see Székely)[104]. It has also been shown by these authors that upon repeated application, tolerance develops primarily to the depressant effects of morphine and only much later to its stimulant action.[105–107] A similar biphasic effect has been observed upon i.c.v. application of morphine,[108] thus the effect is centrally mediated. The specificity of the effect is proven with the ineffectivity of dextrorphan, whereas, levorphanol induces a similar biphasic action.[107] Examining separate brain areas, monophasic facilitation was seen upon injecting morphine into the posterior hypothalamus and ventral tegmental area, whereas its injection into the periaqueductal central gray and locus coeruleus only induced facilitation.[108] Since, as discussed in Chapter 6, the ventral tegmental area and lateral hypothalamus belong to the rewarding circuitry of the brain, these data indicate two points:

TABLE 4
Rewarding and Aversive Effects in Conditioned Place Preference Assay (CPP) of Some Prototypic Ligands of μ-, δ-, and κ-Receptors (in Rats)

Drug examined[a]	Receptor involved	Affective state elicited	Ref.
Morphine and other classical narcotic analgesics	μ-Agonist	Preference	79, 81 82, 83 84
Morphine into various areas of mesolimbic dopamine system	μ-Agonist	Preference	75, 85
Naloxone	Nonselective opiate receptor antagonist	Aversion	81, 86
CTOP[b] i.c.v.	Selective μ-receptor antagonist	Aversion	88
ICI 174864[c] i.c.v.	Selective δ-receptor antagonist	No effect	88
Nor-binaltorphimine	Selective κ-receptor antagonist	No effect	88
Morphine in very low (0.05 mg/kg) i.p. dose	μ-agonist	Aversion	89
Methyl- (quaternary) naltrexone	Nonselective peripherally acting antagonist	Preference	89
Leu-enkephalin i.c.v.	δ-and μ-receptor agonist	No effect	95
(D-Ala^2)-Met-enkephalin i.c.v.	μ- and δ-receptor agonist	Preference	91
β-Endorphin i.c.v.	Agonists of μ-, δ-, and ε-(?)-receptors	Preference	92, 93
DAGO[d]	μ-Agonist	Preference	93
DPDPE[e]	δ-Agonist	Preference	93
U-50488H, tifluadom, bremazocine E-2078[f]	κ-Agonists	Aversion	84, 88
Dynorphin$_{1\text{-}17}$ i.c.v.	κ-Agonist	Preference	95

[a] Drug administered per os, i.p., or s.c. unless indicated otherwise;
[b] CTOP = D-Pen-Cys-Tyr-D-Trp-Orn-Thr-Pen-Thr-NH_2.
[c] ICI 174834 = Allyl^2-Tyr-Aib-Aib-Phe-Leu-OH.
[d] DAGO = (D-Ala^2,NMePhe^4,Gly^5)-enkephalin-ol.
[e] DPDPE = (D-Pen^2,D-Pen^5)-enkephalin.
[f] E-2078 = A synthetic dynorphin analog.

- Facilitation is really the characteristic effect of rewarding mechanisms.
- Facilitation of ICSS seems to be mediated by the rewarding brain areas mentioned above and in Chapter 6.

Subsequently, Kornetsky and co-workers demonstrated that not the eventual alteration of the rate of responding is the most characteristic effect of opiates in this experimental paradigm.[109–112] More informative is the determination of the stimulation threshold, i.e., the intensity of current occasioning lever pressing. Morphine decreases the threshold of reinforcing current and increases the threshold eliciting escape reaction if the electrodes are located in nonrewarding areas, i.e., it elevates the "aversive" threshold.[109,110] Thus, it is more than probable that the general motor depressant/stimulant effect obscures its specific action, i.e., if the rate of lever pressing is examined instead of the threshold of rewarding current occasioning ICSS behavior. Actually, there is no tolerance development if the threshold is examined[111] and measurement of the threshold of rewarding current is a generally used method to determine the eventual hedonic character of drugs under study.[112]

The pharmacological specificity of the effect is proven by stereoselectivity (levorphanol is effective, but dextrorphan is not)[107] and the reversal of the effect of morphine by opiate antagonists,[113–115] though naloxone per se does not affect ICSS.[113]

There are practically no data on the effects of κ-agonists in this paradigm. In a study of Holtzman,[113] cyclazocine and nalorphine, similarly to morphine, decreased the rate of ICSS. Since, as discussed above, in this paradigm the threshold and not the rate is the real measure of specific aversive or rewarding drug effects, further data are needed. In the only study where the stimulation threshold was examined[112] the effects of cyclazocine and nalorphine were equivocal (Table 5).

IV. DISCRIMINATIVE STIMULUS EFFECTS OF μ-, δ- and κ-AGONISTS

The drug discrimination procedure is a powerful tool to classify drugs according to their subjective effects.[116] Based on this concept, drugs belonging to the same pharmacological class are supposed to show stimulus generalization when animals are trained to discriminate between two operant manipulanda. Using this technique,[116] pressing one of the levers is "rewarded" (mostly with food) if the animals are pretreated with the pharmaca under study and the other one is rewarded after saline pretreatment. Injecting the "experimental" drug, the subjects are supposed to select between the drug and saline-associated levers according to identification of the subjective effect of the unknown pharmacon with the former

TABLE 5
Modulation of Intracranial Self-Stimulation (ICSS) by Some Prototypic Ligands of μ-, δ-, and κ-Receptors (in Rats)

Drug examined[a]	Receptor involved	Effect on ICSS	Ref.
Morphine	μ-Agonist	Suppression of the rate of ICSS	102, 113
Morphine	μ-Agonist	Initial suppression followed by facilitation of the rate of ICSS	105, 106 107
Levorphanol	μ-Agonist	Biphasic as above	107
Dextrorphan (i.p. or i.c.v.)	Inactive enantiomer of levorphanol	No effect	
Morphine i.c.v.	μ-Agonist	Biphasic effect as above	108
Into the VTA[a] or PH[b]		Facilitation of the rate of ICSS	
Into the PAG[c]		Suppression of the rate of ICSS	
Cyclazocine, pentazocine	μ-Agonist-antagonist	Suppression of the rate of ICSS	113
Naloxone	Nonselective opiate antagonist	No effect on the rate of ICSS	113
Naltrexone	Nonselective opiate antagonist	No effect on the rate of ICSS	107
Morphine	μ-Agonist	Lowering of the threshold current occasioning ICSS	109, 112
Cyclazocine, nalorphine pentazocine	μ-Agonist-antagonist	Equivocal effect on the threshold current	109, 112
Naloxone	Nonselective opiate antagonist	Elevation of the threshold current occasioning ICSS	109, 112 115

[a] drugs given i.p., s.c., or per os unless stated otherwise.
[b] VTA = ventral tegmental area.
[c] PH = posterior hypothalamus.
[d] PAG = periaqueductal central gray.

or later one. As discussed below, this procedure has provided further evidence of the divergent motivational properties of μ- and κ-agonists, respectively.

Monkeys trained to discriminate between etorphine (a very potent, pure, morphinelike opiate agonist) and saline emit drug-appropriate responses after pretreatment with other narcotic agonists such as codeine, levorphanol, fentanyl, meperidine, etc., but they make a saline-appropriate choice after cyclazocine, (ethyl)ketocyclazocine, pentazocine, or (±)-N-allylnormetazocine.[54,117] The role of central opiate receptors is indicated, among others, by abolition of the discriminative effect of narcotic agonists upon pretreatment with naltrexone, but not after naltrexone methylbromide, i.e., the quaternary analog which does not cross the blood-brain barrier.[54] The pharmacological specificity is also shown by stereoselectivity. Levorphanol-trained animals select the ''saline lever'' after dextrorphan, i.e., its optically inactive isomer.[54]

On the other hand, the lack of stimulus generalization between μ- and κ-agonists is symmetrical, i.e., monkeys trained to discriminate between ethylketocyclazocine and saline fail to emit drug-appropriate response after morphine, etorphine, codeine, or levorphanol.[46,54,68] Nevertheless, naltrexone also blocks the stimulus effect of ethylketocyclazocine,[54,68] proving that both classes of these drugs act on opiate receptors. All the more, since naltrexone generates saline-appropriate responses whether the monkeys are trained to discriminate etorphine or ethylketocyclazocine from saline.[54,68,118]

Narcotic agonists are similarly discriminated from antagonists and other classes of psychoactive pharmaca in rats.[119,120] Furthermore, bremazocine and tifluadom, contrary to morphine and fentanyl, induce drug-appropriate responses in rats trained to discriminate U-50488H from saline.[121] However, in subjects trained with morphine, these κ-agonists (tifluadom and bremazocine) elicit responses on the saline-appropriate lever. Nevertheless, in rats the situation seems to be a bit more equivocal. Rats trained to discriminate fentanyl from saline show partial generalization to (keto)cyclazocine, nalorphine, (±)-N-allylnormetazocine, but they also partially antagonize the effect of fentanyl.[122] At first look simultaneous occurrence of partial generalization and antagonism are not surprising, since these drugs, apart from their κ-agonist property, are known to be agonist/antagonists of μ-receptors.[4–6,8] Nevertheless, an important point of this study of Colpaert and Janssen[122] has not generated real interest. Actually, in this seminal paper it was demonstrated unequivocally that in some subjects these drugs induced stimulus generalization, but did not antagonize the effect of fentanyl, whereas other individuals behaved in just the opposite way in both aspects.[122,] Thus, it may be inferred that these drugs are perceived as narcotics in some individuals, but not necessarily in others. As discussed in Chapters 10 and 11, the same drugs of abuse (opiates and other ones as well) seem to elicit different subjective states in different individuals. Obviously, the interindividual differences in drug-induced subjective states may also account for interindividual differences in propensity to drug abuse.

Further data also indicate that there are species differences in the discriminative stimulus properties of these κ-agonist/μ-agonist/antagonist drugs (cyclazocine, ketocyclazocine, pentazocine, N-allylnormetazocine). Namely, in pigeons the (−)-isomers of all these drugs generalize to fentanyl, whereas their (+)-isomers generalize to bremazocine.[123] An important point, because these drugs are supposed to interact also with the putative σ-receptors,[4] and a careful review of the literature by Musacchio[124] revealed that the dysphoric, i.e., κ- and/or σ-like action of these drugs used in humans in racemic forms, reside in the (+)-isomer.[124]

Returning to the original question, the above reviewed apparent contradictions are obviously due to the lack of sufficient specificity of these drugs. Namely, examining U-50488H, the first really selective κ-ligand,[38] it becomes evident that there is no stimulus generalization between the μ- and κ-agonists in either monkeys or rats.[46,47,125,126] The same applies to spiradoline,[127] a recently synthesized, even more selective κ-agonist.[128] Moreover, a recent careful study[129] revealed that even in the case of U-50488H, κ- and μ-agonists only induce unequivocal stimulus discrimination if the training doses of drugs are high enough.

A related issue is whether or not μ- and κ-agonists modulate the stimulus effect of each other. This issue is very important, just for the reason mentioned in the first section of this chapter. Collectively, these data on stimulus discrimination indicate that the subjective states brought about by the two groups of drugs are different. However, the human studies reviewed above[3,9,10] indicate that the subjective states induced by μ- and κ-agonists are diametrically opposite. In the latter case coadministration of drugs belonging to the two different classes is expected to weaken (abolish) the discriminative stimulus effect of each other. The issue was addressed only in one laboratory. Negus et al.[121] found that in rats morphine, fentanyl, and buprenorphine reduced U-50488H-appropriate responding, but neither U-50488H nor tifluadom modified the drug-appropriate responding in subjects trained to discriminate morphine from saline. On the other hand, in squirrel monkeys neither group of drugs modified the stimulus discriminative effect of the other one.[130] Based on these data the authors suppose that under certain conditions there is a "...perceptual masking"[121] between the μ- and κ-agonists, or "...the effects mediated by different opioid receptors converge on a single experimental endpoint".[130] Actually, both propositions might be correct.

Probably the point is that the stimulus discrimination procedure does not provide reliable information on the eventual hedonic or anhedonic character of drugs under study. In this paradigm only one point is confirmed: that they induce a different emotional state. Dysphoria and euphoria are the antipodes motivationally of each other. However, it does not necessarily mean that they are generated by excitation and blockade of the same receptor population or brought about by two different kinds of receptors coupled to each other as antagonists. From the physiological point of view it is probably enough to draw the most parsimonious conclusion: the μ-receptor-elicited hedonic state induces repetitive-approach behavior, whereas the κ-receptor-mediated dysphoria results in avoidance/inhibitory behavior towards the stimuli generating the opposite emotional states.

REFERENCES

1. **Goldstein, A., Tachibana, S., Lowney, L. I., Hunkapiller, M., and Hood, L.,** Dynorphin-(1–13), an extraordinarily potent opioid peptide, *Proc. Natl. Acad. Sci. U.S.A.*, 76, 6666, 1979.
2. **Chavkin C., James, I. F., and Goldstein, A.,** Dynorphin is a specific endogenous ligand of κ opioid receptor, *Science*, 215, 413, 1982.
3. **Haertzen, C. A.,** Subjective effects of narcotic antagonists cyclazocine and nalorphine on the Addiction Research Center Inventory (ARCI), *Psychopharmacology*, 18, 366, 1970.
4. **Martin, W. R., Eades, C. G., Thompson, J. A., Huppler, R. E., and Gilbert, P. E.,** The effects of morphine and nalorphine-like drugs in nondependent and morphine-dependent chronic spinal dog, *J. Pharmacol. Exp. Ther.*, 197, 517, 1976.
5. **Gilbert, P. E. and Martin, W. R.,** The effects of morphine-like drugs in the nondependent, morphine-dependent and cyclazocine-dependent chronic spinal dog, *J. Pharmacol. Exp. Ther.*, 198, 66, 1976.
6. **Reynolds, E. F., Ed.,** *MARTINDALE The Extra Pharmacopoeia,* Pharmaceutical Press, London, 1989, 837 and 844.
7. **Reynolds, E. F., Ed.,** *MARTINDALE The Extra Pharmacopoeia,* Pharmaceutical Press, London, 1989, 855.
8. **Jaffe, J. H. and Martin, W. H.,** Opioid analgesics and antagonists, in *The Pharmacological Basis of Therapeutics,* Goodman Gilman, A., Rall, T. W., Nies, A. S., and Taylor, P., Eds., Pergamon Press, New York, 1990, 485.
9. **Kumor, K. M., Haertzen, C. A., Johnson, R. E., Kocher, T., and Jasinski, D.,** Human psychopharmacology of ketocyclazocine as compared with cyclazocine, morphine and placebo, *J. Pharmacol. Exp. Ther.*, 238, 960, 1986.
10. **Pfeiffer, A., Brantl, V., Herz, A., and Emrich, H. M.,** Psychotomimesis mediated by κ opiate receptors, *Science*, 233, 774, 1986.
11. **Hambrook, J. M., Morgan, B. A., Rance, M. J., and Smith, C. F. C.,** Mode of deactivation of the enkephalins by rat and human plasma and rat brain homogenates, *Nature*, 262, 782, 1976.

12. **Cornford, E. M., Braun, L. D., Crane, P. D., and Oldendorf, W. H.,** Blood-brain barrier restriction of peptides and the low uptake of enkephalins, *Endocrinology*, 103, 1297, 1978.
13. **Rapaport, S. I., Klee, W. A., Pettigrew, K. D., and Ohno, K.,** Entry of opioid peptides into the central nervous system, *Science*, 207, 84, 1980.
14. **Banks, W. A. and Kastin, A. J.,** Peptide transport systems for opiates across the blood-brain barrier, *Am. J. Physiol.*, 259, E1–E10, 1990.
15. **Graffenried von, B., del Pozo, E., Roubicek, J., Krebs, E., Pöldinger, W., Burmeister, P., and Kerp, L.,** Effects of the synthetic enkephalin analogue FK 33–824 in man, *Nature*, 272, 729, 1978.
16. **Stubbs, W. A., Delitala, G., Jones, A., Jeffcoate, W. J., Edwards, C. R. W., Ratter, S. J., Besser, G. M., Bloom, S. R., and Alberti, K. G. M. M.,** Hormonal and metabolic responses to an enkephalin analogue in normal man, *Lancet*, 2, 1225, 1978.
17. **Krebs, E. and Roubicek, J.,** EEG and clinical profile of a synthetic analogue of methionine-enkephalin, *Pharmakopsychiatr. Neuro-Psychopharmakol.*, 12, 86, 1979.
18. **Nedopil, N. and Rüther, E.,** Effects of methionine-enkephalin analogue FK 33–824 on psychotic symptoms, *Pharmakopsychiatr. Neuro-Psychopharmakol.*, 12, 277, 1979.
19. **Jorgensen, A., Fog, R., and Veilis, B.,** Synthetic enkephalin analogue in treatment of schizophrenia, *Lancet*, 1, 935, 1979.
20. **Földes, J., Török, K., Székely, J. I., Borvendég, J., Karczag, I., Tolna, J., Marosfi, S., Váradi, A., Gara, Á., Rónai, A. Z., and Szilágyi, G.,** Human tolerability studies with D-Met^2, Pro^5-enkephalinamide, *Life Sci.*, 33 (Suppl. I.), 769, 1983.
21. **Székely, J. I., Török, K., Karczag, I., Tolna, J., and Till, M.,** Effects of D-Met^2,Pro^5-enkephalinamide on pain tolerance and cognitive functions in man, *Psychopharmacology*, 89, 409, 1986.
22. **Johnson, R. E. and Jasinski, D. R.,** Opioid profile of BW942C in man, *NIDA Res. Monogr.*, 76, 81, 1987.
23. **Calimlim, J. F., Wardell, W. M., Sriwatanakul, K., Lasagna, L., and Cox, C.,** Analgesic efficacy of parenteral metkephamid acetate in treatment of postoperative pain, *Lancet*, 1, 1374, 1980.
24. **Stacher, G., Bauer, P., Steinringer, H., Schreiber, E., and Schmierer, G.,** Effects of the synthetic enkephalin analogue FK 33–824 on pain threshold and pain tolerance in man, *Pain*, 7, 159, 1979.
25. **Bloomfield, S. S., Barden, T. P., and Mitchell, J.,** Metkephamid and meperidine analgesia after episiotomy, *Clin. Pharmacol. Ther.*, 34, 240, 1983.
26. **Catlin, D. H., Hui, K. K., Loh, H. H., and Li, C. H.,** Pharmacologic activity of β-endorphin in man, *Commun. Psychopharmacol.*, 1, 439, 1977.
27. **Frederickson, R. C. A., Smithwick, E. L., and Henry, D. P.,** Opioid peptides as brain neurotransmitters with therapeutic potential, in *Neuropeptides and Neural Transmission,* Marsan, C. J. and Traczyk, W. Z., Eds., Raven Press, New York, 1980, 227.
28. **Stacher, G., Steinringer, H., and Schmierer, G.,** Stimulatory effects of the synthetic enkephalin analogue FK 33–824 on colonic motor activity antagonized by naloxone, *Hepato-Gastroenterol.*, 28, 110, 1981.

29. **Gaillard, R. C., Grossman, A., Smith, R., Rees, L. H., and Besser, G. M.,** The effects of a Met-enkephalin analogue on ACTH, β-LPH, β-endorphin and Met-enkephalin in patients with adrenocortical disease, *Clin. Endocrinol.*, 14, 471, 1981.
30. **Allolio, B., Winkelmann, W., Hipp, F. X., Kaulen, D., and Mies, R.,** Effects of a Met-enkephalin analog on adrenocorticotropin (ACTH), growth hormone, and prolactin in patients with ACTH hypersecretion, *J. Clin. Endocrinol. Metab.*, 55, 1, 1981.
31. **Pasanisi, F., Sloan, L., and Rubin, P. C.,** Cardiovascular properties of metkephamid, a delta opioid receptor agonist in man, *Clin. Sci.*, 68, 209, 1985.
32. **Ryan, J., Leighton, J., Kirksey, D. F., and McMahon, B.,**Evaluation of an enkephalin analog in man with castor oil-induced diarrhea, *Clin. Pharmacol.*, 39, 40, 1986.
33. **Boucher, B. A., Foster, T. S., Phillips, B. A., Fleck, R. J., Kirksey, D. F., and Cato, A. E.,** The acute effects of the enkephalin analogue BW 942C in humans, *J. Clin. Pharmacol.*, 27, 151, 1987.
34. **Giugliano, D., Quatraro, A., Consoli, G., Ceriello, A., Torella, R., and D'Onofrio, F.,** Inhibitory effect of enkephalin on insulin secretion in healthy subjects and in non insulin-dependent diabetic subjects, *Metab. Clin. Exp.*, 36, 286, 1987.
35. **Vaupel, D. B., Cone, E. J., Johnson, R. E., and Su, T.-P.,** Kappa opioid partial agonist activity of the enkephalin-like pentapeptide BW942C based on urination and *in vitro* studies, *J. Pharmacol. Exp. Ther.*, 252, 225, 1990.
36. **Lord, J. A. H., Waterfield, A. A., Hughes, J., and Kosterlitz, H. W.,** Endogenous opioid peptides: multiple agonists and receptors, *Nature*, 267, 495, 1977.
37. **Illés, P.,** Modulation of transmitter and hormone release by multiple neuronal opioid receptors, *Rev. Physiol. Biochem. Pharmacol.*, 112, 139, 1989.
38. **Voigtlander, von, P. F., Lahti, R. A., and Ludens, J. H.,** U-50,488H: a selective and structurally novel non-mu (kappa) opioid agonist, *J. Pharmacol. Exp. Ther.*, 224, 7, 1983.
39. **Gmerek, D. E., Dykstra, L. A., and Woods, J. H.,** Kappa opioids in rhesus monkeys. III. Dependence associated with chronic administration, *J. Pharmacol. Exp. Ther.*, 242, 428, 1987.
40. **Kay, D. C., Eisenstein, R. B., and Jasinski, D. R.,** Morphine effects on human REM state, waking state and NREM sleep, *Psychopharmacology*, 14, 404, 1969.
41. **Kay, D. C.,** Human sleep during chronic morphine intoxication, *Psychopharmacology*, 44, 117, 1975.
42. **Kay, D. C., Pickworth, W. B., and Neidert, G. L.,** Morphine-like insomnia from heroin in nondependent human addicts, *Br. J. Clin. Pharmacol.*, 11, 159, 1981.
43. **Khazan, N., Weeks, J. R., and Schroeder, L. A.,** Electroencephalographic, electromyographic and behavioral correlates during a cycle of self-maintaining morphine addiction in the rat, *J. Pharmacol. Exp. Ther.*, 155, 521, 1967.

44. **Khazan, N. and Sawyer, C. H.,** Mechanisms of paradoxical sleep as revealed by neurophysiological and pharmacological approaches in the rabbit, *Psychopharmacology*, 5, 457, 1964.
45. **Echols, S. D. and Ewett, R. E.,** Effects of morphine on sleep in the cat, *Psychopharmacology*, 24, 435, 1972.
46. **Katz, J. L., Woods, J. H., Winger, G. D., and Jacobson, A. E.,** Compounds of novel structure having kappa-agonist behavioral effects in rhesus monkeys, *Life Sci.*, 31, 2375, 1982.
47. **Tang, A. H. and Collins, R. J.,** Behavioral effects of a novel kappa opioid analgesic, U-50,488H in rats and rhesus monkeys, *Psychopharmacology*, 85, 309, 1985.
48. **Cowan, A., Zhu, X. Z., Mosberg, H. I., Omnaas, J. R., and Porreca, F.,** Direct dependence studies in rats with agents selective for different types of opioid receptors, *J.Pharmacol. Exp. Ther.*, 246, 950, 1988.
49. **Saelens, J. K., Granat, R. F., and Sawyer, W. K.,** The mouse jumping test: a simple screening method to estimate physical dependence capacity of analgesics, *Arch. Int. Pharmacodyn.*, 190, 213, 1971.
50. **Weeks, J. R.,** Experimental morphine addiction: Method for automatic intravenous injections in unrestrained rats, *Science*, 138, 143, 1962.
51. **Weeks, J. R. and Collins, R. J.,** Factors affecting voluntary morphine intake in self-maintained addicted rats, *Psychopharmacology*, 6, 267, 1964.
52. **Thompson, T. and Schuster, C. R.,** Morphine self-administration, food reinforced, and avoidance behaviors in rhesus monkeys, *Psychopharmacology*, 5, 87, 1964.
53. **Deneau G., Yanagita, T., and Seevers, M. H.,** Self-administration of psychoactive substances by the monkey, *Psychopharmacology*, 16, 30, 1969.
54. **Woods, J. H., Young, A. M., and Herling, S.,** Classification of narcotics on the basis of their reinforcing, discriminative, and antagonist effects in rhesus monkeys, *Fed. Proc. Fed. Am. Soc. Exp. Biol.*, 41, 221, 1982.
55. **Davis, W. M. and Nichols, J. R.,** A technique for self-injection of drugs in the study of reinforcement, *J. Exp. Anal. Behav.*, 6, 233, 1963.
56. **Smith, S. G., Werner, T. E., and Davis, W. M.,** Technique for intragastric delivery of solution: application for self-administration of morphine and alcohol by rats, *Physiol. Psychol.*, 3, 220, 1975.
57. **Jaffe, A. B., Sharpe, L. G., and Jaffe, J. H.,** Rats self-administer sufentanil in aerosol form, *Psychopharmacology*, 99, 289, 1989.
58. **Amit, Z., Brown, Z. W., and Sklar, L. S.,** Intraventricular self-administration of morphine in naive laboratory rats, *Psychopharmacology*, 48, 291, 1976.
59. **Wikler, A., Martin, W. R., Pescor, F. T., and Eades, C. G.,** Factors regulating oral consumption of an opioid (etonitazene) by morphine addicted rats, *Psychopharmacology*, 5, 55, 1963.
60. **Stolerman, I. P. and Kumar, R.,** Preference for morphine in rats: validation of an experimental model of dependence, *Psychopharmacology*, 17, 137, 1970.
61. **Amit, Z., Corcoran, M. E., Amir, S., and Urca, G.,** Ventral hypothalamic lesions block the consumption of morphine in rats, *Life Sci.*, 13, 805, 1973.

62. **Schuster, C. R. and Woods, J. H.,** The conditioned reinforcing effects of stimuli associated with morphine reinforcement, *Int. J. Addict.*, 3, 223, 1968.
63. **Harrigan, S. E. and Downs, D. A.,** Self-administration of heroin, acetylmethadol, morphine, and methadone in rhesus monkeys, *Life Sci.*, 22, 619, 1978.
64. **Mucha, R. F.,** Taste aversion involving central opioid antagonism is potentiated in morphine-dependent rats, *Life Sci.*, 45, 671, 1989.
65. **Tulunay, F. C. and Takemori, A. E.,** The increased efficacy of narcotic antagonists induced by various narcotic analgesics, *J. Pharmacol. Exp. Ther.*, 190, 395, 1974.
66. **Tulunay, F. C. and Takemori, A. E.,** Further studies on the alteration of analgesic receptor-antagonist interaction induced by morphine, *J. Pharmacol. Exp. Ther.*, 190, 401, 1974.
67. **Woods, J. H.,** Narcotic-reinforced responding: a rapid evaluation procedure, *Drug Alcohol. Dep.*, 5, 223, 1980.
68. **Hein, D. W., Young, A. M., Herling, S., and Woods, J. H.,** Pharmacological analysis of the discriminative stimulus characteristics of ethylketazocine in the rhesus monkey, *J. Pharmacol. Exp. Ther.*, 218, 715, 1981.
69. **Young, G. A. and Khazan, N.,** Self-administration of ketocyclazocine and ethylketocyclazocine by the rat, *Pharmacol. Biochem. Behav.*, 19, 711, 1983.
70. **Khazan, N., Young, G. A., and Calligaro, D.,** Self-administration of dynorphin-[1–13] and D-Ala^2-dynorphin-[1–13] (kappa opioid agonists) in morphine (mu opioid agonist)-dependent rats, *Life Sci.*, 33, (Suppl. I), 559, 1983.
71. **Belluzzi, J. D. and Stein, L.,** Enkephalin may mediate euphoria and drive-reduction reward, *Nature*, 266, 556, 1977.
72. **Rónai, A. Z.,** Detection of endogenous ligands of opiate receptors, in *Opioid Peptides,* Vol. III, Opiate Receptors and their Ligands, Székely, J. I. and Rónai, A. Z., Eds., CRC Press, Boca Raton, FL, 1983, 43.
73. **Rónai, A. Z., Székely, J. I., Bajusz, S., and Berzétei, I.,** The *in vitro* pharmacology of D-Met^2,Pro^5-enkephalinamide, *J. Pharmacol. Exp. Ther.*, 33, 534, 1981.
74. **Van Ree, J. M., Smyth, D. G., and Colpaert, F. C.,** Dependence creating properties of lipotropin C-fragment (β-endorphin): evidence for its internal control of behavior, *Life Sci.*, 24, 195, 1979.
75. **Phillips, A. G. and Le Piane, F. G.,** Reward produced by microinjection of D-Ala-2-Met-enkephalinamide into the ventral tegmental area, *Behav. Brain Res.*, 5, 225, 1982.
76. **Olds, M. E. and Williams, K. N.,** Self-administration of D-Ala-2-Met-enkephalinamide at hypothalamic self-stimulation sites, *Brain Res.*, 194, 155, 1980.
77. **Goeders, N. E., Lane, J. D., and Smith, J. E.,** Self-administration of methionine-enkephalin into the nucleus accumbens, *Pharmacol. Biochem. Behav.*, 20, 451, 1984.
78. **Stevens, K. E., Shiotsu, G., Belluzzi, J. D., and Stein, L.,** Dynorphin A: self-administration in CA3 hippocampal field in the rat, *Soc. Neurosci. Abstr.*, 15, 293, 1989.

79. **Rossi, N. A. and Reid, L. D.,** Affective states associated with morphine injections, *Physiol. Psychol.*, 4, 269, 1976.
80. **Katz, R. J. and Gormezano, G.,** A rapid inexpensive technique for assessing the reinforcing effects of opiate drugs, *Pharmacol. Biochem. Behav.*, 11, 231, 1979.
81. **Mucha, R. F., Van der Kooy, D., O'Shaugnessy, M., and Bucenieks, P.,** Drug reinforcement studies by use of place conditioning in rat, *Brain Res.*, 243, 91, 1982.
82. **Mucha, R. F. and Iversen, S. D.,** Reinforcing properties of morphine and naloxone revealed by conditioned place preferences: a procedural examination, *Psychopharmacology*, 82, 241, 1984.
83. **Sherman, J. E., Pickman, C., Rice, A., Liebeskind, J. C., and Holman, E. W.,** Rewarding and aversive effects of morphine: temporal and pharmacological properties, *Pharmacol. Biochem. Behav.*, 13, 501, 1980.
84. **Mucha, R. F. and Herz, A.,** Motivational properties of kappa and mu opioid receptor agonists studied with place and taste preference conditioning, *Psychopharmacology*, 86, 274, 1985.
85. **Van der Kooy, D., Mucha, R. F., O'Shaugnessy, M., and Bucenieks, P.,** Reinforcing effects of brain microinjections of morphine revealed by conditioned place preference, *Brain Res.*, 243, 107, 1982.
86. **Stolerman, I. P., Pilcher, C. W., and D'Mello, G. D.,** Steroespecific aversion property of narcotic antagonists in morphine-free rats, *Life Sci.*, 22, 1755, 1978.
87. **Mucha, R. F., Millan, M. J., and Herz, A.,** Aversive properties of naloxone in non-dependent (naive) rats may involve blockade of central β-endorphin, *Psychopharmacology*, 86, 281, 1985.
88. **Bals-Kubik, R., Herz, A., and Shippenberg, T. S.,** Evidence that aversive effects of opioid antagonists and κ-agonists are centrally mediated, *Psychopharmacology*, 98, 203, 1989.
89. **Bechara, A. and Van der Kooy, D.,** Opposite motivational effects of endogenous opioids in brain and periphery, *Nature*, 314, 535, 1985.
90. **Dymshitz, J. and Lieblich, I.,** Opiate reinforcement and naloxone aversion as revealed by place preference paradigm in two strains of rats, *Psychopharmacology*, 92, 473, 1987.
91. **Stapleton, J. M., Lind, M. D., Merriman, V. J., Bozarth, M.A., and Reid, L. D.,** Affective consequences and subsequent effects on morphine self-administration of d-ala^2-methionine enkephalin, *Physiol. Psychol.*, 7, 146, 1979.
92. **Amalric, M., Cline, E. J., Martinez, J. L., Jr., Bloom, F. E., and Koob, G. F.,** Rewarding properties of β-endorphin as measured by conditioned place preference, *Psychopharmacology*, 91, 14, 1987.
93. **Bals-Kubik, R., Shippenberg, T. S., and Herz, A.,** Involvement of central μ and δ opioid receptors in mediating the reinforcing effect of β-endorphin in the rat, *Eur. J. Pharmacol.*, 175, 63, 1990.
94. **Sweep (Fred), C. G. J., Wiegant, V. M., De Vry, J., and Van Ree, J. M.,** β-Endorphin in brain limbic structures as neurochemical correlate of psychic dependence on drugs, *Life Sci.*, 44, 1133, 1989.
95. **Iwamoto, E. T.,** Dynorphin A [1–17] induces "reward" in rats in the place conditioning paradigm, *Life Sci.*, 43, 503, 1988.

96. **Iwamoto, E. T.,** Characterization of dynorphin A place preference in rats, *NIDA Res. Monogr.*, 95, 308, 1989.
97. **Stevens, K. E. and Stein, L.,** Mu receptors may mediate hippocampal endogenous opioid reinforcement, *Soc. Neurosci. Abstr.*, 15, 36, 1989.
98. **Carboni, E., Acquas, E., Leone, P., and Di Chiara, G.,** $5HT_3$ receptor antagonists block morphine- and nicotine-induced but not amphetamine-induced reward, *Psychopharmacology*, 97, 175, 1989.
99. **Higgins, G. A., Nguyen, P., and Sellers, E. M.,** Blockade of morphine place conditioning by the CCK_A receptor antagonist devazepide, *Eur. J. Pharmacol.*, 197, 229, 1991.
100. **Olds, J. and Milner, P.**, Positive reinforcement produced by electrical stimulation of the septal area and other regions of the brain, *J. Comp. Physiol. Psychol.*, 47, 419, 1954.
101. **Delgado, J. M. R., Roberts, W. W., and Miller, N. E.,** Learning motivated by electrical stimulation of the brain, *Am. J. Physiol.*, 179, 587, 1954.
102. **Olds, J. and Travis, R. P.,** Effects of chlorpromazine, mepbrobamate, pentobarbital and morphine on self-stimulation, *J. Pharmacol. Exp. Ther.*, 128, 397, 1960.
103. **Adams, W. J., Lorens, S. A., and Mitchell, C. L.,** Morphine enhances lateral hypothalamic self-stimulation in the rat, *Proc. Soc. Exp. Biol. (N.Y.)*, 140, 770, 1972.
104. **Székely, J. I.,** The most characteristic in vivo effects of opiates: behavioral effects, in *Opioid Peptides*, Vol. I. Research Methods, Székely, J. I. and Rónai, A. Z., Eds., CRC Press, Boca Raton, FL, 1982, 45.
105. **Lorens, S. A. and Mitchell, C. L.**, Influence of morphine on lateral hypothalamic self-stimulation in the rat, *Psychopharmacology*, 32, 271, 1973.
106. **Glick, S. D. and Rapaport, G.,** Tolerance to the facilitatory effect of morphine on self-stimulation of the median forebrain bundle in rats, *Res. Commun. Chem. Pathol. Pharmacol.*, 9, 647, 1974.
107. **Weibel, S. L. and Wolf, H. H.,** Opiate modification of intracranial self-stimulation in the rat, *Pharmacol. Biochem. Behav.*, 10, 71, 1979.
108. **Broekkamp, C. L., Van den Bogaard, J. H., Heijnen, H. J., Rops, R. H., Cools, A. R., and Van Rossum, J. M.,** Separation of inhibiting and stimulating effects of morphine on self-stimulation behavior by intracerebral microinjections, *Eur. J. Pharmacol.*, 36, 443, 1976.
109. **Marcus, R. and Kornetsky, C.,** Negative and positive intracranial reinforcement thresholds: effects of morphine, *Psychopharmacology*, 38, 1, 1974.
110. **Izenwasser, S. E. and Kornetsky, C.,** Pharmacological effects of morphine on brain-stimulation reward, *Psychopharmacology*, 93, 136, 1987.
111. **Esposito, R. U. and Kornetsky, C.,** Morphine lowering self-stimulation thresholds: lack of tolerance with long-term administration, *Science*, 195, 189, 1977.
112. **Kornetsky, C., Esposito, R. U., McLean, S., and Jacobsen, J. O.,** Intracranial self-stimulation thresholds: a model for the hedonic effects of drugs of abuse, *Arch. Gen. Psychiat.*, 36, 291, 1979.
113. **Holtzman, S. G.,** Comparison of the effects of morphine, pentazocine, cyclazocine and amphetamine on intracranial self-stimulation in the rat, *Psychopharmacology*, 46, 223, 1976.

114. **Trujillo, K. A., Belluzzi, J. D., and Stein, L.,** Effects of opiate antagonists and their quaternary analogues on nucleus accumbens self-stimulation, *Behav. Brain Res.*, 33, 181, 1989.
115. **Trujillo, K. A., Belluzzi, J. D., and Stein, L.,** Naloxone suppression of self-stimulation is independent of response difficulty, *Pharmacol. Biochem. Behav.*, 33, 147, 1989.
116. **Overton, D. A.,** Similarities and differences between behavioral control by drug-produced stimuli and by sensory stimuli, in *Transduction Mechanisms of Drug Stimuli*, Colpaert, F. C. and Balster, R. L., Eds., Springer-Verlag, Berlin, 1988, 176.
117. **Herling, S. and Woods, J. H.,** Discriminative stimulus effect of etorphine in rhesus monkeys, *Psychopharmacology*, 72, 265, 1981.
118. **Valentino, R. J., Herling, S., Woods, J. H., Medzihradsky, F., and Merz, H.,** Evidence for the central mediation of the discriminative stimulus effect of narcotic agonists and antagonists, *J. Pharmacol. Exp. Ther.*, 217, 652, 1981.
119. **Shannon, H. E. and Holtzman, S. G.,** Evaluation of the discriminative effects of morphine in rats, *J. Pharmacol. Exp. Ther.*, 198, 54, 1976.
120. **Colpaert, F. C.,** Discriminative stimulus properties of narcotic analgesic drugs, *Pharmacol. Biochem. Behav.*, 9, 863, 1978.
121. **Negus, S. S., Picker, M. J., and Dykstra, L. A.,** Interactions between mu and kappa opioid agonists in the rat discrimination procedure, *Psychopharmacology*, 102, 465, 1990.
122. **Colpaert, F. C. and Janssen, P. A. J.,** Agonist and antagonist effects of prototype opiate drugs in rats discriminating fentanyl from saline: characteristics of partial generalization, *J. Pharmacol. Exp. Ther.*, 229, 193, 1984.
123. **Picker, M. J., Negus, S. S., and Dykstra, L. A.,** Opiate-like discriminative stimulus properties of benzomorphans in the pigeon: stereospecificity and differential substitution patterns, *Life Sci.*, 45, 1637, 1989.
124. **Musacchio, J. M.,** The psychotomimetic effects of opiates and the sigma receptor, *Neuropsychopharmacology*, 3, 191, 1990.
125. **Tang, A. H. and Code, R. A.,** Discriminative stimulus properties of nalorphine in rhesus monkeys, *J. Pharmacol. Exp. Ther.*, 227, 563, 1983.
126. **Dykstra, L. A., Gmerek, D. E., Winger, G., and Woods, J. H.,** Kappa opioids in rheusus monkeys. I. Diuresis, sedation, analgesia and discriminative stimulus effects, *J. Pharmacol. Exp. Ther.*, 242, 413, 1987.
127. **Holtzman, S. G., Cook, L., and Steinfels, G. F.,** Discriminative stimulus effects of spiradoline, a kappa-opioid agonist, *Psychopharmacology*, 105, 447, 1991.
128. **Meecham, K. G., Boyle, S. J., Hunter, J. C., and Hughes, J.,** An *in vitro* profile of activity for the (+) and (−) enantiomers of spiradoline and PD 117302, *Eur. J. Pharmacol.*, 173, 151, 1989.
129. **Picker, M. J., Doty, P., Negus, S. S., Mattox, S. R., and Dykstra, L. A.,** Discriminative stimulus properties of U-50,488H and morphine: effects of training dose on stimulus substitution patterns produced by mu and kappa opioid agonists, *J. Pharmacol. Exp. Ther.*, 254, 13, 1990.
130. **Negus, S. S., Picker, M. J., and Dykstra, L. A.,** Interaction between the discriminative stimulus effects of mu and kappa opioid agonists in the squirrel monkey, *J. Pharmacol. Exp.Ther.*, 256, 149, 1991.

5 The Role of Dopamine in the Reinforcing Effect of Opiates

As discussed in Section III of this chapter, it was proposed already in the early 1980s that the rewarding action of opiates is brought about by activation of the dopamine system. However, others held just the opposite view: opiate agonists and psychostimulants were supposed to exert their reinforcing/euphoric effects by separate neural systems (Section III). Thus the idea that the euphorigenic action of opiates is brought about by dopaminergic mechanisms is not new, neither is the majority of data which partly confirm partly contradict this postulate. Nevertheless, in view of the opposite motivational effects of μ- and κ-opiates (discussed in Chapter 4), even relatively old observations may obtain new meaning. Accordingly, in the subsequent sections the following questions will be addressed: is dopamine involved in the hedonic state elicited by intracranial electrical self-stimulation (ICSS), i.e., in an experimental model in which the reward is induced by mobilization of ''unspecified'' endogenous neurotransmitters and not by extraneous abused drugs (Section I). If the answer is yes, a next logical question is whether or not dopamine is also involved in the rewarding potency of psychostimulants like amphetamine and cocaine. Namely, these agents are known to enhance the dopaminergic tone by enhancing its release from and/or inhibiting its reuptake into the presynaptic terminals (see Section II and Chapter 7). Since the related data almost unequivocally support this proposition (Section II), it is reasonable to determine from the literature whether or not dopamine is also involved in mediation of the euphorigenic action of μ-opiates (Section III). As discussed there and in Chapters 6 and 7, the majority of related data speak for the existence of uniform or at least partly overlapping reward mechanisms mediating the euphoria induced by μ-opiates and psychostimulants, respectively. Therefore, the next questions are whether or not opiate antagonists modify the rewarding action of psychostimulants (Chapter 4) and in general how the two main groups of abused substances interact with each other (Section V). Finally, in view of the dysphoric action of κ-opiates (Chapter 4), it is logical to examine whether the anhedonic states induced by κ-opioids or withdrawal of μ-opioids are related to hypoactivity of the dopaminergic system (Section VI). A further logical question as to whether or not dopamine is also involved in the rewarding effect of other abused drugs is addressed in a separate chapter (Chapter 10).

I. THE ROLE OF DOPAMINE IN INTRACRANIAL SELF-STIMULATION (ICSS)

Since during ICSS the hedonic state is generated by activating natural "physiological" euphorigenic mechanisms, its pharmacologic manipulation may help to reveal the identity of the underlying neurotransmitters and the receptor populations mediating it. Since, generally, areas in the vicinity of the medial forebrain bundle are stimulated, catecholamines have been the candidates since the very beginning. However, the eventual role of norepinephrine as the main substrate can be excluded, since the drug-induced inhibition of this behavior does not correlate with central adrenergic blockade.[1] On the other hand, in rats, selective dopamine receptor blockers such as chlorpromazine[2] and pimozide[1,3,4] have been shown to decrease the rate of ICSS, whereas amphetamine, a dopamine-releasing agent, increases the rate of self-stimulation.[5,6] Cocaine, the other prototypic dopamine-releasing psychostimulant, also increases the rate of self-stimulation.[7,8] However, as discussed in Chapter 4, in this paradigm not the rate, but rather the reward threshold, i.e., the minimal intensity of current needed to maintain ICSS is the reliable paramater. (Obviously, the eventual changes in the rate of ICSS may always be secondary to aspecific changes in the general motor behavior.) Anyhow, neuroleptics such as chlorpromazine[9] and pimozide[4,10] elevate the reward threshold, whereas amphetamine[9–11] decreases the intensity of the minimal reinforcing current. Similar data were obtained in monkeys: ICSS is facilitated by amphetamine and inhibited by chlorpromazine.[12] Cocaine also lowers the rewarding threshold in rats.[13–15] In rats, apomorphine in low dosage, known to preferentially activate the presynaptic receptors, hence to reduce dopamine release, induces increase in the stimulation current threshold, whereas higher postsynaptically acting doses induce facilitation.[16] Depletion of the brain dopamine with 6-hydroxydopamine results in suppression of ICSS[17–19] and elimination of the facilitatory effect of amphetamine.[20] However, selective depletion of the brain norepinephrine content did not significantly modify the ICSS behavior.[21] These pharmacological data unequivocally prove that an increase in the central dopaminergic tone is rewarding, whereas suppression of the dopamine receptors induces the opposite effect.

Nevertheless, subsequent data indicate that not the whole dopamine system is involved in mediation of dopamine euphoria. Implanting the stimulating electrodes into the ventral tegmental area (VTA), amphetamine facilitated the ICSS if injected into the nucleus accumbens, but it induced just depression upon its administration into the neostriatum.[22] Crystalline application of amphetamine or dopamine into the ventral anterior striatum or the neighboring nucleus accumbens facilitated ICSS (via electrodes placed into the lateral hypothalamus), and destruction by 6-hydroxydopamine of the same area inhibited the behavior.[23] However, in the same study more dorsal or posterior application of these drugs were ineffective.[23] Destruction of the A_9 and A_{10} dopamine-containing neurons by 6-hydroxydopamine injection into the VTA and substantia nigra resulted in persistent loss of ICSS using electrodes aimed, as usual, at the lateral hypothalamus.[24] However, in another study,[25] selective lesion of the VTA (where the cell bodies of A_{10} neurons are located) had no profound effect, whereas destruction of the substantia nigra did suppress ICSS. Finally, Phillips and Fibiger[26] demonstrated that destruction (by 6-hydroxydopamine) of the ascending dopamine system resulted in a permanent loss of ICSS from the VTA. A transient inhibition of ICSS was recorded if the electrodes were located in the n. accumbens or prefrontal cortex. In addition, an ultimate loss of the amphetamine-induced facilitation was seen wherever the electrodes were located.[26] Probably it is pertinent to mention here that electrical stimulation of the VTA results in an amphetaminelike stimulus effect.[27] Namely, upon training rats to press one of the levers in an operant conditioning box for food reward and rewarding the lever presses on one or the other lever, depending on presession treatment with amphetamine or saline, the cueing effect of the former was enhanced if VTA was stimulated in the active drug session.[27] Consequently, electrical stimulation of the VTA and amphetamine might induce similar subjective states.

Collectively, these works, discounting one,[25] indicate that the rewarding effect of psychostimulants, known to act by enhancing dopamine release as amphetamine, is linked to the mesolimbic dopamine system, i.e., to the A_{10} neurons located in the VTA and projecting into the n. accumbens, septum, medial prefrontal cortex, etc. (for details see Chapter 6). Some of the controversies in the above mentioned studies might be due to the poor delineation of n. accumbens from the striatum and eventually to the close location of A_9 and A_{10} cell bodies in the ventral tegmentum.

A further important issue is whether D_1 or D_2 or both subtypes of dopamine receptors are involved in the rewarding action of ICSS. The related data are, for the time being, controversial. In some studies,[28–30] SCH 23390, a highly selective inhibitor of dopamine$_1$ receptors, inhibited ICSS, whereas sulpiride, a pure antagonist of D_2 receptors, had no effect. However, in another study,[31] just metoclopramide, a close pharmacological and structural analog of sulpiride, suppressed ICSS. Furthermore, according to Gallistel and Davis,[32] the relative affinities of various D_2 receptor ligands to their binding sites are closely correlated with their potencies to suppress ICSS.

Thus, for the time being, the subtype of dopamine receptors involved in the rewarding effect of ICSS cannot be specified. On the other hand, the cardinal role of dopamine receptors in the phenomenon is sufficiently proven (see Table 1).

II. DOPAMINERGIC MECHANISMS IN THE REWARDING EFFECT OF PSYCHOSTIMULANTS

The role of dopamine in the rewarding action of psychostimulants (see Table 2) is also unequivocally demonstrated in paradigms where animals have free choice between these drugs and saline. After the pioneering works made with morphine[33,34] it was demonstrated unequivocally that the widely abused psychostimulants such as amphetamine, metamphetamine, and cocaine also are readily **self-administered** by monkeys[35–37] and rats[38–41] in the usual setting, i.e., just as morphine and other classical μ-preferring opiates (see also Chapter 4). Self-administration (SA) can also be established using other psychostimulants such as pipradol and phenmetrazine.[42] Psychostimulants are known to act by dopamine release and/or by inhibition of its reuptake, resulting in higher dopamine concentration at the postsynaptic dopamine receptors.[43–45] Nevertheless, the effect of these psychostimulants is not restricted to dopamine, they inhibit the reuptake of other biogenic amines too, primarily that of norepinephrine.[46,47] Therefore, it is important from a theoretical point of view that GBR 12909, a selective uptake blocker of dopamine, also induces SA behavior in monkeys.[48] In light of these findings it is not surprising that apomorphine, a potent direct dopamine receptor stimulant,[49,50] is also self-administered in rats.[51]

Several other observations provide indirect evidence for the role of dopamine in mediation of the rewarding action of psychostimulants. If animals, trained to self-administer cocaine or amphetamine, are deprived of psychostimulants, i.e., the drugs are substituted by saline infusion, first an extinction like increase can be seen in the rate of responding, which is followed by cessation of lever pressing.[38] Actually, the same alteration in responding, i.e., an initial facilitation followed by discontinuation, can be recorded if rats or monkeys working for amphetamine or cocaine receive a dopamine receptor blocking agent such as chlorpromazine,[52,53] haloperidol,[54,55] butaclamol,[56] pimozide,[57–59] sulpiride, and metoclopramide.[59] In some of these experiments only the "initial" facilitation could be recorded when these drugs were applied acutely (mostly in a single session). Furthermore, in these studies small doses of neuroleptics increased, whereas higher doses decreased the response rate. The most parsimonious explanation of these findings is that partial dopamine receptor blockade, induced by small doses of neuroleptics, can be overcome by higher doses of the dopamine-releasing agents. However, complete omission of reward results in an initial repetition of unreinforced responses followed by cessation of responding. Nevertheless, the question arises whether the increase in re-

TABLE 1
The Role of Dopamine in the Maintenance of Intracranial Self-Stimulation (ICSS)

Species examined	Experimental paradigm	Effect observed	Ref.
Rat	Effect of chlorpromazine on the rate of ICSS	Decrease	2
	Effect of pimozide on the rate of ICSS	Decrease	1,3,4
	Effect of amphetamine on the rate of ICSS	Increase	5,6
	Effect of cocaine on the rate of ICSS	Increase	7,8
	Effect of chlorpromazine on the reward threshold	Increase	9
	Effect of pimozide on the reward threshold	Increase	4,10
	Effect of amphetamine on the reward threshold	Decrease	9,10,11
	Effect of cocaine on the reward threshold	Decrease	13,14,15
Monkey	Modulation of the reward threshold by		12
	Chlorpromazine	Increase	
	Amphetamine	Decrease	
Rat	Effect of apomorphine on the reward threshold		16
	In low doses	Increase	
	In high doses	Decrease	
	Effect of depletion of brain dopamine (by 6-OHDA[a]) on the rate of ICSS	Suppression	17,18 19
	Effect of depletion by 6-OHDA of brain dopamine on the facilitating effect of amphetamine	Elimination	20
	Modulation of the rate of ICSS by microinjecting amphetamine into		22
	N. accumbens	Facilitation	
	Neostriatum	Inhibition	
	Modulation of the rate of ICSS by microinjecting: amphetamine or dopamine into		23
	N. accumbens	Facilitation	
	Ventral-anterior striatum	Facilitation	
	Dorsal/posterior striatum	No effect	
	6-OHDA into		
	Ventral striatum	Inhibition	
	Posterior striatum	No effect	
	Effect of destruction of ventral tegmental area (A_{10} neurons) by 6-OHDA on the rate if ICSS	Inhibition	24
		No effect	25
	Effect of destruction of the ascending mesolimbic dopamine fibers on amphetamine-induced facilitation of ICSS	Abolition	26

[a] 6-OHDA = 6-hydroxydopamine.

TABLE 2
Dopaminergic Mechanisms in the Rewarding Effect of Psychostimulants

Species examined	Experimental paradigm	Effect observed	Ref.
Rat	Effect of withdrawal of rewarding (met)-amphetamine, cocaine on responding in self-administration paradigm (SA)	Initial increase in rate of responding followed by its cessation	38
Rat	Effect of decreasing the unit dose of Cocaine (Met)amphetamine on responding in SA paradigm	Increase in response rate resulting in stable amount of reinforcing drug over time	41,60
Rat and Monkey	Effect of various dopamine receptor blockers: chlorpromazine, haloperidol, butaclamol, pimozide, metoclopramide, sulpiride on responding for cocaine or amphetamine in SA paradigm	Initial increase in response rate followed by its cessation	52–59
Rat	Effect of haloperidol on responding for cocaine on progressive ratio schedule until it is maintained in SA paradigm	Decrease in the maximal response/reinforcement ratio	65
Monkey	Effect of depletion of catecholamines by reserpine or AMT[a] on responding for cocaine in SA paradigm	Increase in response rate	42
Rat and Monkey	Effect of α- and β-adrenoceptor blockers: phenoxybenzamine, phentolamine, propranolol on responding for cocaine, amphetamine (SA paradigm)	No consistent effect	42, 56–58
Rat	Effect of destruction of dopamine neurons in the VTA[b] or its terminals in n. accumbens by 6-OHDA[c] on responding for cocaine (SA paradigm)	Disruption	63,65
Healthy males	Effect of AMT on the euphoric effect of amphetamine	Abolition	66
Rat	Effect of α-flupentixol, haloperidol, pimozide in conditioned place preference assay (CPP) maintained by		68,70
	Amphetamine	Abolition	
	Cocaine	No effect	
Rat	Modulation of cocaine maintained CPP by		68
	Haloperidol	No effect	
	Pimozide	No effect	
	6-OHDA into the N. accumbens	No effect	
Rat	Effect of 6-OHDA into the n. accumbens on CPP maintained by apomorphine	No effect	69

TABLE 2 (continued)
Dopaminergic Mechanisms in the Rewarding Effect of Psychostimulants

Species examined	Experimental paradigm	Effect observed	Ref.
Rat	Modulation of responding for cocaine in SA paradigm by		74
	SCH 23390, a selective $dopamine_1$ receptor and	Increase in response rate	
	Spiperone, a selective $dopamine_2$ receptor blocker	No effect	
Rat	Modulation of responding for cocaine in SA paradigm by SCH 23390, a selective $dopamine_1$ receptor and Spiperone, a selective $dopamine_2$ receptor blocker	Increase in response rate after both	75
Rat	Effect of SCH 23390 on amphetamine-maintained CPP	Abolition	79
Rat	Capability of selective $dopamine_2$ receptor agonists as quinpirole and bromocriptine to maintain CPP	Yes	77

[a] AMT = α-methyl-*p*-tyrosine.
[b] VTA = ventral tegmental area.
[c] 6-OHDA = 6-hydroxydopamine.

sponding after these neuroleptics really reflects the omission of reward, or if it is due to dopamine receptor blockade, or to something else. On the other hand, it is well established that neuroleptics per se only inhibit the conditional reactions. Thus the facilitation can only be due to a process which supersedes their depressant effect. Actually, in the same studies[35,58] reduction of the unit dose of cocaine, injected as reward, induced a similar increase in response rate, indicating that increase in response rate is a manifestation of lack of reinforcement.

As an interesting corollary, it has been repeatedly demonstrated that reduction of the unit dose of rewarding injection results in a higher rate of responding, whereas increasing the unit dose results in decelerating of the response rate whether psychostimulants[41,60] or opiates[61] are used as reinforcers. Furthermore, if rats received supplemental drug infusion, i.e., cocaine or heroin, which was not dependent on operant responding, the response rate decreased, resulting in an approximately constant amount of drug intake on an hourly basis.[62] Thus, the animals self-regulate the amount of drug ingested, i.e., its blood level.

Depletion of the brain catecholamines by reserpine or α-methyl-*p*-tyrosine (AMT) also induced an initial "withdrawal-like" increase in responding for cocaine,[52] hence the possibility arises that other catecholamines are also involved in the reinforcing effect of psychostimulants. However, selective blockade of the α- andor β-adrenoreceptors by phenoxybenzamine, phentolamine, or propranolol failed to elicit the characteristic biphasic effect seen after dopamine receptor blockers.[52,56–58] Thus, their effect could only be due to depletion of dopamine.

Finally, disruption of lever pressing for cocaine was seen after destruction of the A_{10} dopamine neurons in the VTA by 6-hydroxydopamine[63] or their terminals in the nucleus accumbens.[64]

Another approach is the determination of the "breaking point" when using the technique of progressive ratio responding: the highest response:reinforcement ratio is determined which still maintains responding. Haloperidol decreased this ratio for cocaine,[65] which also indicates that combination of the two drugs results in weakening of the reward.

In the only related human study of Gunne et al.,[66] amphetamine-induced euphoria in volunteers was blocked both by AMT and pimozide, but not by phenoxybenzamine and propranolol. Thus dopamine mediates the rewarding effects of psychostimulants in humans, too.

Similar data have been obtained using another paradigm, **conditioned place preference** (CPP), whereupon different contextual stimuli (environments having distinctive features) are associated with the drug under study and saline, then a free choice is given. In this assay, too, amphetamine,[67] cocaine,[68] and even apomorphine[69] behave as positive reinforcers, i.e., trained rats select the environment associated with these direct or indirect dopaminergic stimulants. The amphetamine-maintained place preference can be reversed by pretreatment with α-flupenthixol,[70] but the cocaine-induced CPP could not be attenuated by either pretreatment with haloperidol or pimozide or by injecting 6-hydroxydopamine into the nucleus accumbens, i.e., a main terminal area of A_{10} dopamine neurons.[68,70] The same type of lesion also failed to attenuate the rewarding effect of apomorphine in CPP.[69] As for the ineffectiveness of 6-hydroxydopamine to abolish the rewarding effect of psychostimulants, it can be surmised that the injection of 6-hydroxydopamine into nucleus accumbens only destroys a part of the mesolimbic system and this intervention actually results in receptor supersensitivity on the postsynaptic side, i.e., exaggerated sensitivity of the remaining elements might explain the negative findings. On the other hand, for the time being it is impossible to explain that in some studies[68,70] strong neuroleptics such as haloperidol and pimozide abolished the rewarding effect of amphetamine, but failed to affect the similar action of cocaine.

As for the subtype of dopamine receptors involved in SA of psychostimulants and their ability to induce CPP, the related data are not unequivocal. Monkeys can be trained to self-administer D_2 receptor agonists such as piribedil and bromocriptine, but SKF 38393, a selective agonist of D_1 receptors, has apparently no reinforcing effect.[71–73] Furthermore, rats and monkeys self-administering cocaine responded to sulpiride and metoclopramide, i.e., to coadministration of D_2 receptor agonists with increased rate of responding as a sign of rewarding deficit.[59,73] SCH-23390, a D_1 receptor blocker, induced a similar enhancement of responding for cocaine in rats,[74] but no similar action was seen in rhesus monkeys.[73] However, in a recent study[75] both SCH 233390 and spiperone, a D_2 receptor blocker, enhanced the responding for cocaine and reduced the breaking point, i.e., the highest response:reinforcement ratio completed in rats self-administering cocaine.

In the CPP paradigm, too, the role of D_2 receptors seems to be more important. Namely, D_2 agonists such as quinpirole and bromocriptine induce place preference, whereas the D_1 agonist SKF 38393 elicits place aversion.[76–78] In addition, SCH 23390, i.e., a $dopamine_1$ receptor blocker, suppresses the amphetamine-maintained CPP.[79]

In this context it is worth mentioning that SCH 23390 blocks the stimulus-discriminative property of amphetamine[80] and cocaine as well.[81] Thus further work is needed to clarify whether only D_1 receptors mediate the reinforcing effect of dopaminergic stimulants in CPP paradigm or both D_1 and D_2 ones are involved.

III. DOPAMINE IN MEDIATION OF THE REWARDING EFFECT OF μ-OPIATES

Several lines of evidence indicate that dopamine somehow contributes to the rewarding effect of classical μ-opiates as well. As reviewed below, the eventual role of dopamine in the reinforcing action of morphinelike drugs was demonstrated in experimental setups also used in examination of the rewarding action of psychostimulants (see Table 3).

Davis and Smith[82] first demonstrated that pretreatment with AMT inhibits SA of both amphetamine and morphine in rats. Since they used animals which were first trained to self-administer these drugs, then the operant reaction was extinguished, no "extinction hyperactivity" was recorded. However, using the same experimental paradigm, haloperidol did elicit in small doses the usual facilitation and in higher amounts the suppression of responding.[83] In several subsequent studies various neuroleptics similarly inhibited the responding for self-administered opiates; haloperidol inhibited SA of morphine,[84] α-flupenthixol,[85] and pimozide,[62] the operant responding for heroin (and cocaine as well). However, in all these studies[62,84,85] the effect of these neuroleptics was more or less dose dependent, i.e., no facilitation of responding was seen if applying small doses. The very fact that α-flupenthixol in a certain dose range enhanced responding for cocaine, but inhibited the heroin-motivated lever-pressing

TABLE 3
Dopamine in Mediation of the Rewarding Effect of μ-Opiates

Species examined	Experimental paradigm	Effect observed	Ref.
Rat	Effect of α-methyl-*p*-tyrosine on self-administration (SA) of morphine	Inhibition	82
	Effect of haloperidol on SA of morphine		83
	In small doses	Increase (in response rate)	
	In high doses	Suppression	
	Effect of haloperidol on SA of morphine	Dose-dependent suppression of responding	84
	Effect of α-flupenthixol on SA of heroin	Suppression in high doses	85
	Effect of pimozide on SA of heroin	Suppression in high doses	62
	Modulation of morphine-maintained conditioned place preference (CPP) by		86
	α-Methyl-para-tyrosine	Abolition	
	Diethyldithiocarbamate	No effect	
	Effect of haloperidol on CPP maintained by morphine	Abolition	79,86 87
	Effect of α-flupenthixol on CPP maintained by morphine	No effect	70
	Effect of 6-hydroxydopamine lesion of A_{10} neuronal axons and haloperidol on CPP maintained by (D-Ala^2, Met^5)-enkephalin	Abolition	89
	Effect of 6-hydroxydopamine lesion of n. accumbens on CPP maintained by heroin	Abolition	87
	Effect of SCH 23390, a selective dopamine$_1$ receptor blocker on CPP maintained by morphine	Abolition	79
	Effect of spiperone and sulpiride, i.e., selective dopamine$_2$ receptor blockers on morphine-maintained CPP	No effect	93
	Effect of SCH 23390 on the aversive effects of naloxone and U-69593, a selective κ-agonist	Abolition	93

prompted Ettenberg et al.[85] to postulate that separate neural systems underlay the rewarding effects of dopamine stimulants and opiates. Thus, the mode of interaction of dopamine antagonists with μ-opiates and psychostimulants is not quite the same: neuroleptics in sufficiently high doses simply interrupt the responding for heroin, while, combining them with psychostimulants, facilitation is seen in small doses and inhibition in higher amounts. Hence the possibility arises that the inhibition by neuroleptics of opiate reward is simply due to their conditional reflex inhibitory or general motor depressant actions.

Nevertheless, the opposite conclusion is dictated also by the data collected using the other appropriate assay, CPP. The obvious advantage of this paradigm is that the outcome of the experiments is not dependent on eventual nonspecific, general motor effects. In the first related study morphine-maintained place preference was abolished by AMT (i.e., depletion of all catecholamines) and i.c.v. 6-hydroxydopamine (i.e., selective depletion of brain dopamine), but not by diethyldithiocarbamate, i.e., by selective depletion of brain norepinephrine.[86] Haloperidol also blocked the place preference induced by morphine.[79,86,87] Of course, the latter finding might be due to place aversive effect of pretreatment, i.e., summation of hedonic and aversive effects of two different drugs. However, haloperidol alone did not induced place aversion.[87,88] In the same paradigm the rewarding effect of heroin could be inhibited not only by naloxone, but also by pretreatment with pimozide which prompted Bozarth and Wise[88] to postulate that "...heroin reward is dependent on a dopaminergic substrate".

Truly, in a challenging study[70] haloperidol and α-flupenthixol failed to antagonize the morphine-induced place preference. However, in this work of Mackey and Van der Kooy,[70] these neuroleptics also failed to reverse the effect of cocaine, but not that of amphetamine. Most probably the contradiction with all the other related data is somehow related to procedural differences. Namely, rats in general prefer the dark environment, and the outcome of these experiments also depends on the way of conditioning, whether the preferred drug is paired with the nonpreferred side or the experimental setup is balanced in this aspect, too.[70,79] Some further data also indicate the involvement of dopamine in the rewarding action of classical opiates and opioid peptides. Thus, Phillips et al.[89] established stable place preference to injection of (D-Ala^2, Met^5)-enkephalin into the VTA. This effect was attenuated by both haloperidol or by 6-hydroxydopamine lesions of the ascending dopamine pathways.[89] 6-hydroxydopamine lesion of the n. accumbens also attenuated the heroin-elicited place preference.[87]

Regarding the dopamine receptor subtype involved, the data point to the participation of dopamine$_1$ ones. Leone and Di Chiara[79] demonstrated the antagonism of morphine- (and amphetamine-) induced CPP by pretreatment with SCH 23390, i.e., a selective blocker of dopamine$_1$ receptors,[90,91] but dopamine$_2$ antagonists such as sulpiride and spiperone did not modify the morphine-maintained place preference.[92] Actually, SCH 23390 induced place aversion if given alone,[92] whereas selective blockers of $dopamine_2$ receptors such as sulpiride and spiperone failed to induce place conditioning.[93] Thus, the theoretical possibility arises that the inhibition of morphine-induced CPP by SCH 23390 is simply due to summation of place-aversive and rewarding effects of two drugs. For this reason it is very important that, according to Shippenberg and Herz,[93] SCH 23390 (but not sulpiride and spiperone) also abolished the aversive effect of both naloxone and U-69593, i.e., those of a non-selective opiate antagonist and a selective κ-agonist, respectively.[94]

The theoretical implications of these data are far reaching. It is well established that dopamine release is enhanced by μ-opiates in the nucleus accumbens, whereas it is inhibited by κ-agonists[95] (for details see Chapter 7). Based on these data the euphorigenic action of μ-agonists might be due to excitation of $dopamine_1$ receptors upon release of endogenous opioids, whereas the aversive action of κ-agonists would be due just to the inhibition of dopamine release. Obviously, neither effect can be elicited if the $dopamine_1$ receptors are suppressed by their selective blocker, SCH 23390, which is, per se, aversive.

IV. MODULATION BY OPIATE ANTAGONISTS OF THE REWARDING ACTION OF PSYCHOSTIMULANTS

If the rewarding effect of μ-agonists is brought about by dopamine-release, their reinforcing action is expected to be abolished by both dopamine and opiate receptor blockers. However, following this line of reasoning the rewarding action of psychostimulants is presumed not to be modified by opiate antagonists since they act directly on the dopamine releasing neurons, i.e., beyond the opioid input to the A_{10} cells. As discussed in the previous section, the majority of related data are in accordance with the first proposition, i.e., dopamine receptor blockade seems to abolish the reinforcing effect of both major classes of abused substances.

However, the other proposition that opiate antagonists would only inhibit the reinforcing effect of opiates, but not that of the psychostimulants, apparently does not hold true, or at least there are some conflicting data in the literature.

In the ICSS procedure the threshold-lowering effect of amphetamine is blocked by naloxone[96] and the similar action of cocaine can also be reversed.[97] However, the antagonistic effect of naloxone was in neither study dose dependent,[96,97] indicating that the action might not be specific or the interaction between opiate antagonists and psychostimulants is rather complex. (In other studies the interaction was examined in relation to the rate of bar pressing. However, as discussed in Chapter 4, the overall rate of bar pressing is not a reliable indicator of reward in this assay. Therefore, these data are not discussed here.)

Even less conclusive data have been obtained in the SA paradigm. In monkeys both nalorphine[98] and naloxone[99] failed to modify the bar pressing for cocaine in doses causing characteristic changes (facilitation in small doses, cessation in high amounts) in the same behavior rewarded by morphine or heroin. Naltrexone also failed to modify the SA of metamphetamine in monkeys.[100] Similarly, in rats naltrexone failed to modify the SA of cocaine in any way.[85] Thus in this paradigm, contrary to ICSS, the initial studies apparently confirm the basic assumption that opioid mechanisms do not modify the reinforcing action of psychostimulants. However, some more recent works indicate that the situation might be not so simple. In rats, naltrexone enhanced the rate of responding for cocaine, which might indicate reinforcement deficit, i.e., participation of opioids in its rewarding action.[101] Nevertheless, the effect of naltrexone was not dose dependent and could only be detected in food-satiated animals, which emit less responses than food-deprived ones.[101] However, in some recent studies just the opposite effect was recorded: both naltrexone and naloxone[55,102] suppressed the responding for cocaine in rats, but only in cases when the unit reinforcing dose of cocaine was relatively low.

Finally, in place preference conditioning paradigm pretreatment with naloxone blocked the amphetamine-induced reward.[103] However, in the same work of Trujillo et al.,[103] naloxone by itself induced aversion, which casts serious doubt on the relevance of this conclusion. Obviously, the alleged antagonism might simply be a summation of the rewarding and aversive effects of amphetamine and naloxone, respectively. In another laboratory[104] the cocaine-maintained place preference gradually tapered off during the three consecutive drug-free test sessions if naloxone was coadministered with cocaine in the conditioning sessions. Therefore, Houdi et al.[104] proposed that cocaine might operate through opioid mechanisms, but they did not comment on their findings that in that very experiment there was visibly no difference between the saline + cocaine- and naloxone + cocaine-treated animals on the first test day (a) and the animals conditioned with naloxone + saline spent more time on the drug-associated side on day 2 and 3 than on day 1 (b). The latter fact indicates that naloxone, per se, is aversive and, upon repeating the drug-free test days, the conditioned aversion dissipates. Thus, for the reviewer the conclusion is different: the dysphoric effect of naloxone (for data see Chapter 4) might obscure the euphorigenic effect of cocaine and amphetamine.

Summarily, opiate antagonists failed to modify in a dose-dependent, consistent, and specific way the rewarding effect of cocaine (and amphetamine) in any of the three paradigms. These negative data basically confirm that the reinforcing action of psychostimulants is by and large independent of the endogenous opioids (see Table 4).

TABLE 4
Modulation by Opiate Antagonists of the Rewarding Effect of Psychostimulants

Species examined	Experimental paradigm	Effect observed	Ref.
Rat	Modulation by naloxone of the intracranial self-stimulation threshold lowering effect of amphetamine and cocaine	Reversal of threshold lowering, but the effect is not dose dependent	96,97
Monkey	Modulation by nalorphine of cocaine SA	No effect	98
Monkey	Modulation by naloxone of cocaine SA	No effect	99
Monkey	Modulation by naltrexone of cocaine SA	Enhanced rate of responding, but only in food-satiated state	101
Rat	Modulation by naloxone and naltrexone of the rate of responding for cocaine	Decrease in rate if the unit dose of cocaine is low	55,102
		No effect at all	85
Monkey	Modulation by naltrexone of metamphetamine SA	No effect	100
Rat	Modulation by naloxone of CPP maintained by amphetamine	Inhibition, but naloxone is aversive by itself	103
Rat	Modulation by naloxone of place preference maintained by cocaine	Inhibition, but naloxone seems to be aversive by itself	104

Note: SA = self-administration; CPP = conditioned place preference.

V. INTERACTION OF PSYCHOSTIMULANTS AND CLASSICAL OPIATES IN GENERATION OF EUPHORIA/REWARD AND ANALGESIA

Quite a few data indicate that the functional changes induced by opioids and psychostimulants might be additive or synergistic (Table 5). Since there is a wealth of such observations and their detailed description might exceed the scope of this monograph, they areonly summarized below.

TABLE 5
Evidence for Positive Interaction Between Classical Opiates and Psychostimulants in Induction of Euphoria/Reward

Line of evidence	Ref.
Psychostimulants and classical opiates induce locomotor hyperactivity if injected systemically or into the ventral tegmental area	See Chapter 7
Naloxone attenuates the locomotor hyperactivity, especially ambulation induced by amphetamine, cocaine, apomorphine (as it does with classical opiates)	104–108
In rodents amphetamine, cocaine, apomorphine, and bromocriptine potentiate the antinociceptive action of opiates	112–118, 120–122
Amphetamine and cocaine enhance opiate analgesia in humans	97,109–111
Apomorphine may suppress some opiate-withdrawal signs in humans and mice	118,123,124
Heroin and amphetamines enhance the euphorigenic action of each other	
In humans:	125–128
In rats:	129–130

A. μ- and δ-Agonists

As discussed in Chapter 7, μ- and δ-agonists injected systemically or into the VTA induce hyperactivity, or, more exactly, intense locomotion. Thus, as for the spontaneous motility, or at least its horizontal component, the effects of classical opiates and psychostimulants are similar.

B. Naloxone

Naloxone attenuates the ambulation induced by apomorphine and amphetamine in rats, but does not modify the stereotypy induced by these dopamine stimulants.[105–108] Naloxone also inhibits (apparently the horizontal component) of cocaine-induced hyperactivity.[104] As discussed in Chapter 7, μ- and δ-opioids induce locomotor hyperactivity via activation of the mesolimbic dopamine system. Furthermore, it is well established that stereotypy is induced by dopaminergic agonists brought about via activation of the nigrostriatal system, i.e., this type of hyperactivity is independent of the mesolimbic dopamine system. The antagonism by naloxone of locomotor hyperactivity (but not stereotypy) elicited by direct (apomorphine) or indirect (amphetamine, cocaine) dopamine receptor stimulation raises the possibility that the locomotor stimulant effects of opioids and psychostimulants might be additive or even synergistic.

C. Coadministration of Amphetamine

It is an old clinical experience that the analgesic effect of opiates is enhanced by amphetamine and their coadministration results in more intense euphoria.[109–111] Cocaine also potentiates the clinical analgesic effect of opiates (for review see Bain and Kornetsky).[97] The combination of heroin with cocaine is so advantageous that a formula containing heroin (recently morphine) and cocaine was introduced still in the last century, called after the hospital where it was invented, ''Brompton Coctail''.

D. Synergistic Effects

In experimental animals, too, amphetamine has been found to potentiate the analgesic effect of opiates.[112–115] Cocaine also enhances the antinociceptive effect of morphine.[116,117] In our laboratory apomorphine has been found to potentiate the analgesic effect of morphine in five different tests in rats and mice as well,[118] contrary to an earlier study[119] where antagonism was seen in tail-flick tests in mice. In a recent detailed study of Morgan,[120] similar synergism was seen using the paw formalin and vocalization (after)-discharge tests, i.e., assays which probably are the best experimental models of real clinical pain. Bromocriptine, a relatively selective stimulant of dopamine$_2$ receptors, also potentiates morphine analgesia.[121] Thus it is not surprising that Franklin[122] proposed that ''...analgesia and reward may be related processes...''. Actually, in his experiments[122] morphine and amphetamine proved to be synergistic in a (formalin-induced) writhing test and destruction by 6-hydroxydopamine of the VTA abolished the analgesic effects of both as seen in rewarding paradigms (see Chapter 6).

E. Apomorphine

In some clinical studies,[123,124] apomorphine was helpful in the treatment of opiate withdrawal. In our study,[118] apomorphine has been found to suppress the naloxone-induced withdrawal jumping in morphine-dependent mice.

F. Amphetamines

In light of the above reviewed data it is not surprising that amphetamines have been found to enhance the euphoriant effect of heroin in opiate abusers.[125–129]

Furthermore, similar data have been obtained under experimental conditions as well: coadminstration of amphetamine and morphine resulted in greater decrease in the reward threshold in the ICSS paradigm than these abused substances alone.[130]

G. Summary

Summarily, two conclusions seem to be warranted:

1. The majority of data seem to confirm the postulate of Bozart and Wise[88] that the euphoriant effect of opiates seems to be mediated by the (mesolimbic) dopamine system.
2. Coadministration of the two apparently different classes of abused substances results in a kind of synergism or addition.

As discussed in Chapter 10, opioid and dopaminergic mechanisms are involved in the rewarding effect of the other types of abused substances (e.g., ethanol, nicotine) as well. Thus, the theoretical possibility arises that an at least partially common circuitry mediates the euphoriant effect of abused substances in general. If so, it is easy to explain why the many drug addicts relatively easily switch from one abused substance to another one or consume them simultaneously.

VI. DRUG WITHDRAWAL ANHEDONIA AND HYPOFUNCTION OF DOPAMINE SYSTEM

Operationally, anhedonia is defined as the inability to gain pleasure from stimuli, situations which generally induce pleasure and hedonia. It is well known that in humans withdrawal of abused substances induces not only craving, but also depression, anxiety, dysphoria, decreased appetite and sexual drive, and sleep disturbances.

Some data indicate that drug withdrawal is accompanied with anhedonia in experimental animals as well. Weight loss and decreased food intake are well-known symptoms of drug withdrawal. Nevertheless, the question might be asked whether this decrease is related to physiological changes due to altered physiological state induced by withdrawal or if it is really a sign of anhedonia. In a carefully designed study, Carroll et al.[131] demonstrated that withdrawal of cocaine after its previous chronic SA results in a significant decrease in (oral) consumption of glucose + saccharin solution. This mixture is highly palatable in rats.[132] Since the food and water intake was simultaneously increased in the same experiment,[131] these data indicate that the reduction of intake was due to decreased palatability of this solution and not to some kind of disturbance in food intake in general. In other words, it is a manifestation of drug withdrawal-induced anhedonia.

Another sign of anhedonia is the elevation of ICSS threshold upon termination of cocaine,[15] amphetamine,[133,134] and lowering the rate of bar pressing upon interruption of amphetamine[135] or morphine[136] administration. Thus, in the state of drug withdrawal pleasure can only be elicited by more intense electrical stimulation of the pleasure centers.

Pentylenetetrazol is an anxiogenic substance which can apparently be discriminated from saline in animals too by the aid of a drug-discrimination procedure. Namely, rats can be trained to press one of two levers after saline and the other one under the effect of pentetrazol for food reinforcement. Withdrawal of abused substances generates a pentetrazol-like stimulus. That is, rats press the pentetrazol-appropriate lever even in its absence if various classes of abused drugs are withdrawn from them such as morphine,[137] cocaine,[138] nicotine,[139] ethanol,[138] and diazepam.[140] Thus, a behavioral equivalent of human anxiety must be present in rodents too upon withdrawal.

It is now well established that naloxone and naltrexone administration to drug-naive animals decreases the spontaneous food[141,142] and water[143] intake. Based on the above discussed data, it is tempting to speculate that hypodipsia and hypophagia are manifestations of the anhedonic state brought about by displacement of endogenous opioids from their receptors upon administration of exogenous antagonists. The role of dopamine in the phenomenon is indicated by the observations that naltrexone-elicited hypodipsia can be potentiated by SCH 23390, i.e., by selective blockade of $dopamine_1$ receptors,[144,145] but not by sulpiride, a $dopamine_2$ receptor blocker. However, in the same study[145] apomorphine also potentiated the naltrexone-induced suppression of drinking. Since apomorphine acts both pre- and postsynaptically[146] its potentiating effect might be due to further inhibition of dopamine release via stimulation of the presynaptic autoreceptors. Finally, it is worth mentioning that depletion of the dopamine in various regions of the mesolimbic dopamine system results in temporary aphagia and adipsia.[147]

Summarily, all these data indicate that the dysphoria and various manifestations of anhedonia seen upon withdrawal of addictive drugs are somehow related to an overall hypoactivity of the mesolimbic dopaminergic system. Thus, direct or indirect stimulation of certain dopaminergic pathways seems to generate euphoria reward, whereas drug withdrawal apparently elicits opposite alterations. Actually, a wealth of data indicate that all kinds of pleasurable stimuli and functional states might be related to activation of the mesolimbic dopamine system.[148] However, not all the related observations confirm such a crude generalization, and the issue exceeds the topic of this monograph. Anyhow, the majority of data reviewed above indicate that the mesolimbic system is activated and suppressed during the administration and withdrawal of addictive drugs, respectively (see Table 6).

TABLE 6
Dopamine Deficiency in Drug-Withdrawal Anhedonia

Species examined	Experimental paradigm	Effect observed	Ref.
Rat	Effect of withdrawal of cocaine on voluntary consumption of palatable glucose + saccharin solution	Decrease	101
	Effect of cocaine withdrawal on the reward threshold of intracranial self-stimulation (ICSS)	Elevation	15
	Effect of amphetamine withdrawal on the reward threshold of ICSS	Elevation	133,134
	Effect of amphetamine withdrawal on the rate of bar pressing for ICSS	Decrease	135
	Effect of morphine withdrawal on the rate of bar pressing for ICSS	Decrease	136
	Drug-discrimination test, differentiation of saline from pentetrazol; lever selection after withdrawal of morphine	Selection of the pentetrazol-appropriate lever	137
	Training as above, effect of cocaine withdrawal	Selection of the pentetrazol-appropriate lever	138
	Training as above, effect of nicotine withdrawal	Selection of the pentetrazol-appropriate lever	139
	Training as above, effect of ethanol withdrawal	Selection of the pentetrazol-appropriate lever	138
	Training as above, effect of diazepam withdrawal	Selection of the pentetrazol-appropriate lever	140

REFERENCES

1. **Zarevics, P., Weidley, E., and Setler, P. E.,** Blockade of intracranial self stimulation by antipsychotic drugs: failure to correlate with central adrenergic blockade, *Psychopharmacology*, 53, 283, 1977.
2. **Olds, J. and Ito, M.,** Effects of chlorpromazine, chlordiazepoxide and pentobarbital on neuronal excitability in the medial forebrain bundle during self-stimulation behavior, *Neuropharmacology*, 12, 1117, 1973.

3. **Fibiger, H. C., Carter, D. A., and Phillips, A. G.,** Decreased intracranial self-stimulation after neuroleptics or 6-hydroxydopamine: evidence for mediation by motor deficits rather than by reduced reward, *Psychopharmacology*, 47, 21, 1976.
4. **Wise, R. A.,** Neuroleptic attenuation of intracranial self-stimulation: reward or performance deficits?, *Life Sci.*, 22, 535, 1978.
5. **Stein, L.,** Self-stimulation of the brain and the central stimulant action of amphetamine, *Fed. Proc. Fed. Am. Soc. Exp. Biol.*, 23, 836, 1964.
6. **Phillips, A. G. and Fibiger, H. G.,** Dopaminergic and noradrenergic substrates of positive reinforcement: differential effects of d- and l-amphetamine, *Science*, 179, 575, 1973.
7. **Crow, T. J.,** Enhancement by cocaine of intra-cranial self-stimulation in the rat, *Life Sci.*, 9, 375, 1970.
8. **Wauquier, A. and Niemegeers, C. J. E.,** Intracranial self-stimulation in rats as a function of various stimulus parameters. V. Influence of cocaine on medial forebrain bundle stimulation with monopolar electrodes, *Psychopharmacology*, 38, 201, 1974.
9. **Stein, L. and Ray, O. S.,** Brain stimulation "thresholds" self-determined in rat, *Psychopharmacology*, 1, 751, 1960.
10. **Zarevics, P. and Setler, P. E.,** Simultaneous rate-independent and rate-dependent assessment of intracranial self-stimulation: evidence for the direct involvement of dopamine in brain reinforcement mechanisms, *Brain Res.*, 169, 499, 1979.
11. **Cassens, G. P. and Mills, A. W.,** Lithium and amphetamine: opposite effects on threshold of intracranial reinforcement, *Psychopharmacology*, 30, 283, 1973.
12. **Spencer, J. and Rezvin, A.,** Amphetamine, chlorpromazine and clonidine effects on self-stimulation in caudate hypothalamus of the squirrel monkey, *Pharmacol. Biochem. Behav.*, 5, 149, 1976.
13. **Esposito, R. U., Motola, A. H. D., and Kornetsky, C.,** Cocaine: acute effects on reinforcement thresholds for self-stimulation behavior to the medial forebrain bundle, *Pharmacol. Biochem. Behav.*, 8, 437, 1978.
14. **Frank, R. A., Martz, S., and Pommering, T.,** The effect of chronic cocaine on self-stimulation train-duration thresholds, *Pharmacol. Biochem. Behav.*, 29, 755, 1988.
15. **Markou, M. A. and Koob, G. F.,** Postcocaine anhedonia. An animal model of cocaine withdrawal, *Neuropsychopharmacology*, 4, 17, 1991.
16. **Leith, N. J.,** Effects of apomorphine on self-stimulation responding: does the drug mimic the current?, *Brain Res.*, 277, 129, 1983.
17. **Breese, G. R., Howard, J., and Leahy, P.,** Effect of 6-hydroxydopamine on electrical self-stimulation of brain, *Br. J. Pharmacol.*, 43, 255, 1971.
18. **Stein, L. and Wise, C.,** Possible etiology of schizophrenia: progressive damage to the noradrenergic reward system by 6-hydroxydopamine, *Science*, 171, 1032, 1971.
19. **Liebman, J. M. and Butcher, L. L.,** Comparative involvement of dopamine and noradrenaline in rate-free self-stimulation of substantia nigra, lateral hypothalamus, and mesolimbic ventral grey, *Naunyn-Schmiedeberg's. Arch. Pharmacol.*, 284, 167, 1974.

20. **Cooper, B. R., Cott, J. M., and Breese, G. R.,** Effects of catecholamine-depleting drugs and amphetamine on self-stimulation of brain following various 6-hydroxydopamine treatments, *Psychopharmacology*, 37, 235, 1974.
21. **Lippa, A. S., Antelman, S. M., Fisher, A. E., and Canfield, D. R.,** Neurochemical modulation of reward: a significant role for dopamine, *Pharmacol. Biochem. Behav.*, 1, 23, 1973.
22. **Broekkamp, C. L. E., Pijnenburg, A. J. J., Cools, A. R., and Van Rossum, J. M.,** The effect of microinjection of amphetamine into the neostriatum and the nucleus accumbens on self-stimulation behaviour, *Psychopharmacology*, 47, 149, 1975.
23. **Neil, D. B., Peay, L. A., and Gold, M. S.,** Identification of a subregion within rat neostriatum for dopaminergic modulation of lateral hypothalamic stimulation, *Brain Res.*, 153, 515, 1978.
24. **Koob, G. F., Fray, P. J., and Iversen, S. D.,** Self-stimulation at the lateral hypothalamus and locus coeruleus after specific unilateral lesions of the dopamine system, *Brain Res.*, 146, 123, 1978.
25. **Carey, R. J.,** Unilateral 6-hydroxydopamine lesions of dopamine neurons produce bilateral self-stimulation deficits, *Behav. Brain. Res.*, 6, 101, 1982.
26. **Phillips, A. G. and Fibiger, H. C.,** The role of dopamine in maintaining intracranial self-stimulation in the ventral tegmentum, nucleus accumbens, and medial prefrontal cortex, *Can. J. Psychol.*, 32, 58, 1978.
27. **Druhan, J. P., Fibiger, H. C., and Phillips, A. G.,** Amphetamine-like stimulus properties produced by electrical stimulation of reward sites in the ventral tegmental area, *Behav. Brain Res.*, 38, 175, 1990.
28. **Ferrer, J. M. R., Sanguinetti, A. M., Vives, F., and Mora, F.,** Effects of agonists and antagonists of D1 and D2 dopamine receptors on self-stimulation of the medial prefrontal cortex of the rat, *Pharmacol. Biochem. Behav.*, 19, 211, 1983.
29. **Nakajima, S. and McKenzie, G. M.,** Reduction of the rewarding effect of brain stimulation by a blockade of dopamine D1 receptor by SCH 23390, *Pharmacol. Biochem. Behav.*, 24, 919, 1986.
30. **Kurumiya, S. and Nakajima, S.,** Dopamine D_1 receptors in the nucleus accumbens: involvement in the reinforcing effect of tegmental stimulation, *Brain Res.*, 448, 1, 1987.
31. **Fenton, H. M. and Liebman, J. M.,** Self-stimulation response decrement patterns differentiate clonidine, baclofen and dopamine antagonists from drugs causing performance deficits, *Pharmacol. Biochem. Behav.*, 17, 1207, 1982.
32. **Gallistel, C. R. and Davis, A. J.,** Affinity for the dopamine D2 receptor predicts neuroleptic potency in blocking the reinforcing effect of MFB stimulation, *Pharmacol. Biochem. Behav.*, 19, 867, 1983.
33. **Weeks, J. R.,** Experimental morphine addiction: method for automatic intravenous injections in unrestrained rats, *Science*, 138, 143, 1962.
34. **Thompson, T. A. and Schuster, C. R.,** Morphine self-administration, food-reinforced, and avoidance behaviors in rhesus monkeys, *Psychopharmacology*, 5, 87, 1964.
35. **Woods, J. H. and Schuster, C. R.,** Reinforcement properties of morphine, cocaine, and SPA as a function of unit dose, *Int. J. Addict.*, 3, 231, 1968.
36. **Denau, G., Yanagita, T., and Seevers, M.,** Self-administration of psychoactive substances by the monkey, *Psychopharmacology*, 16, 30, 1969.

37. **Goldberg, S. R., Woods, J. H., and Schuster, C. R.,** Nalorphine-induced changes in morphine self-administration in rhesus monkeys, *J. Pharmacol. Exp. Ther.*, 176, 464, 1971.
38. **Pickens, R.,** Self-administration of stimulants by rats, *Int. J. Pharmacol.*, 3, 215, 1968.
39. **Pickens, R. and Harris, W. C.,** Self-administration of d-amphetamine by rats, *Psychopharmacology*, 12, 158, 1968.
40. **Pickens, R., Meisch, R., and McGuire, L. E.,** Metamphetamine reinforcement in rats, *Psychon. Sci.*, 8, 371, 1967.
41. **Pickens, R. and Thompson, T.,** Cocaine-reinforced behavior in rats: effects of reinforcement magnitude and fixed ratio size, *J. Pharmacol. Exp. Ther.*, 161, 122, 1968.
42. **Wilson, M. C. and Schuster, C. R.,** Aminergic influences on intravenous cocaine self-administration by rhesus monkeys, *Pharmacol. Biochem. Behav.*, 2, 563, 1974.
43. **Ritz, M. C., Lamb, R. J., Goldberg, S. R., and Kuhar, M. J.,** Cocaine receptors on dopamine transporters are related to self-administration of cocaine, *Science*, 237, 1219, 1987.
44. **Jaffe, J. H. and Martin, W. H.,** Opioid analgesics and antagonists, in *The Pharmacological Basis of Therapeutics,* Goodman Gilman, A., Rall, T. W., Nies, A. S., and Taylor, P., Eds., Pergamon Press, New York, 1990, 485.
45. **Hoffman, B. B. and Lefkowitz, R. J.,** Catecholamines and sympathomimetic drugs, in *The Pharmacological Basis of Therapeutics,* Goodman Gilman, A., Rall, T. W., Nies, A. S., and Taylor, P., Eds., Pergamon Press, New York, 1990, 187.
46. **Moore, K. E., Carr, L. A., and Dominic, J. A.,** Functional significance of amphetamine-induced release of brain catecholamines, in *Amphetamines and Related Compounds,* Costa, E. and Garattini, S., Eds., Raven Press, New York, 1970, 371.
47. **Wise, C. D. and Stein, L.,** Amphetamine: facilitation of behavior by augmented release of norepinephrine from the median forebrain bundle, in *Amphetamines and Related Compounds,* Costa, E. and Garattini, S., Eds., Raven Press, New York, 1970, 463.
48. **Howell, L. L. and Byrd, L. D.,** Characterization of the effects of cocaine and GBR 12909, a dopamine uptake inhibitor, on behavior in the squirrel monkey, *J. Pharmacol. Exp. Ther.*, 258, 178, 1991.
49. **Andén, N., Rubenson, A., Fuxe, K., and Hökfelt, T.,** Evidence for dopamine receptor stimulation by apomorphine, *J. Pharm. Pharmacol.*, 19, 627, 1967.
50. **Ernst, A. M.,** Mode of action of apomorphine and dexamphetamine on gnawing compulsion in rats, *Psychopharmacology*, 10, 316, 1967.
51. **Baxter, B. L., Gluckman, M. I., Stein, L., and Scerni, R. A.,** Self-injection of apomorphine in the rat: positive reinforcement by a dopamine receptor stimulant, *Pharmacol. Biochem. Behav.*, 2, 387, 1974.
52. **Wilson, M. C. and Schuster, C. R.,** The effects of chlorpromazine on psychostimulant self-administration in the rhesus monkey, *Psychopharmacology*, 26, 115, 1972.
53. **Risner, M. E. and Jones, B. E.,** Role of noradrenergic and dopaminergic processes in amphetamine self-administration, *Pharmacol. Biochem. Behav.*, 5, 477, 1976.

54. **Davis, W. M. and Smith, S. G.,** Effect of haloperidol on (+)-amphetamine self-administration, *J. Pharm. Pharmacol.*, 27, 540, 1975.
55. **De Vry, J., Donselaar, I., and Van Ree, J. M.,** Food deprivation and acquisition of intravenous cocaine self-administration in rats: effect of naltrexone and haloperidol, *J. Pharmacol. Exp. Ther.*, 251, 735, 1989.
56. **Yokel, R. A. and Wise, R. A.,** Attenuation of intravenous amphetamine reinforcement by central dopamine blockade in rats, *Psychopharmacology*, 48, 311, 1976.
57. **Yokel, R. A. and Wise, R. A.,** Increased lever pressing for amphetamine after pimozide in rats: implication for a dopamine theory of reward, *Science*, 187, 547, 1975.
58. **De Wit and Wise, R. A.,** Blockade of cocaine reinforcement in rats with the dopamine blocker pimozide but not the noradrenergic blockers phentolamine and phenoxybenzamine, *Can. J. Psychol.*, 31, 195, 1977.
59. **Roberts, D. C. S. and Vickers, G.,** Atypical neuroleptics increase self-administration of cocaine: an evaluation of a behavioral screen for antipsychotic activity, *Psychopharmacology*, 82, 135, 1984.
60. **Yokel, R. A. and Pickens, R.,** Self-administration of optical isomers of amphetamine and methylamphetamine by rats, *J. Pharmacol. Exp. Ther.*, 187, 27, 1973.
61. **Glick, D., Cox, R. S., and Crane, A. M.,** Changes in morphine self-administration and morphine dependence after lesions of the caudate nuclei in rats, *Psychopharmacology*, 41, 219, 1975.
62. **Gerber, G. J. and Wise, R. A.,** Pharmacological regulation of intravenous cocaine and heroin self-administration in rats: a variable dose paradigm, *Pharmacol. Biochem. Behav.*, 32, 527, 1989.
63. **Roberts, D. C. S. and Koob, G. F.,** Disruption of cocaine self-administration following 6-hydroxydopamine lesions of the ventral tegmental area in rats, *Pharmacol. Biochem. Behav.*, 17, 901, 1982.
64. **Roberts, D. C. S., Koob, G. F., Klonoff, P., and Fibiger, H. C.,** Extinction and recovery of cocaine self-administration following 6-hydroxydopamine lesions of the nucleus accumbens, *Pharmacol. Biochem. Behav.*, 12, 781, 1980.
65. **Roberts, D. C. S., Loh, E. A., and Vickers, G.,** Self-administration of cocaine on a progressive ratio schedule in rats: dose-response relationship and effect of haloperidol pretreatment, *Psychopharmacology*, 97, 535, 1989.
66. **Gunne, L. M., Änggård, E., and Jönsson, L. E.,** Clinical trials with amphetamine-blocking drugs, *Psychiat., Neurol., Neurochir.*, 75, 225, 1972.
67. **Sherman, J. E., Roberts, T., Roskam, S. E., and Holman, E. W.,** Temporal properties of the rewarding and aversive effects of amphetamine in rats, *Pharmacol. Biochem. Behav.*, 13, 597, 1980.
68. **Spyraki, C., Fibiger, H. C., and Phillips, A. G.,** Cocaine-induced place preference conditioning: lack of effects of neuroleptics and 6-hydroxydopamine lesions, *Brain Res.*, 253, 195, 1982.
69. **Van der Kooy, D., Swerdlow, N. R., and Koob, G. F.,** Paradoxical reinforcing properties of apomorphine: effects of nucleus accumbens and area postrema lesions, *Brain Res.*, 259, 111, 1983.
70. **Mackey, W. B. and Van der Kooy, D.,** Neuroleptics block the positive reinforcing effects of amphetamine but not of morphine as measured by place conditioning, *Pharmacol. Biochem. Behav.*, 22, 101, 1985.

71. **Yokel, R. A. and Wise, R. A.,** Amphetamine-type reinforcement by dopamine agonists in the rat, *Psychopharmacology*, 58, 289, 1978.
72. **Woolverton, W. L., Goldberg, L. I., and Ginos, J. Z.,** Intravenous self-administration of dopamine receptor agonists by rhesus monkeys, *J. Pharmacol. Exp. Ther.*, 230, 678, 1984.
73. **Woolverton, W. L.,** Effects of D_1 and D_2 dopamine antagonists on the self-administration of cocaine and piribedil by rhesus monkeys, *Pharmacol. Biochem. Behav.*, 24, 531, 1986.
74. **Koob, G. F., Le, H. T., and Creese, I.,** The D_1 dopamine receptor antagonist SCH 23390 increases cocaine self-administration in the rat, *Neurosci. Lett.*, 79, 315, 1987.
75. **Hubner, C. B. and Moreton, J. E.,** Effects of selective D1 and D2 dopamine antagonists on cocaine self-administration in the rat, *Psychopharmacology*, 105, 151, 1991.
76. **Gilbert, D. B., Dembski, J. E., Stein, L., and Belluzzi, J. D.,** Dopamine and reward: conditioned place preference induced by dopamine D_2 receptor agonist, *Soc. Neurosci. Abstr.*, 12, 938, 1986.
77. **Hoffman, D. C., Dickson, P. R., and Beninger, R. J.,** The dopamine D2 receptor agonists quinpirole and bromocriptine produce conditioned place preferences, *Prog. Neuropsychopharmacol. Biol. Psychiat.*, 12, 315, 1988.
78. **Beninger, R. J., Hoffman, D. C., and Mazurski, E. J.,** Receptor subtype specific dopaminergic agents and conditioned behavior, *Neurosci. Biobehav. Rev.*, 13, 113, 1989.
79. **Leone, P. and Di Chiara, G.,** Blockade of D-1 receptors by SCH 23390 antagonizes morphine- and d-amphetamine-induced place preference conditioning, *Eur. J. Pharmacol.*, 135, 251, 1987.
80. **Nielsen, E. B. and Jepsen S. A.,** Antagonism of amphetamine cue by both classical and atypical antipsychotic drugs, *Eur. J. Pharmacol.*, 111, 167, 1985.
81. **Kleven, M. S., Anthony E. W., Goldberg, L. I., and Woolverton, W. L.,** Blockade of the discriminative stimulus effect of cocaine in rhesus monkeys with the D1 dopamine antagonist SCH 23390, *Psychopharmacology*, 95, 427, 1988.
82. **Davis, W. M. and Smith, S. G.,** Alpha-methyltyrosine to prevent self-administration of morphine and amphetamine, *Curr. Ther. Res.*, 14, 814, 1972.
83. **Smith, S. G. and Davis, W. M.,** Haloperidol effects on morphine self-administration: testing for pharmacological modification of the primary reinforcement mechanism, *Psychol. Rev.*, 23, 215, 1973.
84. **Glick, S. D. and Cox, R. D.,** Dopaminergic and cholinergic influences on morphine self-administration in rats, *Res. Commun. Chem. Pathol. Pharmacol.*, 12, 17, 1975.
85. **Ettenberg, A., Pettit, H. O., Bloom, F. E., and Koob, G. F.,** Heroin and cocaine intravenous self-administration in rats: mediation by separate neural systems, *Psychopharmacology*, 78, 204, 1982.
86. **Schwartz, A. S. and Marchok, P. L.,** Depression of morphine-seeking behavior by dopamine inhibition, *Nature*, 248, 257, 1974.
87. **Spyraki, C., Fibiger, H. C., and Phillips, A. G.,** Attenuation of heroin reward in rats by disruption of the mesolimbic dopamine system, *Psychopharmacology*, 79, 278, 1983.

88. **Bozarth, M. A. and Wise, R. A.,** Heroin reward is dependent on a dopaminergic substrate, *Life Sci.*, 29, 1881, 1981.
89. **Phillips, A. G., LePiane, F. G., and Fibiger, H. C.,** Dopaminergic mediation of reward produced by direct injection of enkephalin into the ventral tegmental area of the rat, *Life Sci.*, 33, 2505, 1983.
90. **Hyttel, J.,** SCH 23390 — the first selective dopamine-1 antagonist, *Eur. J. Pharmacol.*, 91, 153, 1983.
91. **Iorio, L. C., Barnett, A., Leitz, F. H., Houser, V. P., and Korduba, C. A.,** SCH 23390, a potential benzazepine antipsychotic with unique interactions on dopaminergic systems, *J. Pharmacol. Exp. Ther.*, 226, 462, 1983.
92. **Shippenberg, T. S. and Herz, A.,** The preference conditioning reveals the involvement of D_1-dopamine receptors in the motivational properties of μ- and κ-opioid agonists, *Brain Res.*, 436, 169, 1987.
93. **Shippenberg, T. S. and Herz, A.,** Motivational effects of opioids: influence of D-1 versus D-2 receptor antagonists, *Eur. J. Pharmacol.*, 151, 233, 1988.
94. **Lahti, R. A., Mickelson, M. M., McCall, J. M., and von Voigtlander, P. F.,** $[^3H]$U-69593, a highly selective ligand for the κ-opioid receptor, *Eur. J. Pharmacol.*, 109, 281, 1985.
95. **Imperato, A. and Di Chiara, G.,** Opposite effects of μ- and κ-opiate agonists on dopamine release in the n. accumbens and caudate in the freely moving rats, *Br. J. Pharmacol.*, 86, 532, 1985.
96. **Esposito, R. U., Perry, W., and Kornetsky, C.,** Effects of d-amphetamine and naloxone on brain stimulation reward, *Psychopharmacology*, 69, 187, 1980.
97. **Bain, G. T. and Kornetsky, C.,** Naloxone attenuation of the effect of cocaine on rewarding brain stimulation, *Life Sci.*, 40, 1119, 1987.
98. **Goldberg, S. R., Woods, J. H., and Schuster, C. R.,** Nalorphine changes in morphine self-administration in rhesus monkeys, *J. Pharmacol. Exp. Ther.*, 176, 464, 1971.
99. **Killian, A. K., Bonese, K., and Schuster, C. R.,** The effects of naloxone on behavior maintained by cocaine and heroin injection in the rhesus monkey, *Drug Alcohol Depend.*, 3, 243, 1978.
100. **Harrigan, S. E. and Downs, D. A.,** Continuous naltrexone effects on morphine self-administration in rhesus monkeys, *J. Pharmacol. Exp. Ther.*, 204, 481, 1978.
101. **Carroll, M. E., Lac, S. T., Walker, M. J., Kragh, R., and Newman, T.,** Effects of naltrexone on intravenous cocaine self-administration in rats during food-deprivation, *J. Pharmacol. Exp. Ther.*, 238, 1, 1986.
102. **Corrigall, W. A. and Coen, K. M.,** Opiate antagonists reduce cocaine but not nicotine self-administration, *Psychopharmacology*, 104, 167, 1991.
103. **Trujillo, K. A., Belluzzi, J. D., and Stein, L.,** Naloxone blockade of place preference conditioning, *Psychopharmacology*, 104, 265, 1991.
104. **Houdi, A. A., Bardo, M. T., and Van Loon, G. R.,** Opioid mediation of cocaine-induced hyperactivity and reinforcement, *Brain Res.*, 497, 195, 1989.
105. **Holtzman, S. G.,** Behavioral effects of separate and combined administration of naloxone and d-amphetamine, *J. Pharmacol. Exp. Ther.*, 189, 51, 1974.

106. **Dettmar, P. W., Cowans, A., and Walter, D. S.,** Naloxone antagonism of behavioral effects of d-amphetamine in mice and rats, *Neuropharmacology*, 17, 1041, 1978.
107. **Haber, S., Hatsukami, T., Berger, P. A., Barchas, J. D.,and Akil, H.,** Naloxone blocks amphetamine-rearing: potential interaction between catecholamines and endorphins, *Prog. Neuropsychopharmacol.*, 2, 425, 1978.
108. **Hitzemann, R., Curell, J., Hom, D., and Loh, H.,** Effects of naloxone on d-amphetamine- and apomorphine-induced behavior, *Neuropharmacology*, 21, 1005, 1982.
109. **Ivy, A. C., Goetzl, F. R., and Burrill, D. Y.,** Morphine-dextroamphetamine analgesia, *War Med.*, 6, 67, 1944.
110. **Handley, C. A.,** A comparison of amphetamine sulphate with other stimulants of the central nervous system in morphine respiratory depression, *Anaesthesiology*, 6, 561, 1945.
111. **Forrest, W. H., Jr., Brown, B. W., Brown, C. R., Defalque, R., Gold, M., Gordon, H. E., James, K. E., Katz, J., Mahler, D. L., Schroff, P., and Teutsch, G.,** Dextroamphetamine with morphine for treatment of postoperative pain, *N. Engl. J. Med.*, 296, 712, 1977.
112. **Goetzl, F. R., Burrill, D. Y., and Ivy, A. C.,** The analgesic effect of morphine alone and in combination with dextroamphetamine, *Proc. Soc. Exp. Biol. Med.*, 55, 248, 1944.
113. **Ahmed, S. S., Abraham, G. J. S., and Assari, M.,** Dose-dependent modification of codeine analgesia by d-amphetamine in albino rats, *Arch. Int. Pharmacodyn.*, 184, 240, 1970.
114. **Dewey, W. L., Harris, L. S., Howes, J. F., and Nuite, J. A.,** The effect of various neurohumoral regulators on the activity of morphine and the narcotic antagonists in the tail-flick and phenylquinone tests, *J. Pharmacol. Exp. Ther.*, 175, 435, 1970.
115. **Richards, R. K.,** A study of the effect of d-amphetamine on the toxicity, analgesic potency and swimming impairment caused by potent analgesics, *Arch. Int. Pharmacodyn.*, 216, 225, 1975.
116. **Vedernikov, Y. P. and Afrikanov, I. I.,** On the role of central adrenergic mechanism in morphine analgesic action, *J. Pharm. Pharmacol.*, 21, 845, 1969.
117. **Sierra, V., Duttaroy, A., Lufty, K., Candido, J., Billings, B., Zito, S. V., and Yoburn, B. C.,** Potentiation of opioid analgesia by cocaine: the role of spinal and supraspinal receptors, *Life Sci.*, 50, 591, 1992.
118. **Dunai-Kovács, Z. and Székely, J. I.,** Effect of apomorphine on the antinociceptive activity of morphine, *Psychopharmacology*, 53, 65, 1977.
119. **Tulunay, F. C., Sparber, S. B., and Takemori, A. E.,** The effect of dopaminergic stimulation and blockade on the nociceptive and antinociceptive responses of mice, *Eur. J. Pharmacol.*, 33, 65, 1975.
120. **Morgan, M. J.,** Opioid-Dopamine Interactions in Analgesia in the Formalin Test, unpublished doctoral thesis, 1989.
121. **Robertson, J., Weston, R., Lewis, M. J., and Barasi, S.,** Evidence for the potentiation of the antinociceptive action of morphine by bromocriptine, *Neuropharmacology*, 20, 1029, 1981.
122. **Franklin, K. B. J.,** Analgesia and the neural substrate of reward, *Neurosci. Biobehav. Rev.*, 13, 149, 1989.

123. **Amitai, M., Dickhaut, H. H., and Hasenkopf, P.,** Behandlung des Entzug-Syndroms ei ''Fixern'' (Rauschmittelmissbrauch Jugendlicher p. Inj.), *Nervenarzt*, 43, 157, 1972.
124. **Ziegler von, B. and Schneider, E.,** Behandlung des Entzugsyndroms mit Apomorphin, *Arzneim.-Forsch.*, 24, 1117, 1974.
125. **Jasinski, D. R. and Nutt, J. G.,** Progress report on the assessment program on the NIMH Addiction Center, in *Report of the Thirty-Fourth Annual Meeting Committee on Problems of Drug Dependence,* NIMH, Michigan, 1972, 442.
126. **Greene, M. H. and Dupont, R. L.,** An outbreak of intravenous amphetamine abuse in heroin addicts, in *Fifth National Conference on Methadone Treatment,* Dupont, R. L., Freeman, S. R., Eds., NAPAN, New York, 1973, 776.
127. **Greene, M. H., Turner, N., and Dupont, R. L.,** Amphetamines in the District of Columbia, *Int. J. Addict.*, 9, 653, 1974.
128. **Chambers, C. D., Taylor, W. J. R., and Walter, P. V.,** Drug abuse in the ambulatory detoxification, in *Methadone: Experiences and Issues*, Chambers, C. O. and Brill, L., Eds., Behavioral Publications, New York, 1973, 203.
129. **Ellinwood, E. H., Eibergen, R. D., and Kilbey, M. M.,** Stimulants: interaction with clinically relevant drugs, *Ann. N.Y. Acad. Sci.*, 281, 393, 1976.
130. **Hubner, C. B., Bain, G. T., and Kornetsky, C.,** The combined effects of morphine and d-amphetamine on the threshold for brain stimulation reward, *Pharmacol. Biochem. Behav.*, 28, 311, 1987.
131. **Carroll, M. E., Lac, S. T., and Nygaard, S. L.,** Concurrently available nondrug reinforcer prevents the acquisition or decreases the maintenance of cocaine-reinforced behavior, *Psychopharmacology*, 97, 23, 1989.
132. **Smith, J. C., Williams, D. P., and Jue, S. S.,** Rapid oral mixing of glucose and saccharin by rats, *Science*, 191, 304, 1976.
133. **Leith, N. J. and Barrett, R. J.,** Amphetamine and the reward system: evidence for tolerance and post-drug depression, *Psychopharmacology*, 46, 19, 1976.
134. **Cassens, G., Actor, C., Kling, M., and Schildkraut, J. J.,** Amphetamine withdrawal: effects on threshold of intracranial reinforcement, *Psychopharmacology*, 73, 318, 1981.
135. **Kokkinidis, L., Zacharko, R. M., and Predy, P. A.,** Post-amphetamine depression of stimulation-responding from the substantia nigra: reversal by tricyclic antidepressants, *Pharmacol. Biochem. Behav.*, 13, 379, 1980.
136. **Schaefer, G. J. and Michael, R. P.,** Morphine withdrawal produces differential effects on the rate of lever pressing for brain self-stimulation in the hypothalamus and midbrain in rats, *Pharmacol. Biochem. Behav.*, 18, 571, 1983.
137. **Emmett-Oglesby, M. W., Harris, C. M., Lane, J., and Lal, H.,** Withdrawal from morphine generalizes to a pentylenetetrazol stimulus, *Neuropeptides*, 5, 37, 1984.
138. **Wood, D. M. and Lal, H.,** Anxiogenic properties of cocaine withdrawal, *Life Sci.*, 41, 1431, 1987.

139. **Harris, C. M., Emmett-Oglesby, M. W., Robinson, N. G., and Lal, H.,** Withdrawal from chronic nicotine substitutes partially for the interoceptive stimulus produced by pentylenetetrazol (PTZ), *Psychopharmacology*, 95, 435, 1986.
140. **Emmett-Oglesby, M. W., Spencer, D. G., Jr., Elmesallamy, F., and Lal, H.,** The pentylenetetrazol model of anxiety detects withdrawal from diazepam in rats, *Life Sci.*, 33,161, 1983.
141. **Holtzman, S. G.,** Behavioral examination of combined and separate administration of naloxone and d-amphetamine, *J. Pharmacol. Exp. Ther.*, 189, 51, 1974.
142. **Brands, B., Thornhill, J. A., Hirst, M., and Gowdey, C. W.,** Suppression of food intake and body weight gain by naloxone in rats, *Life Sci.*, 24, 1773, 1979.
143. **Brown D. R. and Holtzman, S. G.,** Suppression of deprivation-induced food and water intake in rats and mice by naloxone, *Pharmacol. Biochem. Behav.*, 11, 567, 1979.
144. **Gilbert, D. B. and Cooper, S. J.,** Effects of dopamine D-1 antagonist SCH 23390 and the D-2 antagonist sulpiride on saline acceptance-rejection in water deprived rats, *Pharmacol. Biochem. Behav.*, 26, 687, 1987.
145. **Ukai, M., Nakayama, S., and Kameyama, T.,** Inhibition of drinking by naltrexone in the rat: interaction with the dopamine D-1 antagonist SCH 23390 and the D-2 antagonist sulpiride, *Pharmacol. Biochem. Behav.*, 32, 651, 1989.
146. **Protais, P., Dubuc, I., and Costentin, J.,** Pharmacological characteristics of dopamine receptors involved in the dual effect of dopamine agonists on yawning behavior in rats, *Eur. J. Pharmacol.*, 94, 271, 1983.
147. **Koob, G. F., Riley, S. J., Smith, S. C., and Robbins, T. W.,** Effects of 6-hydroxydopamine lesions of the nucleus accumbens and olfactory tubercle on feeding, locomotor activity, and amphetamine anorexia in the rat, *J. Comp. Physiol. Psychol.*, 92, 917, 1978.
148. **Wise, R. A.,** The dopamine synapse and the notion of ''pleasure centers'' in the brain, *Trends Neurosci.*, 3, 91, 1980.

6 Elements of a Uniform Brain Reward Circuitry

I. INTRODUCTION

During the last decades intensive experimental analysis of the rewarding mechanism of abused substances and other manipulations resulting in "reinforcement" has led to a yet much contested and poorly defined concept of a uniform "reward circuitry". Olds[1] first observed that the same hypothalamic sites mediate the positive reinforcing effect of electrical intracranial self-stimulation (ICSS) and self-administration of morphine into this brain area. Simultaneously, Stein and Belluzzi[2] also proposed that morphine and enkephalins directly activate the neural pathways subserving self-stimulation behavior. Subsequently, Wise and Bozarth[3] suggested that the brain centers mediating four different rewarding manipulations such as direct injection of amphetamine into the nucleus accumbens (NA), microinjections of opioids into the ventral tegmental area (VTA), and electrical (self)-stimulation of VTA and lateral hypothalamus (LH) are interconnected with each other. Thus they concluded that "...it would seem unlikely, at best, that reward messages from all sources would ultimately funnel through a common "pleasure center" or "reward circuit...".[3] They suppose the existence of a "...variety of independent systems responsive to a variety of rewards...".[3] Nevertheless, they state that "...it is not clear how many independent reward circuits exist in the brain...". All reward sites in the brain may ultimately converge on a common efferent pathway".[3]

Actually, some years later Koob and Bloom[4] (1988) outlined a common reward pathway. According to their proposal, NA is the centerpiece of the rewarding circuitry receiving afferents not only from the VTA, but also from the limbic and olfactory cortices. Besides sending recurrent projection to the VTA (as discussed in Chapter 7), NA would send its main efferent projection to the ventral pallidum (VP) and from the latter structure the next links would be the pathways to the pedunculopontine nucleus (PPN), the dorsal medial thalamus (DMT), and the (medial) prefrontal cortex (PFC).[4]

According to this model, activation of the VTA-NA-VP circuitry might explain the rewarding, motor, and other common properties of opioids, cocaine and other psychostimulants, and ethanol.[4] Subsequently, Franklin[5] and Morgan[6] stressed another point. Namely, as discussed in Chapter 5, not only opioids, but also psychostimulants exert an analgesic effect (for review see also Franklin[5] and Morgan),[6] i.e., they induce a state when the pain sensation per se is retained, but it loses its aversive character. Thus clinical analgesia and the rewarding effect of the main types of abused substances might be brought about by a common neural mechanisms. Of the many evidences for the commonality between drug reward and analgesia only three are listed below:

1. As discussed in Chapters 4 and 5, dopamine receptor blockers suppress the rewarding effects of opioids, psychostimulants, and their analgesic action as well (for review also see Székely).[7]
2. Dopamine receptor stimulants and dopamine uptake blockers, i.e., psychostimulants, potentiate the analgesic effect of opioids in animal experiments and clinical settings as well (see Chapter 5).
3. In rats, destruction by 6-hydroxydopamine (6-OHDA) of VTA abolishes the analgesic effect of morphine and amphetamine as well.[5] 1989). On the other hand, as reviewed in this chapter, all kinds of rewarding experimental manipulations are dependent on the integrity of VTA (and its afferent and/or efferent projections).

In this chapter the experimental findings are reviewed in the light of this proposition, i.e., an answer is sought to the question: is there a uniform "hedonic (rewarding) circuitry" in the brain mediating the reinforcing effects of abused substances?

II. AFFERENT AND EFFERENT PROJECTIONS OF A_{10} DOPAMINE NEURONS; A PROPOSED REWARDING CIRCUITRY

It has been demonstrated primarily by functional studies and not in morphological ones that VTA gets *afferents* from the (posterior)-lateral part of the hypothalamus (LH) and periaqueductal central grey (PAG).[8] The former connection runs in the medial forebrain bundle (MFB).[9] It is an important point since, in experiments on ICSS, the stimulating electrodes are generally inserted into the hypothalamic segment of the MFB.[10] Actually, electrical stimulation is rewarding (in rats) if the stimulating electrodes are located in a region of the medial brain stem that extends about 4 mm in anterior-posterior axis from the VTA.[9] Some data indicate that the VTA receives afferents from the amygdala as well.[11]

The abundance of afferent projections to and efferent pathways from the PAG is well established. The VTA receives fibres from the PAG and the dorsal raphe areas as well.[12,13] On the other hand, the PAG also sends fibres to the other end of A_{10} neurons: its connections with the NA are verified by morphological studies.[14,15] Thus the VTA, the PAG, the LH, and certain raphe nuclei are interconnected with each other.

The primary *efferents* of the VTA are actually the axons of the A_{10} neurons having their perikarya in the mesencephalon. In the meantime the original findings of Fuxe[16] and Ungerstedt[17] on the monoamine pathways of the brain have been extended and studied in detail in relation to the efferent projections of A_{10} dopamine neurons as well. According to these studies,[18–21] the ascending projections include the NA; the PFC; the cingular, entorhinal, and piriform cortices; the medial part of the lateral septum; the interstitial nucleus of the stria terminalis; the olfactory tubercle; the anteromedial part of the caudate nucleus; and the amygdaloid complex. In these works,[18–21] descending connections were detected to the lateral habenula, the medial hypothalamus, the locus coeruleus, the dorsal raphe, and the cerebellum.

Some studies[22,23] indicate that besides the NA the neocortex, specifically the frontal cortex is the main efferent projection area of A_{10} neurons.

It is worth mentioning here that nearly all the areas known to induce ICSS belong to these projections of A_{10} neurons (or their afferent projections), but not all the projection areas maintain self-stimulation behavior. Besides the LH (actually not an important projection of A_{10} neurons, rather an input to them) the VTA itself seems to be the most appropriate area to induce ICSS.[24] Nevertheless, apparently ICSS can be elicited from many other areas receiving A_{10} projections, including the midbrain central grey, the bed nucleus of the stria terminalis, the globus pallidus (GP),[2] the NA and the PFC (vide infra).

Furthermore, the NA gets afferents not only from the VTA. Efferent fibers from the hippocampus, the amygdala, and the cingulate cortex converge with those arising from the VTA.[25,26] On the other hand, certain (but not all) efferent projections of the NA seem to constitute the next link in the rewarding circuitry (vide infra). Thus the NA is another focal point of the rewarding circuitry where the afferents from rewarding areas such as the VTA and the LH converge with efferents from the main limbic areas.

The exit from the NA, at least a part of its efferents, represents the subsequent link in the rewarding circuitry. The next neuron is a (GABA)- and enkephalin-releasing pathway leading from the NA into the VP including the substantia innominata (SI), GP, and lateral preoptic area (LPA).[27–30] Some studies indicate that this pathway exerts a tonic GABA-ergic inhibitory effect on the structures connected to it. Namely, the inhibition of unit activity in the GP induced by electrical stimulation of the NA was attenuated by iontophoretic application of picrotoxin, a GABA antagonist, into the GP.[28] Furthermore, the locomotor stimulation induced by dopamine injection into the NA was abolished by an antagonist of GABA catabolism, ethanolamine *O*-sulfate, administered into the VP.[31]

Several further data also indicate that the activity of these inhibitory GABA-releasing neurons projecting to the VP, SI, the LPA is suppressed by dopamine if liberated within the NA. Namely, it is well established[32,33] that microinjection of dopamine itself or dopamine-releasing amphetamine into the NA results in locomotor hyperactivity. This hyperactivity can be inhibited not only by dopamine antagonists, but also by GABA, whereas picrotoxin enhances the effect.[34] Furthermore, the enhancement of apomorphine-induced locomotor facilitation induced by 6-OHDA injection into the NA, i.e., by dopaminergic denervation-elicited supersensitivity, could be reversed by lesioning of the SI or LPA (electrolytically or by local administration of ibotenic acid) or by injecting muscimol, a GABA agonist, into these regions.[35–37] Thus the GABA-releasing inhibitory pathway emerging from the NA is tonically active, and this inhibition can be overcome by dopamine at its origin still in the NA. There are enkephalinergic axon terminals in the VP too,[30] indicating that they might also be involved in the inhibitory effect. Injecting (D-Ala^2,NMePhe^4,Gly^5-ol)-enkephalin (DAGO), a μ-selective opioid, into the NA also results in locomotor hyperactivity which can be reversed by injecting muscimol, a $GABA_A$ agonist into the VP.[38] In a related electrophysiological study[39] the spontaneous firing rate of VP units was enhanced upon injecting dopamine into the NA, and iontophoretic application of picrotoxin or naloxone also resulted in facilitation. Thus both the GABA- and enkephalin-releasing axons seem to tonically inhibit the activity of the neurons in the VP.

Beyond the VP the rewarding circuitry seems to spread out. Namely, the efferent projections of VP include cholinergic fibers leading to the PFC[40] and others reaching the PPN[41] and the DMT.[42,43]

Hence the question arises whether or not all these structures are involved in the rewarding circuitry (beyond the VP). The data available are scarce and contradictory. In one study,[44] lesioning with kainic acid of the PPN lead to attenuation of the locomotor facilitation induced by intra-NA injection of amphetamine. However, in another laboratory,[45] electrolytic lesion of the DMT, but not PPN and PFC, abolished the hyperlocomotion to apomorphine in rats made hypersensitive by 6-OHDA injections into the NA. Furthermore, the locomotor hyperactivity induced by intra-VP-injected picrotoxin (i.e., temporary abolition of the inhibitory effect of GABA-ergic NA-VP projection) could be reversed by electrolytic lesion of the DMT.[45]

The above reviewed data are are summarized in graphic form in Figure 1. Of course, this model has numerous limitations; thus it should not be looked at otherwise as a brief summary of experimental data available for the time being. Upon reading this model, caution is indicated for multiple reasons:

- Each class of psychotropic drugs affects various brain regions simultaneously, resulting in diverse subjective, autonomic, motor, and

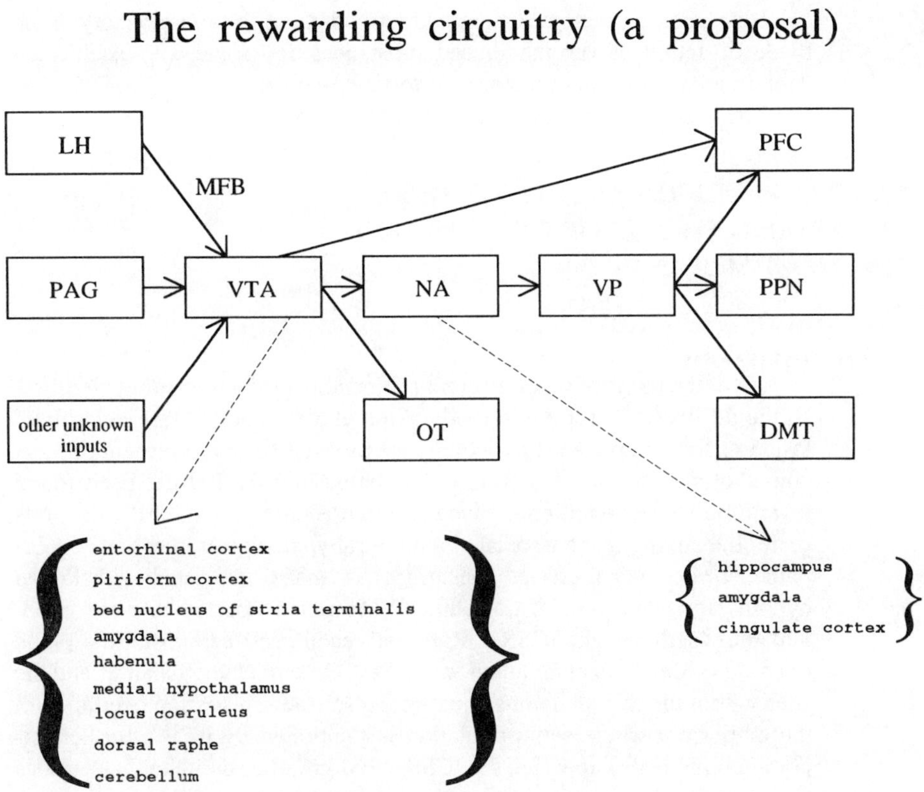

FIGURE 1. Preliminary outline of a uniform reward circuitry in the brain; solid lines indicate the pathways proven to be involved in the rewarding effect of opioids or psychostimulants; broken lines indicate afferent and efferent connections of the VTA and NA which might modulate the activity of the proposed circuitry. Abbreviations: DMT: dorsal medial thalamus; LH: lateral hypothalamus; MFB: medial forebrain bundle; NA: nucleus accumbens; OT: olfactory tubercle; PAG: periaqueductal central grey; PFC: (medial) prefrontal cortex; PPN: pedunculopontine nucleus; VP: ventral pallidum; VTA: ventral tegmental area.

other effects. The data reviewed in the next section (and other related chapters) only indicate that abused subtances, especially opioids and psychostimulants, among other brain regions activate one or more elements of the presumed rewarding circuitry as well.

- Our present knowledge is rather limited; most data referring to the participation of these regions in drug effects are relatively new.
- Various classes of abused subtances seem to activate the rewarding circuitry at different points.
- As reviewed in the introduction to this chapter, most authors doubt whether or not the euphoria-inducing effect of various kinds of exogenous substances (abused drugs) and endogenous neurohormones inducing a similar action would act in the same way.

Nevertheless, there are more and more data on the commonality in the mode of action of various abused substances and some well-established experimental manipulations proven to be rewarding.

These findings are summarized in the following section.

III. BRAIN LOCI RESPONSIVE TO THE REWARDING EFFECTS OF OPIOIDS AND PSYCHOSTIMULANTS

A. Hypothalamus

As in the cases of other afferent projections of the rewarding circuitry, all the positive data refer to opioids. Amit et al.[46] reported that electrolytic lesion of the ventrolateral hypothalamus blocked the consumption of morphine solution in rats. Injection of morphine into the LH has been found rewarding in the conditioned place preference paradigm (CPPP), i.e., rats prefer the environment associated with intrahypothalamic injection of morphine.[47] Mice too preferred the arm of Y maze associated with lateral hypothalamic injection of morphine.[48] Morphine injected into the hypothalamus facilitates the ICSS.[49] Rats also readily self-administer morphine or (D-Ala^2,Met^5)-enkephalinamide) (DALA) into the hypothalamus and the sites within the hypothalamus occasioning self-administration overlap with the ones from where self-stimulation is facilitated by it.[50] A further important point is that applying the ICSS paradigm, the stimulating electrodes are most frequently implanted into the LH to reach the MFB.

Psychostimulants aparently do not affect these inputs to the rewarding circuitry. For example, bilateral lesions of the dorsal and ventral ascending "noradrenergic" systems, i.e., pathways, which are known to contain dopaminergic fibers and pass via the hypothalamus to the VTA, failed to modify the i.v. self-administration of cocaine in rats.[51]

B. Periaqueductal Central Grey (PAG)

Some data indicate that the PAG also is a part of the rewarding circuitry, though it is better known for mediation of the analgesic effect of opiates and many symptoms of opiate withdrawal. Rats can be trained to self-stimulate this area and this behavior is suppressed by i.c. naloxone,[2] indicating the participation of opioid mechanisms. Both rats[47] and mice[48] show preference in CPPP paradigm for the environment associated with morphine administration into the PAG. Coadministration of naloxone abolishes the rewarding effect of morphine,[48] proving that opiate receptors mediate this effect. Furthermore, local injection of methylnaloxone into this region induces conditioned place aversion[52] and increase in the rate of lever pressing for i.v. heroin.[53] The latter finding[53] points to reinforcement deficit.

Summarily, these data indicate that areas sending afferents to the A_{10} neurons induce preference via opioid mechanisms. No data are available about the eventual similar effects of psychostimulants. In other words, if A_{10} neurons are activated transsynaptically, the origin is an opioid-mediated pathway.

C. Ventral Tegmental Area (VTA)

Bozarth and Wise[54] demonstrated first that rats can be trained to press the lever in a standard operant chamber if depression was rewarded by microinjection of morphine (0.1 μg) into the VTA. In the same study[54] i.p. naloxone blocked the acquisition of this response, proving that the rewarding action of morphine in the VTA is mediated by opiate receptors. Furthermore, a compensatory increase in the rate of lever pressing was observed if either diallyl-normorphinium bromide[55] or methylnaloxonium chloride[56] was injected locally into the VTA, i.e., opiate antagonists not supposed to spread into other brain areas. Thus the VTA is involved in the rewarding effect of systemically administered opiates as well. The same conclusion is confirmed by another experiment[57] where injection of morphine into the VTA reinstated the responding in rats which were first trained to press a lever for i.v. heroin reward, then the response was extinguished by withdrawing the reward. The importance of VTA in reinforcement is substantiated by the finding that, of morphine, as small an amount as 50 ng was enough to maintain its self-administration into this area.[58]

The role of the VTA in reward has also been demonstrated in CPPP. Rats prefer the environment associated with intra-VTA application of minute amount of morphine[59] or DALA.[60] (Systemic naloxone abolished the preference in both studies.)[59,60] It was demonstrated in a careful subsequent study that the loci in the VTA maintaining CPPP to locally injected morphine correspond well with the distribution of A_{10} cells in the mesencephalon.[61]

The participation of the VTA in reward has been demonstrated in ICSS paradigm as well. Broekkamp et al.[62] first observed that intra-VTA injection of morphine facilitated the electrical self-stimulation of the brain with electrode placement in the LH. DALA induced a similar effect, and the action of both opioids could be reversed by systemic (i.p.) injection of naloxone.[62] The opiate receptor subtype involved was examined in one study.[63] Applying them into the VTA, both morphine and a δ-receptor-preferring analog (D-Pen^2,D-Pen^5)-enkephalin (DPDPE), dose dependently decreased threshold of hypothalamic self-stimulation but the κ-selective U-50488H induced no change. Thus, in the VTA μ- and δ-, but not κ-receptor might mediate the rewarding effect of opioids.

Nevertheless, the next step in reinforcement is dopamine. Namely, the CPPP induced by injection of DALA into the VTA is abolished by both systemic haloperidol, a nonselective dopamine receptor blocker, or by a 6-OHDA lesion of the ascending dopaminergic system.[64] Furthermore, lesioning by 6-OHDA of the VTA also abolished the facilitation induced by systemic morphine of hypothalamic self-stimulation.[65]

The eventual participation of VTA in the rewarding effect of psychostimulants is a controversial issue. In one study,[57] intra-VTA injection of morphine reinstated the previously extinguished cocaine self-administration behavior. Thus at the level of VTA the reinforcing mechanisms of opiates and psychostimulants seem to converge. However, in another laboratory,[65] 6-OHDA lesions of the VTA blocked the morphine, but not the amphetamine-induced facilitation of hypothalamic self-stimulation, indicating that the reinforcing effect of the latter is independent of the VTA. Nevertheless, a different group reported that 6-OHDA lesions of the VTA disrupted the i.v. self-injection of cocaine.[66] However, others found that VTA lesions enhanced the rate of responding for amphetamine in an operant acquisition paradigm[67] and speeded up the process of acquisition of amphetamine self-administration behavior.[68] The conflicting data can be explained by supposing that lesioning of the A_{10} neurons is generally incomplete. If so, the loss of A_{10} neurons will induce two opposite processes:

1. Less dopamine will be released from their terminals, leading to reduction in reinforcing capacity.
2. The postsynaptic dopamine receptors might become hypersensitive, generating an exaggerated response to the dopamine released from the unimpaired axon terminals as proposed by Le Moal et al.[68]

Some data indicate that the VTA is under a tonic inhibitory control maintained by a pathway connecting amygdala with the VTA.[11,69,70] Accordingly, the more rapid acquisition of amphetamine self-administration behavior recorded after 6-OHDA lesion of the amygdala[71] has been attributed to elimination of this inhibitory pathway. Namely, after 6-OHDA lesion of the amygdala, not only the self-administration of amphetamine is enhanced, but also the dopamine turnover in the NA is facilitated and the amphetamine-indced locomotor hyperactivity is augmented as well.[72]

D. Nucleus accumbens

Morphine[73] and Met-enkephalin[74] are readily self-administered into the NA and, as predictable, naloxone abolishes the behavior after an initial, extinction-like facilitation.[74] Local microinjection of quaternary methylnaloxone enhances the rate of response for i.v. heroin,[53,56] indicating that direct activation of the opiate receptors in the NA is involved in the reinforcing effect of systemically administered opiates. In two studies, local, i.e., intra-NA injection of 6-OHDA, i.e., selective and local destruction of the dopamine-releasing terminals, failed to significantly modify the i.v. self-adminstration of heroin[75] and morphine,[76] though in the same studies the self-administration of cocaine and amphetamine was attenuated. On the other hand, lesioning of the NA by kainic acid, i.e., destroying the cell bodies located there, resulted in attentuation of response for i.v. morphine[77] and heroin as well.[78]

These findings are not easy to interpret (vide infra), but they also are in partial contradiction to some other findings collected using the two other main experimental paradigms of drug reward. Rats show place preference for environment associated with intra-NA injection of morphine[47] and place aversion was seen with intra-accumbens injection of methylnaloxone.[52] 6-OHDA treatment of this structure also attenuates the place preference induced by i.p. heroin.[79]

Intra-accumbens injections of both μ-agonists such as morphine and DAGO and δ-selective ones such as DPDPE facilitate ICSS (with electrodes in the LH).[80] Finally, a long series of experiments prove that ICSS is inhibited with opiate antagonists such as naloxone and naltrexone if the stimulating electrode is located in the NA.[81–84]

Summarily, these data indicate that

1. NA is involved in the rewarding effect of opiates whether administered locally or systemically.
2. Opiates might retain their rewarding potency even after selective lesioning of the dopamine axon terminals in the NA in the self-administration paradigm,[75,76] but not in CPPP.[79] Of course, opiates might activate A_{10} units which have axon terminals bypassing the NA. However, if opiates are capable of mediating reward without the participation of the NA, why is their rewarding effect attenuated by kainic acid lesion of the NA? Furthermore, how can opiate antagonists, injected into the NA, attenuate the rewarding effect of opioids in all the paradigms examined if this structure is not indispensable for their rewarding effect? For the time being, these data[75,76] cannot be explained.

The data are nearly unequivocal in relation to psychostimulants. Rats readily self-administer amphetamine into the NA.[85] Intra-accumbens injection of methylnaloxone only induced selective increase in responding for heroin, but not for cocaine, indicating that opiate receptors are not involved in the rewarding effect of psychostimulants.[53] On the other hand, destroying with 6-OHDA the dopamine-releasing terminals in the NA resulted, in all related studies, in extinction of responding for i.v. amphetamine and cocaine,[75,76,86–88] and lesioning of the cell bodies by kainic acid elicits a similar effect.[78] Furthermore, place preference could be established by injecting amphetamine into the NA, but not into the caudate nucleus.[89] Depletion by 6-OHDA of the dopamine in the NA abolished the place preference for amphetamine,[90] but not for cocaine in the same laboratory.[91] Discounting the latest finding, all these data fit into a simple concept. As discussed elsewhere, psychostimulants exert their rewarding effect by releasing dopamine from the terminals of A_{10} (and A_9) cells. Apparently, the ones projecting to the NA are indispensable for this effect.

E. Ventral Pallidum

In the only related study, Hubner and Koob[92] found the i.v. self-administration of both cocaine and heroin attenuated upon destruction of the VP with ibotenic acid (in rats). This finding that self-administration of both heroin and cocaine was attentuated points to the existence of a NA-VP pathway participating in the rewarding effects of both opioids and psychostimulants. Since only attenuation and no complete abolition of responding was recorded,[92] it is not clarified yet whether NA-VP is the only efferent (rewarding) pathway exiting the NA or simply the destruction of the VP was not complete.

F. Medial Prefrontal Cortex

The eventual role of the PFC in the reinforcing of opiates has hardly been examined. In the only related study, ICSS of the PFC was not modified by local application of morphine or naloxone.[93] Thus the eventual participation of the PFC in the rewarding effect of opioids is questionable.

However, the abundance of VTA-PFC projections[22,94] and the data below indicate that this cortical area might play a significant role in the rewarding action of psychostimulants. Firstly, Phillips et al.[95] demonstrated that rhesus monkeys can be trained to self-administer *d*-amphetamine into the orbitofrontal cortex after establishing ICSS behavior by placing the stimulating electrodes into the same area. Interestingly, in the same study the monkeys failed to self-administer amphetamine into the NA.[95] In agreement with this finding, rats could be trained to self-administer cocaine into the medial PFC, but not to the VTA or the NA.[96] These studies indicate that, eventually, the PFC might mediate the rewarding effect of psychostimulants. Furthermore, Goeders and Smith[96] proposed that the suppression by NA lesions of cocaine and amphetamine self-administration was actually due to accidental destruction of the fibers passing nearby from the VTA to the PFC.

However, as discussed above, rats could be trained to the intra-NA administration of amphetamine in another laboratory.[85] Furthermore, the findings obtained in experiments using the technique of selecting cortical lesions are contradictory. In one study, 6-OHDA lesions of the PFC failed to modify the previously established i.v. cocaine self-administration behavior in rats.[97] But in a recent study[98] the same type of 6-OHDA lesion of the PFC did induce supersensitivity to the reinforcing effect of i.v.-injected cocaine, i.e., it shifted the dose-effect curve to the left. The situation is further complicated by the observations that destruction of the PFC enhances the dopamine turnover in the NA and the striatum,[99] indicating that the PFC exerts a tonic inhibitory feedback effect on the DA turnover. Thus further studies are needed first to repeat the more than 10-year-old observations and to clarify whether PFC only contributes to the rewarding effects of opiates or if it is capable of doing it without participation of the NA. In either case a further question to be answered is whether or not the PFC might induce euphoria without participation of the VP and the structures attached to its output.

G. Hippocampus

There is a set of experiments which should probably be reviewed separately, since they apparently contradict all the other experimental data with regard to both the structures involved in reward mechanisms and especially the findings of others on the subtype specificity of opiate receptors mediating reinforcement.

In one laboratory, rats could be easily trained to self-administer dynorphin-A and U-50488H both into the lateral ventricle and specifically into the CA 3 field of hippocampus.[100,101] The specificity of the rewarding effect of these κ-ligands was confirmed by the ability of i.c.v.-injected dynorphin_{1-17} to induce CPPP as opposed to the lack of efficacy of dynorphin_{2-17}, i.e., a close structural analog without any significant κ-agonist potency.[102] Subsequently, the same group demonstrated that, using hippocampal slice preparation, the CA 3 cells could be conditioned by dynorphin microinjection, i.e., by rewarding the initially spontaneous bursts of their firing with iontophoretic administration of dynorphin A, the burst activity became more and more frequent.[101]

If these data are corroborated, the structures involved in the rewarding areas will have to be redefined and the generality of findings on the aversive effects of κ-agonists should also be questioned, or at least with respect to the hippocampus.

REFERENCES

1. **Olds, M. E.,** Hypothalamic substrate for the positive reinforcing properties of morphine in the rat, *Brain Res.*, 168, 351, 1979.
2. **Stein, L. and Belluzzi, J. D.,** Brain endorphins: possible role in reward and memory formation, *Fed. Proc. Fed. Am. Soc. Exp. Biol.*, 38, 2468, 1979.
3. **Wise, R. A. and Bozarth, M. A.,** Brain reward circuitry: four circuit elements "wired" in apparent series, *Brain Res. Bull.*, 12, 203, 1984.

4. **Koob, G. F. and Bloom, F. E.,** Cellular and molecular mechanisms of drug dependence, *Science*, 242, 715, 1988.
5. **Franklin, K. B.,** Analgesia and the neural substrate of reward, *Neurosci. Biobehav. Rev.*, 13, 149, 1989.
6. **Morgan, M. J.,** Opioid-Dopamine Interactions in Analgesia in the Formalin Test, Thesis, McGill University, Montreal, 1989.
7. **Székely, J. I.,** The most characteristic in vivo effects of opiates. VI. Drug interaction studies, in *Opioid Peptides,* Vol. I. Research Methods, Székely, J. I. and Rónai, A. Z., Eds., CRC Press, Boca Raton, FL, 1982, 53.
8. **Boye, S. and Rompré, P.-P.,** Evidence for direct axonal link between reward related neurons in the ventral tegmental area-posterior hypothalamus and the medial mesencephalon, *Soc. Neurosci. Abstr.*, 13, 1544, 1987.
9. **Rompré, P.-P. and Wise, R. A.,** Opioid neuroleptic interaction in brainstem self-stimulation, *Brain Res.*, 477, 144, 1989.
10. **Olds, J.**, Hypothalamic substrates of reward, *Physiol. Rev.*, 42, 554, 1962.
11. **Kelley, A. E., Domesick, V. B., and Nauta, W. J. H.,** The amygdalostriatal projection in the rat: an anatomical study by anterograde and retrograde tracing methods, *Neuroscience*, 7, 615, 1982.
12. **Phillipson, O. T.,** Afferent projections to the ventral tegmental area of Tsai and interfascicular nucleus: a horseradish peroxidase study in the rat, *J. Comp. Neurol.*, 187, 117, 1979.
13. **Oades, R. D. and Halliday, G. M.,** Ventral tegmental (A10) system: neurobiology. I. Anatomy and connectivity, *Brain Res. Rev.*, 12, 117, 1987.
14. **Li, Y. Q., Rao, Z. R., and Shi, J. W.,** Serotonergic projection from the midbrain periaqueductal grey to nucleus accumbens in the rat, *Neurosci. Lett.*, 98, 276, 1989.
15. **Li, Y. Q., Rao, Z. R., and Shi, J. W.,** Collateral projections from the midbrain periaqueductal gray to the nucleus raphe magnus and nucleus accumbens in the rat. A fluorescent double labelling study, *Neurosci. Lett.*, 117, 285, 1990.
16. **Fuxe, K.,** Evidence for the existence of monamine neurons in the central nervous system. IV. Distribution of monoamine nerve terminals in the central nervous system, *Acta Physiol. Scand.*, 247, (Suppl.) 39, 1965.
17. **Ungersted, U.,** Stereotaxic mapping of the monoamine pathways in the rat brain, *Acta Physiol. Scand. Suppl.*, 367, 1, 1971.
18. **Versteeg, D. M. G., Van der Gugten, J., De Jong, W., and Palkovits, M.,** Regional concentrations of noradrenaline and dopamine in the rat brain, *Brain Res.*, 113, 563, 1976.
19. **Fallin, J. H. and Moore, R. Y.,** Catecholamine innervation of the basal forebrain. IV. Topography of the dopamine projection to the basal forebrain and neostriatum, *J. Comp. Neurol.*, 180, 545, 1978.
20. **Moore, R. Y. and Bloom, F. E.,** Central catecholamine neuron systems: anatomy and physiology of the dopamine systems, *Annu. Rev. Neurosci.*, 1, 129, 1978.
21. **Simon, H., Le Moal, M., and Calas, A.,** Efferents and afferents of the ventral tegmental A10 region studied after local injection of [^{3}H]leucine and horseradish peroxidase, *Brain Res.*, 178, 17, 1979.

22. **Fuxe, K., Hökfelt, T., Johansson, O., Jonsson, G., Lidbrink, P., and Ljungdahl, A.,** The origin of the dopamine nerve terminals in limbic and frontal cortex: evidence for meso-cortico dopamine neurons, *Brain Res.*, 82, 349, 1974.
23. **Lindvall, O. and Björklund, A.,** The organization of the ascending catecholamine neuron system in the rat brain as revealed by the glyolxylic acid fluorescence method, *Acta Physiol. Scand.*, 412, (Suppl.) 1, 1974.
24. **Corbett, D. and Wise, R. A.,** Intracranial self-stimulation in relation to the ascending dopaminergic systems of the midbrain: a moveable electrode mapping study, *Brain Res.*, 185, 1, 1980.
25. **Swanson, L. W. and Cowan, W. M.,** Autoradiographic study of organization of the efferent connections of the hippocampal formation in the rat, *J. Comp. Neurol.*, 172, 49, 1977.
26. **DeFrance, J. F., Marchand, J. E., Stanley, J. C., Sikes, R. W., and Chronister, R. B.,** Convergence of excitatory amygdaloid and hippocampal inputs in the nucleus accumbens septi, *Brain Res.*, 185, 183, 1980.
27. **Nauta, W. J. H., Smith, G. P., Faull, R. L. M., and Domesick, V. B.,** Efferent connections and nigral afferents of the nucleus accumbens septi in the rat, *Neuroscience*, 3, 385, 1978.
28. **Jones, D. L. and Mogenson, G. J.,** Nucleus accumbens to globus pallidus GABA projection subserving ambulatory activity, *Am. J. Physiol.*, 238, 65, 1980.
29. **Mogenson, G. J., Swanson, L. W., and Wu, M.,** Neural projections from nucleus accumbens to globus pallidus, substantia innominata, and lateral preoptic-lateral hypothalmic areas: an anatomical and electrophysiological investigation in the rat, *J. Neurosci.*, 3, 189, 1983.
30. **Zahm, D. S., Zaborsky, L., Alones, V. E., and Heimer, L.,** Evidence for the coexistence of glutamate decarboxylase and met-enkephalin immunoreactivities in axon terminals of rat ventral pallidum, *Brain Res.*, 325, 317, 1985.
31. **Pycock, C. and Horton, R.,** Evidence for an accumbens-pallidal pathway in the rat and its possible GABAergic control, *Brain Res.*, 110, 629, 1976.
32. **Costall, B. and Naylor, R. J.,** Comparison of circling models for the detection of antiparkinson activity, *Psychopharmacology*, 45, 57, 1975.
33. **Pijnenburg, A. J. J., Honig, W. M. M., and Van Rossum, J. M.,** Inhibition of d-amphetamine-induced locomotor activity by injections of haloperidol into the nucleus accumbens of the rat, *Psychopharmacology*, 41, 87, 1975.
34. **Jones, D. L., Mogenson, G. J., and Wu, M.,** Injections of dopaminergic, cholinergic, serotonergic and gabaergic drugs into the nucleus accumbens: effects on locomotor activity in the rat, *Neuropharmacology*, 20, 29, 1981.
35. **Swerdlow, N. R. and Koob, G. F.,** The neural substrates of apomorphine-stimulated locomotor activity following denervation of the nucelus accumbens, *Life Sci.*, 35, 2537, 1984.
36. **Swerdlow, N. R., Swanson, L. W., and Koob, G. F.,** Substantia innominata: a critical link in the behavioral expression of mesolimbic dopamine stimulation in the rat, *Neurosci. Lett.*, 50, 19, 1984.

37. **Swerdlow, N. R., Swanson, L. W., and Koob, G. F.,** Electrolytic lesions of the substantia innominata and lateral preoptic area attenuate the ''supersensitive'' locomotor response to apomorphine resulting from denervation of the nucleus accumbens, *Brain Res.*, 306, 141, 1984.
38. **Austin, M. C. and Kalivas, P. W.,** Blockade of enkephalinergic and GABA mediated locomotion in the nucleus accumbens by muscimol in the ventral pallidum, *Jpn. J. Pharmacol.*, 50, 487, 1989.
39. **Yang, C. R. and Mogenson, G. J.,** Ventral pallidal neuronal responses to dopamine receptor stimulation in the nucleus accumbens, *Brain Res.*, 489, 237, 1989.
40. **Divac, I., Kosmal, A., Björklund, A., and Lindval, O.,** Subcortical projections to the prefrontal cortex in the rat as revealed by the horseradish peroxidase technique, *Neuroscience*, 3, 785, 1978.
41. **Swanson, L. W.,** An autoradiographic study of the efferent connections of the preoptic region in the rat, *J. Comp. Neurol.*, 167, 227, 1976.
42. **Vives, F., Gayoso, M. J., Osorio, C., and Mora, F.,** Afferent pathways to the points of self-stimulation in the medial prefrontal cortex of the rat as revealed by the horseradish peroxidase technique, *Behav. Brain. Res.*, 8, 23, 1983.
43. **Young, W. S., Alheid, G. F., and Heimer, L.,** The ventral pallidal projection to the mediodorsal thalamus: a study with fluorescent retrograde tracers and immunohistofluorescence, *J. Neurosci.*, 4, 1626, 1984.
44. **Brudzynski, S. M. and Mogenson, G. J.,** Association of the mesencephalic locomotor region with locomotor activity induced by injections of amphetamine into the nucleus accumbens, *Brain Res.*, 334, 77, 1985.
45. **Swerdlow, N. R. and Koob, G. F.,** Lesions of the dorsomedial nucleus of the thalamus, medial prefrontal cortex and pedunculopontine nucleus: effects on locomotor activity mediated by nucleus accumbens-ventral pallidal circuitry, *Brain Res.*, 412, 233, 1987.
46. **Amit, Z., Corcoran, M. E., Amir, S., and Urca, G.,** Ventral hypothalamic lesions block the consumption of morphine in rats, *Life Sci.*, 13, 805, 1973.
47. **Van der Kooy, D., Mucha, R. F., O'Shaughnessy, M., and Bucenieks, P.,** Reinforcing effects of brain microinjections of morphine revealed by conditioned place preference, *Brain Res.*, 243, 107, 1982.
48. **Cazala, P.,** Dose-dependent effects of morphine differentiate self-administration elicited from lateral hypothalamus and mesencephalic gray area in mice, *Brain Res.*, 527, 280, 1990.
49. **Broekkamp, C. L., Van den Bogaard, J. H., Jeijnen, H. J., Rops, R. H., Cools, A. R., and Van Rossum, J. M.,** Separation of inhibitory and stimulating effects of morphine on self-stimulation behavior by intracerebral microinjection, *Eur. J. Pharmacol.*, 36, 443, 1976.
50. **Olds, M. E. and Williams, K. N.,** Self-administration of D-Ala^2-Met-enkephalinamide at hypothalamic self-stimulation sites, *Brain Res.*, 194, 155, 1980.
51. **Roberts, D. C. S., Corcoran, M. E., and Fibiger, H. C.,** On the role of ascending catecholaminergic systems in intravenous self-administration of cocaine, *Pharmacol. Biochem. Behav.*, 6, 615, 1977.

52. **Koob, G. F., Stinus, L., Le Moal, M., and Bloom, F. E.,** Opponent process theory of motivation: neurobiological evidence from studies of opiate dependence, *Neurosci. Biobehav. Rev.*, 13, 135, 1989.
53. **Corrigal, W. A. and Vaccarino, F. J.,** Antagonist treatment in nucleus accumbens or periaqueductal grey affects heroin self-administration, *Pharmacol. Biochem. Behav.*, 30, 443, 1988.
54. **Bozarth, M. A. and Wise, R. A.,** Intracranial self-administration of morphine into the ventral tegmental area in rats, *Life Sci.*, 28, 551, 1981.
55. **Britt, M. D. and Wise, R. A.,** Ventral tegmental site of opiate reward: antagonism by a hydrophylic opiate receptor blocker, *Brain Res.*, 258, 105, 1983.
56. **Vaccarino, F. J., Bloom, F. E., and Koob, G. F.,** Blockade of nucleus accumbens opiate receptors attenuates intravenous heroin reward in the rat, *Psychopharmacology*, 86, 37, 1985.
57. **Stewart, J.,** Reinstatement of heroin and cocaine self-administration behavior in the rat by intracerebral application of morphine into the ventral tegmental area, *Pharmacol. Biochem. Behav.*, 20, 917, 1984.
58. **Welzl, H., Kuhn, G., and Huston, J. P.,** Self-administration of small amounts of morphine through glass micropipettes into the ventral tegmental area of the rat, *Neuropharmacology*, 28, 1017, 1989.
59. **Phillips, A. G. and Le Piane, F. G.,** Reinforcing effects of morphine microinjected into the ventral tegmental area, *Pharmacol. Biochem. Behav.*, 12, 965, 1980.
60. **Phillips, A. G. and Le Piane, F. G.,** Reward produced by microinjection of (D-Ala^2),Met^5-enkephalinamide into the ventral tegmental area, *Behav. Brain Res.*, 5, 225, 1982.
61. **Bozarth, M. A.,** Neuroanatomical boundaries of reward-relevant opiate-receptor field in the ventral tegmental area as mapped by the conditioned place preference method in rats, *Brain Res.*, 414, 77, 1987.
62. **Broekkamp, C. L., Phillips, A. G., and Cools, A. R.,** Facilitation of self-stimulation behavior following intracerebral microinjections of opioids into the ventral tegmental area, *Pharmacol. Biochem. Behav.*, 11, 289, 1979.
63. **Jenck, F., Gratton, A., and Wise, R. A.,** Opioid receptor subtypes associated with ventral tegmental facilitation of lateral hypothalamic brain stimulation reward, *Brain Res.*, 423, 34, 1987.
64. **Phillips, A. G., LePiane, F. G., and Fibiger, H. C.,** Dopaminergic mediation of reward produced by direct injection of enkephalin into the ventral tegmental area of the rat, *Life Sci.*, 33, 2505, 1983.
65. **Hand, T. H. and Franklin, K. B. J.,** 6-OHDA lesions of the ventral tegmental area block morphine-induced but not amphetamine-induced facilitation of self-stimulation, *Brain Res.*, 328, 233, 1985.
66. **Roberts, D. S. and Koob, G. F.,** Disruption of cocaine self-administration following 6-hydroxydopamine lesions of the ventral tegmental area in rats, *Pharmacol. Biochem. Behav.*, 17, 901, 1982.
67. **Demenière, J. M., Simon, H., Herman, J. P. and Le Loal, M.,** 6-Hydroxydopamine lesion of the dopamine mesocorticolimbic cell bodies increases (+)-amphetamine self-administration, *Psychopharmacology*, 83, 281, 1984.

68. **Le Moal, M., Stinus, L., and Simon, H.,** Increased sensitivity to (+)-amphetamine self-administration following meso-cortico-limbic dopamine neuron destruction, *Nature*, 280, 156, 1979.
69. **Maeda, H. and Mogenson, G. J.,** Electrophysiological responses of neurons of ventral tegmental area to electrical stimulation of amygdala and lateral septum, *Neuroscience*, 6, 367, 1981.
70. **Louilot, A., Simon, H., Taghzouti, K., and Le Moal, M.,** Modulation of dopaminergic activity in the nucleus accumbens following facilitation or blockade of the dopaminergic transmission in the amygdala: a study by in vivo differential pulse voltametry, *Brain Res.*, 346, 141,1985.
71. **Demenière, J. M., Piazza, P. V., Le Moal, M., and Simon, H.,** Experimental approach to individual vulnerability to psychostimulant addiction, *Neurosci. Biobehav. Rev.*, 13, 141, 1989.
72. **Simon, H., Taghzouti, K., Gozlan, H., Studler, J. M., Louilot, A., Hervé, D., Glowinski, J., Tassin, J.P., and Le Moal, M.,** Lesions of dopaminergic terminals in the amygdala produces enhanced locomotor response to d-amphetamine and opposite changes in dopaminergic activity in the frontal cortex and nucleus accumbens, *Brain Res.*, 447, 335, 1988.
73. **Olds, M. E.,** Reinforcing effects of morphine in the nucleus accumbens, *Brain Res.*, 237, 429, 1982.
74. **Goeders, N. E., Lane, J. D., and Smith, J. E.,** Self-administration of methionine enkephalin into the nucleus accumbens, *Pharmacol. Biochem. Behav.*, 20, 451, 1984.
75. **Pettit, H. O., Ettenberg, A., Bloom, F. E., and Koob, G. F.,** Destruction of dopamine in the nucleus accumbens selectively attenuates cocaine but not heroin self-administration in rats, *Psychopharmacology*, 84, 167, 1984.
76. **Dworkin, S. I., Gucrin, G. F., Co, C., Goeders, N. E., and Smith, J. E.,** Lack of an effect of 6-hydroxydopamine lesions of the nucleus accumbens on intravenous morphine self-administration, *Pharmacol. Biochem. Behav.*, 30, 1051, 1988.
77. **Dworkin, S. I., Guerin, G. F., Goeders, N. E., and Smith, J. E.,** Kainic acid lesions of the nucleus accumbens selectively attenuate morphine self-administration, *Pharmacol. Biochem. Behav.*, 29, 175, 1988.
78. **Zito, K. A., Vickers, G., and Roberts, D. S.,** Disruption of cocaine and heroin self-administration following kainic acid lesions of the nucleus accumbens, *Pharmacol. Biochem. Behav.*, 23, 1029, 1985.
79. **Spyraki, C., Fibiger, H. C., and Phillips, A. G.,** Attenuation of heroin reward in rats by disruption of the mesolimbic dopamine system, *Psychopharmacology*, 79, 278, 1983.
80. **Wise, R. A.,** Opiate reward: sites and substrates, *Neurosci. Biobehav. Rev.*, 13, 129, 1989.
81. **Stapleton, J. M., Merriman, V. J., Coogle, C. L., Gelbard, S. D., and Reid, L. D.,** Naloxone reduces pressing for intracranial stimulation sites in the periaqueductal gray area, accumbens nucleus, substantia nigra, and lateral hypothalamus, *Physiol. Psychol.*, 7, 427, 1979.
82. **Katz, R. J.,** Identification of a novel class of central reward sites showing delayed and cummulative response to opiate blockade, *Pharmacol. Biochem. Behav.*, 15, 131, 1981.

83. **West, T. E. G. and Wise, R. A.,** Effects of naltrexone on nucleus accumbens, lateral hypothalamic and ventral tegmental self-stimulation rate-frequency functions, *Brain Res.*, 462, 126, 1988.
84. **Trujillo, K. A., Belluzzi, J. D., and Stein, L.,** Opiate agonists and self-stimulation: extinction-like response patterns suggest selective reward deficit, *Brain Res.*, 492, 15, 1989.
85. **Hoebel, B. G., Monaco, A. P., Hernandez, L., Aulisi, E., Stanley, B. G., and Lénárd, L.,** Self-injection of amphetamine directly into the brain, *Psychopharmacology*, 81, 158, 1983.
86. **Roberts, D. C. S., Corcoran, M. E., and Fibiger, H. C.,** On the role of ascending catecholamine systems in intravenous self-administration of cocaine, *Pharmacol. Biochem. Behav.*, 6, 615, 1977.
87. **Lyness, W. H., Friedle, N. M., and Moore, K. E.,** Destruction of dopaminergic nerve terminals in nucleus accumbens: effect on d-amphetamine self-administration, *Pharmacol. Biochem. Behav.*, 11, 553, 1979.
88. **Roberts, D. C. S., Koob, G. F., Klonoff, P., and Fibiger, H. C.,** Extinction and recovery of cocaine self-administration following 6-hydroxydopamine lesions of the nucleus accumbens, *Pharmacol. Biochem. Behav.*, 12, 781, 1980.
89. **Carr, G. D. and White, N. M.,** Conditioned place-preference from intra-accumbens but not intra-caudate amphetamine injections, *Life Sci.*, 33, 2551, 1983.
90. **Spyraki, C., Fibiger, H. C., and Phillips, A. G.,** Dopaminergic substrates of amphetamine-induced place preference conditioning, *Brain Res.*, 253, 185, 1982.
91. **Spyraki, C., Fibiger, H. C., and Phillips, A. G.,** Cocaine-induced place preference conditioning: lack of effects of neuroleptics and 6-hydroxydopamine lesions, *Brain Res.*, 253, 195, 1982.
92. **Hubner, C. B. and Koob, G. F.,** The ventral pallidum plays a role in mediating cocaine and heroin self-administration in the rat, *Brain Res.*, 508, 20, 1990.
93. **Shaw, S. G., Vives, F., and Mora, F.,** Opioid peptides and self-stimulation of the medial prefrontal cortex in the rat, *Psychopharmacology*, 83, 288, 1984.
94. **Lindvall, O., Björklund, A., Moore, R. Y., and Stenevi, U.,** Mesencephalic dopamine neurones projecting to neocortex, *Brain Res.*, 81, 325, 1974.
95. **Phillips, A. G., Mora, F., and Rolls, E. T.,** Intracranial self-administration of amphetamine by rhesus monkeys, *Neurosci. Lett.*, 24, 81, 1981.
96. **Goeders, N. E. and Smith, J. E.,** Cortical dopaminergic involvement in cocaine-reinforcement, *Science*, 221, 773, 1983.
97. **Martin-Iverson, M. T., Szostak, C., and Fibiger, H. C.,** 6-Hydroxydopamine lesions of the medial prefrontal cortex fail to influence intravenous self-administration of cocaine, *Psychopharmacology*, 88, 310, 1986.
98. **Schenk, S., Horger, B. A., Peltier, R., and Shelton, K.,** Supersensitivity to reinforcing effects of cocaine following 6-hydroxydopamine lesions to the medial prefontal cortex in rats, *Brain Res.*, 543, 227, 1991.

99. **Carter, C. J. and Pycock, C. J.,** Behavioral and biochemical effects of dopamine and noradrenaline depletion within the medial prefrontal cortex of the rat, *Brain Res.*, 192, 163, 1980.
100. **Stevens, K. E., Shiotsu, G., Belluzzi, J. D., and Stein, L.,** Dynorphin A: self-administration in CA3 hippocampal field in the rat, *Soc. Neurosci. Abstr.*, 14, 293, 1988.
101. **Stein, L. and Belluzzi, J. D.,** Cellular investigations of behavioral reinforcement, *Neurosci. Biobehav. Rev.*, 13, 69, 1989.
102. **Iwamoto, E. T.,** Dynorphin $A_{[1-17]}$ place preference in rats: opioid-specific and dopaminergic, *Soc. Neurosci. Abstr.*, 14, 293, 1988.

7 Sensitization and the Pathology of Addiction

I. THE CONCEPT OF ADDICTION

Addictive drugs are rewarding *per definitionem*. However, the nature of their rewarding effect is a much debated issue.

For a long time it was supposed that the rewarding character of addictive drugs is due to their ability to alleviate the withdrawal symptoms in subjects physically dependent on them.[1,2] Obviously, for physically dependent addicts, abolition of the aversive subjective feelings associated with withdrawal of the drug to which they are addicted can be homologous to the effects of food or water in subjects deprived of them. Therefore, their reinforcing effect might be considered as a kind of "drive reduction" where the dysphoric character of the withdrawal state is the motivating factor.

Nevertheless, numerous clinical observations contradict this simplistic interpretation. The dependence theory of addiction does not explain why the drug self-administration habits get established in originally nondependent subjects. This theory also fails to explain the high rate of relapse in addicts successfully detoxicated earlier.[3-5] A further basic argument against such a simplistic explanation is the lack of serious physical dependence symptoms upon withdrawal of several main classes of abused substances, e.g., amphetamine, cocaine, nicotine, caffeine, etc., contrary to opiates, alcohol, barbiturates, etc.[6-9] Withdrawal of the latter drugs does induce intense craving and aversion, but the accompanying autonomic reactions, i.e., the most easily measurable manifestations of addiction are much milder than in the case of classical opiates.[9] Furthermore, it is well established clinically that "passive dependence", i.e., the use of opiates in chronic pain patients only very rarely induces craving and physical dependence.

Thus, a second possibility arises that drugs of abuse are rewarding *per se*. Together, these hypotheses indicate that drugs may serve as "positive" or "negative" reinforcers as well. According to Werner et al.,[10] "...negative reinforcement occurs when a physically dependent subject, by making an appropriate response, receives drug and so escapes from or avoids the noxious stimuli associated with the withdrawal state. Positive reinforcement is inferred to occur when a response which causes drug to be delivered is emitted at levels indicative of reinforcement but...in the absence of physical dependence...''.

The majority of related experimental data also are at variance with the dependence theory. Truly, if rats are trained to self-administer abused substances, the animals respond in a way which incurs relatively stable interinjection time, i.e., stable blood level independently of the reinforcement schedule employed.[11] This finding[11] can be interpreted in favor of ''dependence-hypothesis''. However, Woods and Schuster[12] showed that morphine was capable of maintaining self-administration behavior in doses smaller than needed to induce physical dependence. In animals operant responding for drugs typically displays inverted U-shaped dose-effect curves.[13] No definite explanation can be given for the decline in response rate upon increasing the amount of unit dose.[13] However, these findings also indicate that the reinforcing property might not be simply explained by alleviation of withdrawal symptoms. Namely, elevation of the (unit) dose might turn the drug aversive, but it does not reduce its ability to suppress withdrawal.

Some further data also indicate that the two phenomena may be independent of each other. For example, rats could be trained to press a lever for morphine injection into the ventral tegmental area (VTA) without seeing withdrawal phenomena at the end of these sessions or upon administration of naloxone.[8] On the other hand, injection of morphine into the periaqueductal grey did not occasion operant responding. Nevertheless, strong withdrawal phenomena were seen upon cessation of its continuous infusion into this region.[14] Actually, morphine is most readily self-injected into brain regions which also induce electrical self-stimulation.[15] Thus the brain regions rendering morphine rewarding and the ones inducing physical dependence must be anatomically, at least partially, distinct.

According to another hypothesis, drugs act as conditioned positive reinforcers.[16,17] Clinically, it is a well-known fact that the environment associated with previous drug consumption renders the subjects particularly vulnerable to substance abuse. For example, visual or olfactory stimuli reminding the addicts of the abused drugs might precipitate especially intense craving and relapse, whereas detoxification is easier with simultaneous changes of the environment.[18] Apparently, the same applies to experimental animals. If drug injection is paired with conditional stimuli (stimulus light or simply the sound of infusion pump), subsequently these stimuli *per se* will maintain responding for a given time.[19,20] Nevertheless, no report claims that responding, originally reinforced by drugs, could be indefinitely maintained by conditional stimuli, and nonreinforced operant behavior is extinguished after an initial ''extinction burst''.[13]

Returning to the withdrawal theory, one of its most significant shortcomings is that it does not account for an important and probably the most peculiar property of drug reinforcement. Namely, it disregards the fact (discussed in detail in the subsequent sections of this chapter) that repeated drug administration results in enhancement of the reinforcing potency of abused substances. Truly, in some clinical works the euphoric effect of opiates seemed to be attenuated after long-term consumption, but even in these studies[21,22] the tolerance was not complete. As discussed below, in well-controlled laboratory studies the rewarding effect of abused substances gradually increases upon its repeated administration or its "voluntary intake", whereas under clinical situation the reinforcing effect of a drug might be obscured by symptoms which are overcome by it. For example, in rats, repeated application of morphine, cocaine, and amphetamine resulted in enhancement of their rewarding effect; using the conditioned place preference assay the animals spent a longer and longer time in the drug-paired environment upon repeated exposure to these abused substances.[23] Furthermore, previous exposure to morphine enhanced the rewarding effects of amphetamine and cocaine, and vice versa.[23] Thus the ultimate substrate of sensitization may be the same whether opiates or psychostimulants are abused.

II. PSYCHOMOTOR STIMULATION AND DOPAMINE RELEASE

In this context it is important to emphasize that all classes of abused substances induce psychomotor stimulation. Not only the nominal ones such as amphetamines or cocaine, but even alcohol, benzodiazepines, etc., i.e., substances generally considered "sedative". (For detailed data see Chapter 10.) Namely, as postulated by Glickman and Schiff,[24] drugs which have a reinforcing effect induce forward locomotion. Forward locomotion is homologous to approach behavior which is evidently induced by rewarding stimuli. Truly, even opiates induce in most species locomotor depression upon their first application. However, their effect is always biphasic, an initial phase of locomotor depression is invariably followed by locomotor stimulation.[25–30] However, upon repeated application the stimulant effect becomes more and more conspicuous, whereas the phase of locomotor depression shortens, i.e., tolerance develops to just the sedative effects of opiates and other substances too, which generally are considered depressant, but induce addiction. That is why Wise and Bozarth[8] postulated that "psychomotor stimulation and positive reinforcement are homologous... they derive from common or at least overlapping neural mechanisms". As discussed elsewhere, the locomotor stimulant effect of opiates is due to dopamine release (see Section V), and other types of abused substances also enhance directly or indirectly the dopamine release in the forebrain (see Chapters 5, 7, and 10).

As for the psychostimulants, i.e., the other main class of abused substances, it is evident by definition that intense locomotor stimulation can be elicited by direct or indirect dopaminergic stimulants. Namely, it is also well established that amphetamines and cocaine inhibit the (re)uptake of dopamine into the presynaptic terminals from where they are released; therefore, their net effect is stimulation of the postsynaptic dopamine receptors.[31–37] This effect can also be observed in various terminal areas of the mesolimbic dopamine neurons such as the nucleus accumbens (NA), the olfactory tubercle, and the medial prefrontal cortex.[38–40] This effect is brought about by binding of cocaine (and other psychostimulants) to the dopamine transporter.[41,42] This dopamine transporter is a glycoprotein, with a molecular weight of about 60,000 to 80,000.[43–45]

In this relation, an important point must be raised. Though it is generally accepted that psychostimulants enhance the dopamine liberation in the forebrain by inhibiting its reuptake, some data indicate that their mode of action might be at least partially different in the nigrostriatal and mesolimbic circuitries. Working on rat brain slices, Missale et al.[46] concluded that "...Na^+-dependent [^{3}H]cocaine binding is not detectable in the nucleus accumbens...", i.e., in the main terminal area of the mesolimbic dopaminergic circuitry. Furthermore, in their study nomifensine inhibited dopamine uptake in the striatum by binding to high- and low-affinity binding sites, but they found only low-affinity binding sites in the NA.[46] However, Boja and Kuhar[47] detected Na^+-dependent cocaine binding in similar densities in the striatum and NA, and mazindol inhibited this binding with similar IC_{50}. Thus they concluded that "...the dopamine transporter is similar in both the striatum and nucleus accumbens...", and they attributed the evident contradiction with the former study[46] to the fact that they used synaptosomal preparation.[47] Subsequent studies indicate that there really are certain differences between the two circuitries, but these are of a quantitative and not a qualitative nature. Namely, recently Marshall et al.[48] reported that the density and affinity of dopamine uptake sites, labeled by [^{3}H]mazindol, are two- to threefold higher in the dorsal striatum than in the NA. A decrease in the density in [^{3}H]mazindol binding was also observed along the dorso-to-ventral axis in the striato-accumbens region by autoradiography.[48] Actually, a similar difference was observed in the densities of dopamine binding sites between the dorsal and ventral striatum by electron microscopic autoradiography.[49] Furthermore, the transient increase in extracellular dopamine concentration, induced by electrical stimulation, has been found to normalize much more slowly in the NA than in the striatum.[50]

Finally, according to a recent study,[51] the molecular weights of dopamine transporters have been found to be different in the striatum and NA, respectively. Thus, besides the basic similarity, there are certain differences in the dopamine reuptake in the two structures. Of course, despite these differences dopamine-releasing abused substances enhance the amount of this transmitter reaching the postsynaptic membranes in both structures. Nevertheless, the differences in the reuptake might explain that the process of sensitization is more conspicuous in the mesolimbic than in the nigrostriatal circuitry *(vide infra)*.

Anyhow, there is still a theoretical possibility that locomotor stimulation and reinforcement are two effects of dopamine which are independent of each other, i.e., these effects might not be causally related. However, Yokel and Pickens[52] demonstrated that of the two isomers of amphetamine the (+) is the more potent one in maintaining self-administration behavior in rats, i.e., just the isomer which is the stronger inhibitor of reuptake of dopamine, whereas they are equiactive in the inhibition of norepinephrine (re)uptake.[53] The sympathomimetic amines act by binding to a dopamine transporter. Thus, the conclusive evidence was provided by Ritz et al.,[54] who found that the potencies of cocainelike drugs to maintain self-administration was closely correlated with their potencies to inhibit +-mazindol binding to the dopamine transporter in rat striatum. Recently, a similar correlation has been demonstrated between binding to the dopamine transporter and the locomotor stimulant activities of some cocaine analogs.[55] The role of dopaminergic stimulation in reward has been also demonstrated in electrophysiological experiments. Nishino et al.[56] demonstrated that the A_{10} dopamine cells fire in a ''reinforcing situation'', e.g., when extending their arm to grab food given as reward during operant conditioning.

All these findings are not surprising in view of the data reviewed in Chapter 5, proving that dopamine mediates the rewarding effects of opiates and psychostimulants alike. Summarily, these findings indicate that drugs of abuse are rewarding *per se,* not due only to alleviation of physical withdrawal symptoms. Furthermore, rewarding drugs induce addiction and locomotor stimulation by dopaminergic stimulation, and this effect seems to become more and more preponderant upon repetitive drug administration/consumption. As reviewed below, just the latter phenomenon, i.e., enhancement of rewarding potency of drugs of abuse might explain the phenomenon of addiction.

III. ACUTE MOTOR EFFECTS OF OPIATES AND PSYCHOSTIMULANTS

The locomotor stimulant effects of amphetamine[57–62] and cocaine[61,62] upon their systemic administration and the role of catecholamines, especially dopamine, in this effect are well established as discussed in Section I. It also is firmly proven that after an initial phase of general motor depression, locomotor stimulation prevails upon systemic application of classical (μ) opiates, especially if given in small doses or repeatedly (for review see Babbini and Davis,[27] Amalric and Koob,[63] Wise and Bozarth,[8] and Székely).[64]

Destruction of the NA abolishes the locomotor stimulant effects of amphetamine and cocaine, especially the increase in ambulatory activity,[65,66] and injection of the dopamine receptor antagonist haloperidol into this area also inhibits the locomotor facilitation induced by systemic amphetamine.[67] However, destruction of the nigrostriatal circuitry suppresses only the stereotypic component of enhanced locomotor activity induced by psychostimulants.[68] Thus the locomotor stimulant effect of psychostimulants (at least one of its main components, forward locomotion, which is linked to a reward-related approach)[24] seems to be mediated by the mesolimbic circuitry. Actually, it is probably not surprising that the locomotor stimulant effect of dopamine-releasing drugs can be abolished by lesioning the main forebrain dopamine-releasing circuitry.

However, it was really an unexpected finding that bilateral injection of morphine into the ventral tegmental area (VTA) induced strong locomotor facilitation.[69] Similar hyperactivity can be elicited by bilateral injection of stabilized enkephalin analogs such as (D-Ala^2,Met^5)-enkephalinamide (DALA), and (D-Ala^2,D-Leu^5)-enkephalin (DADLE) into the VTA.[70–72] β-Endorphin also induces hyperactivity if injected bilaterally into the VTA.[73] Similarly to the effects of psychostimulants, of the various kinds of spontaneous motor activities of rats, enkephalins specifically facilitate forward locomotion and rearing, but inhibited grooming.[70–73] Ambulation and not stereotypy was the prevailing component of locomotion even upon injection of DALA and DADLE into the globus pallidus.[72] Thus opiates seem to preferentially activate mesolimbic circuitry arising from the A_{10} dopamine neurons, but not the nigrostriatal pathway originating from the A_9 dopamine cell bodies in the neighboring substantia nigra.

As emphasized by Broekkamp et al.[70] and confirmed by all the other studies reviewed here, the locomotor hyperactivity induced by intra-VTA injection of opioids is monophasic, i.e., no initial behavioral depression can be seen as upon systemic injection. Thus the depressant effect of opioids is obviously mediated by other brain structures (for review see Székely).[64] Consequently, opiates might simultaneously activate the rewarding mesolimbic circuitry and other structures inducing motor depression.

The hyperactivity elicited by injection of opioids into the VTA can be abolished by both naloxone and neuroleptics such as α-flupenthixol and haloperidol.[69–73] Intra-VTA DALA also potentiates the locomotor hyperactivity induced by systemic amphetamine.[71] Destruction by 6-hydroxydopamine (6-OHDA) of NA abolished the hyperactivity induced by DALA or β-endorphin infusion into the VTA,[71,73] just as in the case of systemically administered amphetamine and cocaine.[65,66] Thus both dopaminergic and opiate receptor-mediated mechanisms are involved in the hyperactivity induced by opioids when injecting them into the VTA. More specifically, opiates apparently activate the mesolimbic dopaminergic circuitry arising from the A_{10} neurons in the VTA. Thus, actually, the behavioral symptoms *per se* indicate that opioids somehow activate the A_{10} neurons in the VTA,[71] a proposal confirmed by electrophysiological data *vide infra*.

Few studies addressed the important issue of which subtype of opiate receptors mediates the motor effect. Latimer et al.[74] compared the effect of a specific μ-receptor agonist, (D-Ala^2,NMePhe^4,Gly^5-ol)-enkephalin (DAGO) and that of a δ-selective one, (D-Pen^2,D-Pen^5)-enkephalin (DPDPE), and the former has been found much more active on a weight basis. Since the facilitation caused by DAGO was significantly attenuated by naloxazine, they concluded that both μ- and δ-receptors, but specifically the μ_1 receptors, might mediate the facilitation.[74]

Furthermore, as discussed in Section III, no sensitization can be induced by κ-agonists. Thus it is reasonable to suppose, though it has not been examined in acute experiments, that κ-agonists are probably incapable of activating the mesolimbic circuitry. Anyhow, the direct confirmation of such a proposition is still missing. However, there is one crucial study. Namely, in all these works opioids were injected bilaterally. Upon unilateral injection contraversive circling has been seen whether morphine,[75] i.e., μ-agonist, or DPDPE, a δ-selective enkephalin analog, was injected.[76] However, U-50488H failed to induce contraversive circling.[76] The contraversive circling induced by unilateral injection of opioids into the VTA is obviously homologous with the locomotor facilitation recorded upon bilateral infusion. (This phenomenon also is reversible by both opiate antagonists and the dopamine receptor antagonist pimozide.)[77] The complete lack of effectiveness of U-50488H indicates that stimulation of κ-receptors in the VTA does not activate the mesolimbic dopamine circuitry.

Another approach in examining the modulation of mesolimbic circuitry by opioids and psychostimulants is their injection into the NA, one of the main terminal areas of A_{10} neurons.

Based on the above reviewed data, i.e., supposing that psychostimulants directly release dopamine from the terminals of A_{10} neurons, whereas opiates activate these cells transsynaptically in the VTA, it is expected that

- Dopamine antagonists injected into the NA depress the stimulant actions of both opioids and psychostimulants.
- Opiate antagonists have no effect if injected into the NA.

However, the related data are not so unequivocal. In good harmony with these propositions, cocaine injected directly into the NA resulted in locomotor activation which could be reversed by the dopamine antagonist *cis*-flupenthixol.[78] Irreversible lesioning of NA by 6-OHDA has been found to attenuate the locomotor facilitation caused by s.c.-injected amphetamine, but the similar effect of heroin was not modified.[66] The same lesion reduced the locomotor facilitation induced by systemic amphetamine, but did not eliminate it completely.[79] On the other hand, this lesion enhanced the locomotor stimulant effect of apomorphine, a direct dopamine receptor agonist, obviously as a manifestation of denervation supersensitivity of postsynaptic receptors.[79,80] Summarily, these findings still can be explained by considering that NA is not the only terminal area of the A_{10} dopamine neurons.

However, the situation is more complex in relation to opioid mechanisms. In a frequently quoted study, Pert and Sivit[81] found that bilateral injection of morphine or (D-Ala^2)-Met-enkephalin into the NA resulted in long-lasting, monophasic locomotor facilitation reversible by naloxone, but not by haloperidol. Consistent with these findings, Met-enkephalin (but not Leu-enkephalin) and DALA have been found to induce strong locomotor facilitation if injected into the NA.[82] This effect of DALA was reversed by naloxone, but not by fluphenazine; nevertheless, the stimulant effects of DALA and also locally injected dopamine have been found additive.[82] In another laboratory,[83] intra-accumbens injection of morphine induced naloxone-sensitive locomotor depression and catalepsy; DADLE caused biphasic, partially naloxone-sensitive modulation of locomotion, whereas MR 2033, a κ-agonist, had no effect at all. Furthermore, local intra-NA injection of methylnaloxonium antagonized the locomotor stimulant effect of s.c.-injected heroin.[63] This finding is not surprising in view of the data on the existence opiate receptors on the dopamine-releasing terminals in the NA.[84,85] Thus opiates might have a dual mode of action: they might activate the dopamine release locally in the NA besides their complex effect at the level of VTA.

A third line of evidence indicates the existence of further independent physiological-pharmacological mechanisms. As demonstrated by Pijnenburg et al.,[86] injection of dopamine or direct dopamine receptor stimulants into the NA results in intense and long-lasting locomotor facilitation, primarily ambulation, but no stereotypy, and these effects are reversible by classical neuroleptics. On the other hand, morphine, injected into the NA, induces naloxone-reversible initial behavioral depression followed by excitation as seen upon systemic application.[87] In our study,[88] morphine and (D-Met^2,Pro^5)-enkephalinamide both potently inhibited the s.c. apomorphine-induced stereotypy and turning behavior in rats with unilateral lesion of substantia nigra. The antiapomorphine effects of these opioids were stronger than those of haloperidol and chlorpromazine.[88] This "third line" of findings is not easy to interpret or to bring into harmony with the previous ones. Most probably they should not. These data are obviously related to events at the level and beyond the postsynaptic dopamine receptors.

Thus the drugs under study modify the activity of the mesolimbic system at least at three different levels: they might act in the VTA, i.e., at the input of the system, in the NA on the dopamine-releasing terminals, and in the same structure postsynaptically.

Several lines of evidence indicate that the effect of opiates within the VTA is rather complex. They represent only one class of the multiple neuromodulators which take part in regulation of the activity of A_{10} neurons. Primarily, electrophysiological data (see Section II) prove that opioids activate the A_{10} dopamine neurons in a rather indirect way. Nevertheless, even a cursory analysis of their behavioral and neurochemical effects is helpful in perceiving the complexity of the local circuitry within the VTA regulating the activity of these cells.

A. GABA Agonists and Antagonistics

In one study, direct local injection of ethanolamine-*o*-sulfate, a GABA (α-aminobutyric acid) agonist, into this area abolished the locomotor activation seen with systemically injected amphetamine or also locally given (D-Ala^2,Met^5)-enkephalin.[89] Furthermore, injection of picrotoxin, a GABA antagonist, into the VTA resulted in locomotor facilitation not reversible by naloxone.[90,91] However, the hyperactivity induced by intra-VTA injection of picrotoxin can be attenuated by administration of spiroperidol, a dopamine receptor blocker, into the NA.[91] Therefore, Mogenson et al.[91] suggested that "...picrotoxin disinhibits the A_{10} dopaminergic neurons projecting to the nucleus accumbens...". These data, in good harmony with the outcome of related electrophysiological studies (see Section IV), indicate that opioids might activate the A_{10} cells by suppressing a tonic GABA-ergic inhibitory input. However, in another laboratory,[92] microinjection of muscimol, a selective agonist of $GABA_A$ receptor, per se, induced locomotor facilitation and baclofen, a $GABA_B$ receptor agonist, had no effect alone, but suppressed the locomotor facilitation induced by locally injected DAGO, muscimol, and systemically administered cocaine and amphetamine. Therefore, Kalivas et al.[92] proposed a model according to the GABA-ergic cells tonically inhibiting the dopamine release bear $GABA_A$ receptors, but there are $GABA_B$ receptors on the soma of A_{10} dopamine cells which mediate a direct inhibitory effect.

B. Neurotensin

Activation of the A_{10} cells[93] and locomotor facilitation can also be elicited by microinjecting neurotensin into the VTA.[94,95] This effect is obviously brought about by activation of the neurotensin receptors located on the soma of A_{10} neurons.[96,97] Its microinjection into this area is rewarding as indicated by conditioned place preference[98] and it induces dopamine release in the NA, but not in the striatum.[99] Nevertheless, its injection into the NA induces behavioral depression[94,100] and antagonizes the amphetamine-induced behavioral stimulation.[101] Thus neurotensin might induce opposite changes upon its injection into the VTA or in the NA, similarly to enkephalins (*vide supra*).

C. Substance P

Substance P is another endogenous neuropeptide in the VTA. It is also rewarding in various behavioral paradigms.[102,103] Its local injection into the VTA also results in locomotor facilitation[90,104] and increased dopamine liberation from the mesolimbic dopamine terminals.[95,105,106]

D. Amino Acid Receptors

Some sporadic data point to the involvement of excitatory amino acid receptors in the complex regulation. For example, L-glutamic acid diethyl ester, a glutamate/quisqualate receptor antagonist, antagonizes the locomotor facilitation induced by systemic cocaine and amphetamine injection, but does not modify the similar effect of caffeine as a proof of the selectivity of the effect.[107]

Thus a rather complex circuitry regulates the activity of A_{10} neurons, and besides the indirect facilitatory effect of opioids these neurons can also be excited directly with substance P and neurotensin. For the time being, there are no data on the eventual role of these endogenous neuropeptides in the rewarding effect of drugs of abuse.

IV. ACUTE ELECTROPHYSIOLOGICAL EFFECTS OF OPIATES AND PSYCHOSTIMULANTS IN THE MESOLIMBIC SYSTEM

The electrophysiological data regarding the modulation by opioids and psychostimulants of the mesolimbic system are very scanty. The few pertinent studies were done mostly in the 1980s and were obviously prompted by the related behavioral studies in the preceding years (see Section III). In several works, all performed in chloralose-anesthetized rats, systemically (i.v.) injected morphine was found to increase the discharges of A_{10} cells in the VTA.[108–110] (In most studies the A_9 cells in the neighboring substantia nigra were modulated in the same way.) Facilitation was also seen upon local microiontophoretic application of morphine and enkephalin analogs as DALA and DADLE.[109]

The facilitatory effect of morphine was dose dependent, and even after high doses, monophasic excitation was seen[110] just as in relation to locomotor stimulation seen after intra-VTA administration of μ- and δ-opioids (see the previous section). Furthermore, the apparently smallest doses of morphine needed for facilitation of A_{10} cells also induced locomotor activation,[108] indicating that the two phenomena, i.e., excitation of A_{10} neurons and spontaneous locomotion, are causally interrelated.

In all related works the systemic morphine-induced facilitation could be reversed by naloxone. In one of these studies, morphine and the enkephalin analogs were also examined upon local microiontophoretic application, and the facilitation could hardly be reversed by naloxone, prompting the authors to propose that the effect of opiates is indirect[109] *(vide infra)*.

In the only related *in vitro* study, morphine, Met- and Leu-enkephalin, and β-endorphin all increased the VTA cell activity upon adding the drugs into the incubating medium.[111] Comparing the sensitivity of neurons in the VTA and the substantia nigra, the former region was found more responsive to morphine, but there were no qualitative differences in responsivity.[111]

TABLE 1
Mesolimbic Dopamine System in the Motor Effects of Opiates and Psychostimulants

Species examined	Experimental paradigm	Observed effect	Ref.
Rat	Effect of bilateral injection of morphine into the VTA[a]	Hyperactivity	69
	Effect of bilateral injection of DALA,[b] DADLE[c] into the VTA	Hyperactivity	70–72
	Effect of bilateral injection of β-endorphin into the VTA	Hyperactivity	73
	Effect of dopamine receptor blockers such as α-flupenthixol, haloperidol, and opiate antagonists on the hyperactivity induced by opioids injected into the VTA	Reversal	69–72
	Effect of intra-VTA DALA on hyperactivity induced by systemic amphetamine	Potentiation	71
	Effect of destruction by 6-OHDA[d] of NA[e] on hyperactivity induced by intra-VTA β-endorphin, DALA	Abolition	71,73
	Effect of destruction by 6,7-OHDA of NA on hyperactivity induced by systemically injected amphetamine, cocaine	Abolition	65,66
	Effect of unilateral injection of morphine, DPDPE[f] into the VTA	Contraversive circling	75–76
	Effect of pimozide and opiate antagonists on the contraversive circling induced by unilateral injection of morphine into the VTA	Blocade	77
	Effect of injection of cocaine into NA	Hyperactivity reversible by *cis*-flupenthixol	78
	Effect of lesion by 6-OHDA of NA on hyperactivity induced by systemically injected		
	Heroin	No effect	66
	Amphetamine	Attenuation	79
	Apomorphine	Potentiation	80
	Effect of intra-NA injection of DALA, (D-Ala[2], Met[5])-enkephalin	Hyperactivity	81,82
	Effect of intra-NA injection of		
	Dopamine	Locomotor stimulation	86
	Morphine	Locomotor depression	83,87

TABLE 1 (continued)
Mesolimbic Dopamine System in the Motor Effects of Opiates and Psychostimulants

Species examined	Experimental paradigm	Observed effect	Ref.
	DALA	Biphasic effect	83
	Morphine and DMPEA[g]	Reversal of stimulation induced by s.c. apomorphine	88

[a] VTA: Ventral tegmental area.
[b] DALA: (D-Ala^2,Met^5)-enkephalinamide.
[c] DADLE: (D-Ala^2,D-Leu^5)-enkephalin.
[d] 6-OHDA: 6-Hydroxydopamine.
[e] NA: Nucleus accumbens.
[f] DPDPE: (D-Pen^2,D-Pen^5)-enkephalin.
[g] DMPEA: (D-Met^2,Pro^5)-enkephalinamide.

κ-Agonists have not been examined. However, there are data on the modulation by opioids of the A_9 neurons in the substantia nigra. They reacted similarly to the A_{10} cells to morphine and μ-preferring enkephalin analogs as DAGO, but U-50488, i.e., a prototypic κ-agonist, had no effect.[112] Thus one might surmise that probably it might also be true for the A_{10}.

A series of data indicate that opioids facilitate the A_{10} cell indirectly. Already in the first related study,[108] some cells in the VTA were found inhibited by morphine and this inhibition was hardly reversible by naloxone. As it subsequently turned out, actually just the naloxone-insensitive inhibition of certain units might lead to activation of the A_{10} cells. Namely, several lines of evidence indicate that the opiate-induced activation of A_{10} neurons is indirect; it is brought about by disinhibition, i.e., suppression of the tonically active GABA-ergic inhibitory input of dopamine cells:

- Upon its iontophoretic application, morphine only facilitated the dopamine neurons, i.e., the ones inhibited by apomorphine and facilitated by haloperidol (obviously due to feedback by autoreceptors located on the perikaryons), but opiates inhibited the closely located nondopaminergic ones.[109]
- Even upon i.v. administration, some cells in the VTA are inhibited by morphine.[108]
- Microiontophoretic application of GABA into the VTA suppresses the activity of dopamine neurons in the VTA.[113,114]

- Electrical stimulation of the NA similarly inhibits the activity of VTA dopamine neurons.[114–116]
- It is morphologically well established that NA sends efferent projections to the VTA.[117,118]

Thus, in accordance with nonelectrophysiological data reviewed in Section III, it is more than probable that opiates facilitate the A_{10} cells by inhibiting adjacent GABA-releasing neurons which tonically inhibit their activity. These GABA-releasing neurons are activated by a feedback loop projecting from the NA into the VTA.

Probably it is reasonable to mention here that morphine infusion into the VTA inhibited some units in the NA, and this inhibition could also be abolished by α-flupenthixol.[119] Thus the dopamine released from the terminals of A_{10} neurons might inhibit certain neurons in the NA. However, some cells in the NA were inhibited by systemically given heroin, even in the presence of α-flupenthixol or local opiate microiontophoretic application of morphine.[119] Thus some of the cells in the NA are inhibited via dopamine release from the terminals of A_{10} neurons, and others directly. These electrophysiological data are in harmony with the data in the previous section (III) on the inhibition[87,88] of spontaneous or apomorphine-activated locomotion by direct injection of opioids into the NA.

Summarily, these data confirm the existence of a dopaminergic VTA-NA projection and a GABA-ergic NA-VTA feedback loop. Of course, the functional interrelationships are much more sophisticated. Namely, as discussed in Chapter 6, NA also sends projecting pathways proximally as part of the rewarding circuitry, and its activity can also be modified locally by both dopaminergic and opioid agents.

As for the effect of psychostimulants, cocaine has been reported to inhibit the VTA dopamine neurons upon systemic application.[120] Several data indicate that cocaine also affects the VTA neurons indirectly as opiates do. Cocaine inhibits the A_{10} neurons *in vitro* in brain slice preparation too, and it powerfully potentiates the similar effect of dopamine.[121] Thus it obviously enhances the dopaminergic autoreceptor-mediated feedback inhibition. Even this effect is in part indirect, namely, ibotenic acid lesions of the NA and hemitranssection of the brain rostral to the VTA both reduce the cocaine-induced feedback inhibition of A_{10} neurons.[120]

Thus, the net effect of opiates and psychostimulants is similar. Opiates facilitate the A_{10} neurons indirectly, but their final effect in the mesolimbic system is dopamine release (in the NA), whereas psychostimulants have a direct dopamine-releasing effect as discussed above. The NA-VTA feedback loop seems to be activated similarly by both main types of these abused drugs.

TABLE 2
Modulation by Opioids and Psychostimulants of the A_{10} Cell Activity (Electrophysiological Data)

Species examined	Experimental paradigm	Observed effect	Ref.
Rat	Effect of i.v. morphine on the activity of A_{10} neurons in the VTA[a]	Increase in firing	108–110
	Effect of microintophoretic application of morphine, DALA, DADLE into the VTA	Increase in firing	109
Mouse	Effect of Met- and Leu-enkephalin, β-endorphin and morphine, adding them to the incubation medium of brain slice preparation	Increase in firing of VTA cells	111
Rat	Effect of opioids on the nondopaminergic (GABA-ergic) cells	Inhibition	109
	Effect of GABA intophoresis on the firing of A_{10} cells in the VTA	Inhibition	113,114
	Effect of electrical stimulation of NA[b] on the activity of A_{10} cells	Inhibition	114–116
	Effect of morphine infusion into the VTA on the activity of NA units	Inhibition	119
	Effect of cocaine (given i.v.) on the activity of A_{10} cells	Inhibition	120
	Effect of cocaine added to the incubation medium of brain slice	Inhibition of VTA dopamine cells; the effect reversible by $dopamine_2$ receptor antagonist sulpiride	121

[a] VTA: Ventral tegmental area.
[b] NA: Nucleus accumbens.

V. ACUTE METABOLIC EFFECTS OF OPIATES AND PSYCHOSTIMULANTS IN THE MESOLIMBIC SYSTEM

Based on elevated tissue levels of homovanillic acid and 3,4-dihydroxyphenylacetic acid, it was established a long time ago that morphine (and other classical opiates) facilitate the dopamine turnover in the nigrostriatal system in rats,[122–125] as do β-endorphin[126] and DALA.[125] In some other dopaminergic systems, such as the tuberoinfundibular[127–129] and tuberohypophyseal neurons,[130] morphine decelerates the turnover. Nevertheless, morphine and etorphine have been found to enhance the turnover in terminal areas of the mesolimbic system, including the NA and olfactory tubercle,[124,131,132] furthermore in the cingulate, pyriform, and frontal cortices.[133]

As for enkephalins, injecting them centrally (i.c.v. or intracisternally), both δ-receptor-preferring ones such as DADLE[134] and (D-Ser^2,Leu^5)-enkephalin-Thr[135] and μ-preferring ones such as (D-Ala^2,Met^5)-enkephalin[136] and FK 33–824 (D-Ala^2,NMePhe^4,Met(O)^5)-enkephalin-ol[135] enhance the turnover in the nigrostriatal system. DALA also increases the dopamine turnover in the NA, septum, and striatum if microinjected into the VTA.[104] Morphine, (D-Ala^2,Met^5)-enkephalin, and the δ-receptor-preferring (D-Ser^2,Leu^5)-enkephalin-Thr all enhance the release of dopamine if injecting them locally into the caudate nucleus in cats.[137] Thus both classical opiates and the μ- and δ-receptor-preferring enkephalin derivatives apparently accelerate the dopamine turnover in both main forebrain circuitries. All these data indicate that the A_9 and A_{10} neurons have enkephalinergic input as postulated by Wood.[125] This input is obviously tonically active, namely since tiorphan, an inhibitor of enkephalinase, also increases the concentrations of dopamine metabolites in the striatum[138] and SCH 32615, another enkephalinase inhibitor, induces the same effect in the NA.[139]

However, as reported by Wood and co-workers,[125,134] κ-agonists such as dynorphin, MR 2034, tifluadom, and U-50488H do not enhance the dopamine turnover by themselves, but, combining the synthetic κ-agonists with morphine, the former antagonized the activation caused by the latter in the nigrostriatal and mesolimbic systems as well.[125] Similar data were reported recently: U-50488 given alone, did not modify the dopamine turnover in various brain regions, but it antagonized the haloperidol-induced increase in its metabolism in the nigrostriatal, mesolimbic (and tuberohypophyseal) neurons, and the acceleration induced by prolactin in the tuberoinfundibular cells.[140]

Since μ- and δ-preferring opioids injected into the VTA induce locomotor facilitation and enhance the firing of A_{10} cells, as discussed above, the increase in dopamine turnover, might be a biochemical manifestation of their activation. Therefore it is not surprising that not only the dopamine turnover but in most related studies, also its release was enhanced. Morphine accelerates the dopaminergic impulse flow in the nigrostriatal system[124,141] and increases the concentration of 3-methoxytyramine, the extraneuronal metabolite of dopamine.[125,134]

However, contrary to the more or less unequivocal data obtained *in vivo,* the data obtained *in vitro* are rather contradictory. In the first related study, Loh et al.[142] claimed that morphine, similarly to Met-enkephalin and β-endorphin, inhibited the potassium-evoked dopamine release from rat striatal slice preparation. These findings have not been confirmed in the subsequent studies.

Morphine was found ineffective in modulation of dopamine release from rat striatum,[143] but in two studies δ-agonists such as DADLE and (D-

Ser^2,Leu^5)-enkephalin-Thr[144] and (D-Thr^2, Leu^5)-enkephalin-Thr (DTLET)[145] and kelatorphan, an enkephalinase inhibitor, all enhanced the basal and stimulation-induced release of dopamine from rat striatal slice preparation, but μ-agonists such as DALA[144] and DAGO[145] had no effect. In the latter study[145] the stimulatory effect of DTLET and kelatorphan was abolished by ICI 154129, a selective antagonist of δ-receptors, indicating that the stimulation was really mediated by δ-receptors. However, in other studies[146,147] δ- and μ-agonists such as Leu-enkephalin, DADLE, DPDPE, and DAGO all failed to significantly modify the basal or K^+ stimulation-induced dopamine release. The contradictory outcome of these studies is difficult to explain.

However, in all related *in vitro* studies, κ-agonists, including dynorphin,[146] U-50488H,[147] ethylketocyclazocine, tifluadom, bremazocine, U 69593, and PD 117302[146] all inhibited the K^+ and electrically induced dopamine release from rat striatal slice or synaptosomal preparation.[147]

Thus, while *in vivo* the μ-and δ-receptor-mediated facilitation of dopamine release is the most evident, *in vitro* just the κ-receptor-mediated inhibition is the most firmly established, at least in the nigrostriatal system.

It is tempting to speculate that the *in vitro* inefficacy of μ- and δ-agonists is in harmony with the previously discussed data indicating that opioids activate transsynaptically the A_{10} neurons. (In an *in vitro* system these connections are obviously impaired.) On the other hand, the unequivocal inhibitory potency of κ-ligands *in vitro* as opposed to the lack of effect *in vivo* might be due to a tonic μ-δ-receptor-mediated excitatory input which can obscure the inhibitory effects of the formers in intact organism.

In the NA the situation is a bit different. Both DAGO and DTLET failed to modify the spontaneous and potassium-induced dopamine liberation in slice preparation.[145] Furthermore, direct injection of DALA into the NA also failed to modify the dopamine release from this structure *in vivo*.[82] Thus the direct, local modulation by enkephalins of dopamine release might not be significant (in contradiction with the data in Sections III and IV).

On the other hand, according to one study,[148] in slice preparations made from the terminal areas of the mesolimbic dopamine system (NA, frontal cortex, and olfactory tubercle) U-50488H strongly inhibited the dopamine release induced by electrical stimulation, and this effect could be reversed by nor-binaltorphimine, a selective blocker of κ-receptors. However, DAGO and DPDPE, i.e., selective inhibitors of μ- and δ-receptors, had no effect.[148] Hence the possibility arises that the A_{10} neurons receive a κ-receptor-mediated inhibitory input at their terminals. For the time being there are no additional data corroborating the finding that the mesolimbic circuitry is inhibited by κ-opioids at the level of NA besides the scattered negative findings about the inability of κ-agonists to induce motor activation or dopamine release if administered directly into the VTA. Nevertheless, whether they act at the level of VTA and/or in the NA, these data[148] supplement our present knowledge on the reciprocal modulation of A_{10} cells by μ-/δ- and κ-opioid mechanisms, respectively.

However, using the powerful technique of *in vivo* microdialysis, Di Chiara and Imperato[149] demonstrated that in unrestrained rats, upon systemic administration μ-agonists such as morphine, methadone, and fentanyl did modify the dopamine release from the NA (and also from the caudate nucleus): they induced a powerful monophasic increase in its release, whereas prototypic κ-agonists such as U-50488H, bremazocine, and tifluadom suppressed the liberation. Not less important is the observation that the minimally effective dose of these classical opiates induced both locomotor facilitation and dopamine release from the NA.[149] The stimulant effect of classical opiates could be abolished by β-funaltrexamine, an irreversible ligand of μ-receptors, whereas naloxone reversed even the inhibitory action of κ-agonists.[149] The same authors demonstrated that other types of abused substances, including cocaine, amphetamine, ethanol, and nicotine, also increased the dopamine release from the NA, whereas psychotropic drugs not abused by humans, such as imipramine, atropine, and diphenhydramine, failed to modify the dopamine release.[150] Thus, these pioneering experiments[149,150] have far-reaching implications: they indicate that liberation of dopamine from the mesolimbic terminals is a common element in the rewarding effects of abused substances whether they are μ-opiates, psychostimulants, or other types of drugs inducing euphoria. Some subsequent studies only confirmed and extended these findings in relation to morphine[151] and other μ- and δ-opioids, as well. Dopamine release could be elicited from the NA by i.c.v. administration of both by μ-selective enkephalin analogs such as DAGO and δ-specific ones such as DPDPE, whereas E-2068, a dynorphin analog, and U-69593, a selective κ-agonist, decreased the release.[152] In this study[152] the dopamine release induced by DAGO could be selectively antagonized by CTOP, a selective μ-antagonist, and the similar effect of DPDPE was found reversible by ICI 174864, i.e., a δ-antagonist. Thus the mesolimbic rewarding dopamine release can be activated by both μ- and δ-receptors.

In line with these observations, β-endorphin has also been reported to enhance the dopamine turnover in and its release from the NA.[153–155] According to the latter study,[155] the dopamine-releasing effect of β-endorphin is stronger in awake than in anesthetized rats, i.e., when the enhanced dopamine liberation is accompanied by locomotor stimulation. Injection of FK 33–824, a μ-selective agonist, into the A_{10} region also increased the dopamine release from the NA and, of course, the locomotor activity as well.[156]

Using the same technique, i.e., measuring the extracellular concentration of dopamine or its metabolites in the NA by the aid of a chronically implanted microdialysis probe, psychostimulants such as cocaine,[157–163] amphetamine,[164] and other dopamine-uptake inhibitors such as nomifensine, Lu17133, Lu19005, GBR 12783,[157,164] GBR 12909, amfonelic acid, and methylphenidate[39] have also been found to enhance release of dopamine in the NA and striatum just as seen with μ- and δ-opioids. Not only passive administration, but voluntary consumption also enhances dopamine release. As was demonstrated first by Pettit and Justice[165] in rats trained to lever-press for cocaine, i.v. self-administration results in a gradual increase in dopamine level in the NA. Nevertheless, letting the animals operate the lever on a continuous reinforcement schedule, the dopamine level stabilized at a higher extracellular lever.[165] Thus the rats seem to titrate the amount of cocaine self-administered in a way to maintain the dopamine level in the NA at a certain, above the baseline, level. However, if the unit dose of cocaine is raised (i.e., the amount given during a single infusion, i.e., as a single reinforcement), the dopamine level in the NA stabilizes at a higher level.[166] The latter findings are difficult to interpret in the light of the generally held belief that animals always titrate the number of lever presses to maintain a stable blood level. It is better to say that, apparently, the unit dose might somehow codetermine the level stabilized and maintained by regular operant responding.

Furthermore, similarly to μ- and δ-preferring opiates also, these direct dopamine uptake inhibitors enhance its release in other terminal areas of the mesolimbic dopaminergic system, including the olfactory tubercle, the medial prefrontal cortex, and the piriform cortex.[39,162] Actually, the data regarding the medial prefrontal cortex are contradictory,[160] but it might be explained by the findings[39] that the dopamine uptake-inhibiting effect of cocaine is rather weak in this area and its alleged effectiveness in this brain region is obscured by simultaneous inhibition of norepinephrine reuptake in this region.[39]

In the above mentioned studies the dopamine-releasing effect of psychostimulants was seen after their systemic (mostly i.v.) administration. Nevertheless, adding cocaine locally to the perfusing medium (in the striatum), the release was enhanced just as seen after systemic application, but not if injected into the substantia nigra.[164] Thus its effect is direct; it must act on the dopamine-releasing terminals. Finally, it is worth recalling that a close temporal correlation was found between the cocaine-induced enhancement of dopamine turnover and locomotor stimulation in an early study.[167]

Probably it is pertinent to mention here that using another relatively new and sophisticated method, *in vivo* voltametry, enhanced dopamine release was detected in the NA upon electrical self-stimulation of the VTA.[168] In addition, i.p. injection of cocaine augmented the self-stimulation-induced dopamine release.[168] These observations provide additional evidence concerning the commonalities in the two apparently different kinds of rewarding mechanisms, addiction to drugs and intracranial electrical self-stimulation (ICSS).

Summarily, all these data confirm the postulate of Wise and Bozarth[8] that the reinforcing effect of various classes of drugs of abuse is mediated by enhancement of dopamine release from its mesolimbic dopaminergic terminals, primarily in the NA.

VI. ENHANCEMENT OF REWARDING POTENCIES OF ABUSED SUBSTANCES UPON REPEATED ADMINISTRATION: ALTERED BEHAVIORAL AND NEUROCHEMICAL EFFECTS

Examining the effect of morphine on spontaneous behavior in rats, Kaymakcalan and Woods[169] first observed that upon repeated administration of morphine its sedative effect gradually disappeared and it was replaced by excitation. Similar observations were made in other laboratories as well,[170,171] but the phenomenon was studied in a comprehensive way first by Babbini and Davis.[27] Morphine in rats in smaller doses induced monophasic locomotor facilitation, but higher doses induced a biphasic response: an initial phase of motor depression was followed by locomotor stimulation. Upon repeated administration, tolerance developed to the sedative effect and the locomotor stimulation became more and more intense.[27] A similar progressive increase in the opiate-induced hyperactivity has been observed in other laboratories, too.[172–174] These findings have never been questioned ever since and nobody has claimed to record tolerance development to the excitatory effect of morphine or its classical derivatives. An interesting auxiliary finding is that sensitization becomes less conspicuous if the animals receive the daily morphine injection in the home cage instead of the laboratory, where the activity is recorded.[175] However, separation of the site of daily injections and observations only weakens, but does not eliminate sensitization.[176] Thus conditioning contributes to the process, but sensitization cannot be equated with a kind of ''learning process''.

As for psychostimulants, based on an initial, very old observation,[177] Post and Rose[178] reported that upon repeated administration the cocaine-induced locomotor facilitation was augmented, including its horizontal and vertical components as well. The lack of tolerance development to the amphetamine-induced locomotor facilitation,[179] and the progressive increase in its efficacy to induce hyperlocomotion and stereotypy[180,181] was described in the same period.

Repeated pretreatment with psychostimulants might enhance the locomotor facilitation induced by opiates and vice versa. For example, in rats repeatedly preexposed to morphine the amphetamine-induced locomotor activation was enhanced.[182] In another study,[183] mice repeatedly

TABLE 3
Modulation by Opioids and Psychostimulants of the Dopamine Turnover in the Mesolimbic System

Species examined	Experimental paradigm	Observed effect	Ref.
Rat	Modulation by morphine of the dopamine turnover in the terminal areas (NA,[a] olfactory tubercle, cingulate, pyriform, and prefrontal cortices) of the mesolimbic system	Facilitation	124,131 133
	Modulation by DALA[b] injected into the VTA[c] of dopamine turnover in the NA and septum	Facilitation	104
	Effect of β-endorphin on dopamine turnover in the NA	Facilitation	153,154
	Effect of SCH 32615, an enkephalinase inhibitor on the dopamine turnover in the NA	Facilitation	139
	Effect of U-50488H on dopamine turnover in the NA	No effect per se, but reversal of haloperidol-induced activation	140
	Effect of DAGO[d] and DTLET[e] on spontaneous and potassium-induced release of dopamine from NA slice preparation *in vitro*	No effect	145
	Effect of DALA injected on the dopamine release locally	No effect	82
	Modulation of electrical stimulation-induced dopamine release from slices prepared from NA, frontal cortex, and olfactory tubercle by		148
	U-50488H	Suppression	
	DAGO	No effect	
	DPDPE[f]	No effect	
	Modulation of dopamine release from the NA as determined by microdialysis by opioids as		
	Morphine, methadone, fentanyl	Increase	149–151
	DAGO	Increase	152
	FK-33824[g] into the VTA	Increase	156
	DPDPE	Increase	152
	U-50488H	Decrease	149,150
	E-2078[h]	Decrease	152
	β-endorphin (i.c.v.)	Increase	152

TABLE 3 (continued)
Modulation by Opioids and Psychostimulants of the Dopamine Turnover in the Mesolimbic System

Species examined	Experimental paradigm	Observed effect	Ref.
	Inhibitors of dopamine (re)uptake as		
	Amphetamine and cocaine	Increase	149,150 158–164
	Nomifensine	Increase	157,164
	Methylphenidate	Increase	39
	Modulation of dopamine release from NA during self-administration of cocaine	Increase	165,166
	Modulation by cocaine and methylphenidate of dopamine (re)uptake *in vitro* in slice preparations made of NA, olfactory tubercle, medial prefrontal cortex	Inhibition	39,162
	Modulation of the dopamine turnover by cocaine in NA, prefrontal, and piriform cortices	Deceleration	162

[a] NA: Nucleus accumbens.
[b] DALA: (D-Ala^2,Met^5)-enkephalinamide.
[c] VTA: Ventral tegmental area.
[d] DAGO: (D-Ala^2,NMePhe^4,Gly^5-ol)-enkephalin.
[e] DTLET: Tyr-D-Thr-Gly-Phe-Leu-Thr.
[f] DPDPE: (D-Pen^2,D-Pen^5)-enkephalin.
[g] FK-33824: (D-Ala^2,NMePhe^4,Met(*O*)^5-ol)-enkephalin.
[h] E-2078: Dynorphin analog.

pretreated with metamphetamine showed an increased sensitivity to the ambulation-increasing effect of morphine as compared to the drug-naive controls. This phenomenon, called cross sensitization, indicates that

- The same pharmacological mechanism might mediate the sensitization induced by the two main groups of abused substances (opioids and psychostimulants).
- The anatomical substrate of this process must be the mesolimbic dopamine system, where the opioid and mesolimbic systems converge.

Sensitization develops to the neurochemical effects of abused substances as well. The first findings were obtained with amphetamine. As observed by Robinson and Becker[184] and Kolta et al.,[185] in rats repeatedly pretreated with amphetamine, a subsequent dose induced not only more intense locomotor facilitation and stereotypy, but also the amphetamine-induced release of dopamine was enhanced from striatal slice if the preparation was made from previously chronically treated animals. Thus sensitization may develop to both the characteristic behavioral and neurochemical effects of certain drugs. According to the latter study[185] the progressive increase in amphetamine-induced behavioral response and dopamine release run parallel. The phenomenon is not specific to amphetamine; namely, in another study,[186] amphetamine pretreatment also potentiated KCl or electrical stimulation-induced dopamine release from striatum *in vitro*. In good harmony with these findings, repeated pretreatment with cocaine also enhanced both the behavioral response (hyperlocomotion) and the neurochemical changes (decrease in the tissue level of dopamine metabolites in the NA as a manifestation of its more rapid turnover) seen upon a subsequent challenge with cocaine.[187]

Nevertheless, this phenomenon, i.e., the enhancement of drug-induced dopamine release, could be examined in a more reliable way only recently using the powerful technique of *in vivo* microdialysis. In rats, pretreatment with metamphetamine augmented both the locomotor facilitation and the increase in dopamine release in the ventral striatum, i.e., NA upon a challenge dose, whereas the baseline value of dopamine liberation did not change.[188] In another study,[189] sensitization with metamphetamine augmented the dopamine content of the striatal dialysates in rats challenged with either metamphetamine or cocaine. Thus it is not surprising that repeated cocaine treatment gradually enhanced the dopamine release in the NA, as measured by local microdialysis, and augmented the locomotor response, and a close correlation was found between the two phenomena.[163] Two studies apparently contradict the previous ones. In one work a decrease in the dopamine metabolites in the perfusate of NA was detected upon repeated cocaine administration, but the dopamine concentration was not measured.[190] Of course, it is difficult to figure out whether the decrease in the dihydroxyphenylacetic level in the NA reflects decreased release or accelerated turnover of dopamine. In another study the concentrations of both dopamine and its metabolites decreased upon repeated cocaine application, but a sensitization like elevation was seen in the anterior caudate putamen.[161] Thus, neither study is really conclusive. Nevertheless, a recent interesting study might explain these contradictions. According to Imperato et al.,[191] chronic cocaine treatment (twice daily 10 mgkg) only temporarily enhanced the dopamine release from the NA in rats; from the fifth day on the cocaine-induced dopamine release dropped below the level recorded in the vehicle-treated rats, though the stereotypy induced by it remained maximal. The drastic decrease in the extracellular concentration of dopamine in the NA in spite of continuing cocaine treatments[191] might explain craving and the urge of cocaine addict subjects and the especially strong addictive potential of cocaine among the abused substances. Of course, these initial findings need to be extended and corroborated. On the other hand, for the time being it cannot be even guessed whether in these subjects the dopamine stores were depleted, the dopamine-producing capacity exhausted, or, for some reason, the negative feedback effects enhanced.

The opposite phenomenon, i.e., drug withdrawal-induced suppression, was demonstrated so far only in few studies. Namely, Acquas et al.[192] and Parson et al.[193] reported on a profound depression of dopamine release in the NA upon morphine withdrawal in dependent rats. Obviously, that is expected by supposing that dopamine release in the mesolimbic system mediates the rewarding effects of abused substances, and their withdrawal leads to dysphoria for lack of the neurohormones maintaining euphoria under the influence of drugs of abuse.

VII. THE SITE OF SENSITIZATION

Upon looking for the sites of sensitization, probably lesioning of the suspected brain areas provides the simplest answer. In one study, electrolytic lesion of either the VTA or the NA attenuated and delayed the onset, but did not completely abolish locomotor excitation seen upon chronic i.p. administration of morphine.[194] This finding is in agreement with the data obtained by other types of interventions *(vide infra)*. Of course, the failure of these lesions to completely eliminate the morphine-induced sensitization might equally be due to limitations of this type of experiment, i.e., it may be that not all the A_{10} neurons were destroyed in this study[194] or, besides these neurons there might be other alternative pathways mediating the effect. In another study, lesioning of the A_{10} neurons by 6-OHDA injection into the VTA or the NA enhanced the locomotor facilitation induced by injection of DALA, β-endorphin, and morphine into the NA.[195] Since 6-OHDA destroys only the presynaptic part of dopaminergic synapses, this ''negative'' finding might actually point to the participation of A_{10} neurons in the process.

Repeated administration of morphine into the VTA results in a progressive increase in its motor excitant as compared to the locomotor facilitation recorded upon acute treatment.[69,175] In animals previously implanted with morphine pellets the behavioral arousal induced by intra-VTA β-endorphin was also augmented.[196] Furthermore, repeated systemic preexposure to either morphine or amphetamine enhanced the locomotor response induced by intra-VTA morphine.[197] Similarly, repeated preexposure to cocaine enhanced the locomotor response to intra-VTA DAGO.[198] Thus sensitization can be elicited via this structure, and, even upon ''systemic'' sensitization with either opiates or psychostimulants, the responsivity to intra-VTA opiates is enhanced. The sensitization induced by i.p. morphine could be reversed not only by systemically given naloxone, but also by its local injection into the VTA.[199] Thus there is little doubt that VTA is (one of) the main site(s) mediating sensitization to opiates.

Actually, the opiate receptor subtypes involved in sensitization were also identified by injecting the substances into this area. As reported by Kalivas et al.,[176] repeated injections of both DALA (a relatively μ-preferring enkephalin analog) and DADLE (a δ-preferring one) induced progressively increasing hyperactivity; however, $dynorphin_{1-13}$ failed to induce sensitization. Thus the process can be initiated by both μ- and δ-receptors, but not by κ-ones.

Kalivas and co-workers[176,200] demonstrated that daily injections of DALA into the VTA resulted in progressive enhancement of locomotor stimulation caused by the enkephalin analog, and the behavioral stimulation induced by systemic amphetamine and intra-VTA neurotensin also was augmented. Thus during sensitization the reactivity of VTA is enhanced whether the sensitizing injections of opiates are administered systemically or locally into this area. Thus the site of both sensitization and cross sensitization might be the VTA.

As for the neurochemical effects of opiates, Kalivas[200] reported that the DALA-induced enhancement of dopamine turnover in the NA also became more and more intense, but there was no change in the baseline levels of dopamine and its metabolites. Subsequently, the same group demonstrated that in sensitized rats not only the dopamine turnover (based on measurement of the tissue concentrations of its metabolites), but also its *in vivo* release (detected by the aid of microdialysis) was progressively enhanced upon repeated administration, and furthermore that the enhancement of dopamine release was closely correlated with augmentation of the locomotor activation.[163] Furthermore, according to this study there are no permanent changes in the dopamine turnover in the NA, but the responsivity of the system is augmented.[200] Subsequently, the same group reported that in animals repeatedly pretreated with morphine, its application induced less intense increase in the dopamine turnover in the VTA, contrary to the NA and prefrontal cortex, where the response was enhanced.[199] Furthermore, if rats were sensitized by peripheral cocaine or morphine, the potassium depolarization-induced dopamine release from the slices made of A_9 and A_{10} areas was also reduced.[201] Thus the above mentioned increased responsivity of the mesolimbic terminal areas is accompanied with decreased reactivity of the VTA.

As for psychostimulants, the situation seems to be symmetrical. Namely, repeated preexposure to intra-VTA amphetamine potentiated the locomotor response induced by peripheral amphetamine or cocaine, though intra-VTA injection in itself did not elicit hyperactivity.[201,202] In addition, repeated preexposure to cocaine augments the facilitation of dopamine turnover in the NA seen in acute experiments.[187]

Contrary to VTA, injection of amphetamine into the NA does not induce behavioral sensitization, though it does elicit locomotor hyperactivity if injected into this region.[202,203] Similarly, repeated administration of morphine and DAGO into the NA also failed to induce behavioral sensitization.[204]

As discussed in Chapter 9, A_{10} neurons and their projections only comprise a part of the rewarding circuitry. These cells have physiological inputs and output and activate further pathways through their outputs. For the time being, almost nothing is known about the eventual role of these adjoining pathways in the process of sensitization. Nevertheless, according to some recent data there is at least one communicating circuitry which seems to be involved in the process. Yoshikawa et al.[205] reported recently that electrolytic lesion of the fimbria fornix abolished the behavioral sensitization induced by repeated metamphetamine injections. In view of the dopaminergic innervation of the hippocampus arising from the VTA and the existence of a glutamatergic input of the NA from the hippocampus, the authors postulate the existence of a VTA-hippocampal-NA circuitry activated during sensitization.[205] As discussed below (Section VIII), some behavioral data are in harmony with this proposition.

Summarily, these findings indicate unequivocally that the main or only site of sensitizing effects of opioids and psychostimulants is the mesolimbic, i.e., VTA-NA system. More exactly,

- The site of sensitization (and cross sensitization) to opiates and psychostimulants is the VTA and not the NA.
- Sensitization is accompanied with enhanced release of dopamine from the NA, but the dopamine turnover is actually slowed down in the VTA.
- Sensitization apparently does not affect the baseline turnover and release of dopamine, but the reactivity to both opiates and psychostimulants becomes augmented as a result of this process.

VIII. THE BIOCHEMICAL BASIS OF SENSITIZATION

A. The Role of Dopamine

Pretreatment with pimozide (injected i.p.), a nonselective blocker of dopamine receptors, suppressed the "conditioned" sensitization induced by intra-VTA morphine,[175] i.e., it inhibited the process (in an experimental arrangement where the sensitization was specific for the testing). Haloperidol also blocked the sensitization to the amphetamine-, metamphetamine-, and cocaine-induced locomotor facilitation[206–208] and amphetamine-

TABLE 4
Ventral Tegmental Area: The Main Site of Sensitization to Opioids and Psychostimulants

Species examined	Experimental paradigm	Observed effect	Ref.
Rat	Effect of repeated administration of morphine into the VTA[a]	The locomotor facilitation progressively increases upon repetition: sensitization	69,175
	The effect of electrolytic lesion of either VTA or NA[b] on sensitization to i.p. morphine-induced locomotor excitation	Attenuation	194
	Effect of previous continuous morphine treatment on locomotor hyperactivity induced by injection of β-endorphin or morphine into the VTA	Enhanced response: cross sensitization between peripheral and intra-VTA-injected opioids	196,197
	Effect of previous repeated cocaine treatments on locomotor activation induced by intra-VTA injection of DAGO[c]	Enhanced response: cross sensitization between peripheral cocaine and intra-VTA opioid	198
	Effect of naloxone given into the VTA on sensitization induced by i.p. morphine	Inhibition	199
	Examining the receptor subtype involved: induction of sensitization by intra-VTA injection of		200
	DALA[d]	Yes	
	DADLE[e]	Yes	
	Dynorphin$_{1\text{-}13}$	No	
	Effect of repeated injections of DALA into the VTA on locomotor facilitation induced by systemic amphetamine or local neurotensin	Potentiation, i.e., cross sensitization	176,200
	Effect of repeated opiate treatments (systemically or locally into the VTA) on dopamine turnover	Enhancement in the NA and prefrontal cortex, but no change or deceleration in the VTA	163,199, 200
	Effect of repeated peripheral cocaine treatments on the dopamine turnover in		187
	The NA	Progressive acceleration	
	In the VTA	Progressive attenuation of the initial acceleration	
	Effect of repeated peripheral cocaine or morphine treatments on the K^+-induced dopamine release from slice preparation of the A_{10} area	Decrease in dopamine release	201

TABLE 4 (continued)
Ventral Tegmental Area: The Main Site of Sensitization to Opioids and Psychostimulants

Species examined	Experimental paradigm	Observed effect	Ref.
	Effect of repeated administration of amphetamine into the VTA on subsequent motor response induced by peripherally given amphetamine or cocaine	Enhancement of locomotor hyperactivity upon their peripheral application through intra-VTA amphetamine per se does not induce hyperactivity	201,202
	Effect of repeated pre-exposure to cocaine on its acute metabolic effect	Augmented facilitation of the dopamine turnover in the NA	187
	Effect of repeated injections of morphine or DAGO into the NA	No sensitization	204
	Effect of repeated amphetamine injections into the NA	No sensitization	202,203
	Effect of electrolytic lesion of the fimbria fornix on sensitization to the motor effects of metamphetamine	Abolition	205

[a] VTA: Ventral tegmental area.
[b] NA: Nucleus accumbens.
[c] DAGO: (D-Ala^2,NMePhe^4,Gly^5-ol)-enkephalin.
[d] DALA: (D-Ala^2,Met^5)-enkephalinamide.
[e] DADLE: (D-Ala^2,D-Leu^5-enkephalin.

induced sensitization to *N*-methyl-DL-aspartic acid-elicited convulsions.[208] These findings (and many others discussed below) indicate that dopamine receptors are needed to induce sensitization. Of course, it is evident to ask which subclass of dopamine receptors is involved in the process. The same authors[209] reported subsequently that pretreatment with i.p. SCH-23390, a selective antagonist of dopamine$_1$ receptors, blocked the sensitization induced by systemically applied amphetamine, but failed to modify the similar action of intra-VTA morphine. On the other hand, intra-VTA injection of SCH-23390 inhibited the sensitization to systemically administered amphetamine.[210] Thus in spite of the above mentioned data on cross sensitization between opioids and dopamine uptake blockers, they might elicit sensitization in at least partially different ways.

On the other hand, sulpiride, a selective inhibitor of dopamine$_2$ receptors, failed to suppress the sensitization to either peripheral amphetamine or central morphine. Furthermore, in this work even pimozide was ineffective in suppression of the sensitization by morphine.[209] (In the latter study the authors attribute the opposite action in their earlier paper[175] to depolarization block.) However, in another laboratory,[211] both SCH-23390, i.e., a selective dopamine$_1$ receptor blocker, and YM-09151, a dopamine$_2$ receptor inhibitor, suppressed the sensitization towards the locomotor facilitation induced by metamphetamine.

These studies point to the indispensable role of dopamine receptors (primarily dopamine$_1$ ones) in the process of sensitization to opiates and psychostimulants, but do not explain by themselves the intricate mechanism of this gradually developing phenomenon.

1. Increased Turnover?

The simplest theoretical possibility is that the dopamine turnover might become permanently increased in the sensitized state. However, upon repeated amphetamine treatment the steady-state turnover of dopamine does not change.[212] Repeated intra-VTA application of DALA also failed to alter the baseline rate of turnover of dopamine in the NA.[200] Thus in the resting state there is no alteration; only the reactivity of the mesolimbic dopamine system is enhanced for some reason. Furthermore, no sensitization could be elicited by repeated intra-VTA application of apamine (an octadecapeptide extracted from the venom of *Apis mellifera*), though it did elicit motor hyperactivity and enhanced dopamine release in the NA.[213] Apamine has been shown to block the C^{2+}-dependent K^+ channels.[214] Thus, lasting depolarization in itself is not enough to induce sensitization. Both the ineffectiveness of lasting depolarization[213] and the lack of any change of steady-state dopamine turnover of A_{10}[212] indicate that sensitization is not an intracellular process; the mechanism of sensitization might be related to alteration(s) at the synaptic level.

2. Down-Regulation of Autoreceptors?

The findings on progressive enhancement of dopamine metabolism in the NA with simultaneous attenuation of its release from the perikarya of A_{10} cells in the VTA upon repeated cocaine[187] and opioids[163,199,200] administration indicates that the autoreceptors on these cells might be gradually downregulated.[188] Actually, it was proposed by Muller and Seeman[215] even before these studies that "...presynaptic subsensitivity is a possible basis for sensitization for long-term dopamine mimetics...". Obviously, downregulation of inhibitory presynaptic autoreceptors might lead to enhanced dopamine release from the axon terminals upon transsynaptic activation of these cells, which is recorded as an incremental enhancement of the motor responses and increase in the amount of dopamine (metabolites) in the NA and other regions containing the nerve endings of A_{10} neurons.

The subsensitivity of dopamine autoreceptors has also been demonstrated in electrophysiological experiments. After repeated cocaine administration the inhibition of A_{10} neurons by i.v.-injected apomorphine or locally iontophoresed dopamine was found attenuated.[216] Since in these experimental arrangements dopamine acts directly on the A_{10} cell bodies, these findings can only be explained by subsensitivity of autoreceptors.

Based on the above mentioned findings that SCH 23390 inhibits the sensitization to systemic amphetamine, but does not modify the same process elicited by intra-VTA morphine,[209] one has to suppose that dopamine uptake inhibitors elicit a subsensitivity of autoreceptors, but opioids do not. Since the formers liberate dopamine by acting directly on the A_{10} neurons, whereas opioids act by inhibiting the GABA-ergic inhibition of their activity (see Section II), this might be true, but for the time being no experimental proof confirms such a proposition.

3. Upregulation of Postsynaptic Receptors?

A further theoretical possibility is an upregulation of the postsynaptic dopamine receptors, which in turn should induce more pronounced behavioral response to dopamine released from the presynaptic side. Actually, Taylor et al.[217] already observed a long time ago that in cocaine-sensitized animals the binding of spiroperidol in the striatal tissue was enhanced and, in conjunction with sensitization to cocaine, the behavioral effects of apomorphine were also enhanced. Similarly, after repeated morphine administration the apomorphine-induced stereotypy is enhanced.[218] These observations[217,218] indicate that postsynaptic dopaminergic hypersensitivity might (also) contribute to the enhanced responses. In this respect it is worth mentioning that after 6-OHDA lesion of the A_{10} neurons in the VTA injection of DALA, morphine, or β-endorphin into the NA elicits more intense locomotor activation than in intact animals.[195,219] The finding is a bit surprising, since opioids are expected to stimulate the A_{10} neurons in the VTA as discussed in the previous sections. However, opiates do induce locomotor hyperactivity even if injected into the NA.[81,82] Actually, after chronic amphetamine pretreatments, its acute injection into the NA induces more intense locomotor facilitation than in drug-naive animals.[220] As discussed in Section VI, there are postsynaptic dopamine receptors in the NA, and these might become supersensitive during sensitization. In another study,[221] lesioning by 6-OHDA of the lateral hypothalamus in neonatal rats, i.e., damaging the ascending dopaminergic system also resulted in potentiation of the hyperactivity elicited by intra-NA injection of DALA and systemically applied apomorphine. This supersensitivity could be abolished by implanting embryonic mesencephalic extract, rich in dopamine cells, into the NA and the striatum.[221] In the corresponding electrophysiological study,[216] repeated cocaine pretreatments enhanced the inhibitory effect of dopamine, iontophoresed into the NA as a sign of postsynaptic receptor supersensitivity.

Of course, the direct evidence of supersensitivity would be the demonstration of enhancement of B_{max} of dopamine binding sites. For the time being, there are not many such studies. Akiyama et al.[222] recorded an increase in the B_{max} of spiroperidol binding sites in the NA and the frontal cortex in conjunction with metamphetamine-induced behavioral hypersensitivity. Subsequently, Goeders and Kuhar[223] (1987) recorded enhanced receptor binding (increase in B_{max}) of sulpiride in the NA. Peris et al.[224] also recorded an increase in the B_{max} of dopamine$_2$ receptors in the NA after chronic cocaine administration. These findings indicate that the downregulation of dopamine$_1$ autoreceptors in the VTA is accompanied with upregulation of dopamine$_2$ receptors postsynaptically in the NA. The ultimate effect of both receptorial changes is enhanced responsivity, more exactly the former one (downregulation of dopamine$_1$ receptors) will only enhance the dopamine release to adequate stimuli, whereas the enhancement of dopamine$_2$ receptors will augment the behavioral response to unit dose of dopamine. Thus the two apparently opposite changes induce functionally the same effect: enhanced reactivity of the mesolimbic dopamine system.

Interestingly, the data referring to the eventual changes of opiate receptors in the striatum are contradictory (see Goeders and Kuhar[223] for review). In cocaine addict humans, the methylspiroperidol binding was found reduced.[225] Anyhow, the eventual lack of upregulation of dopamine receptors in the nigrostriatal system might explain that behavioral sensitization is a phenomenon specific for the mesolimbic dopamine system.

4. Depletion of Dopamine?

Based on the alleged subsensitivity of dopamine autoreceptors, dopamine receptor stimulants are expected to induce more intense dopamine release in the NA in sensitized animals than in drug-naive ones. However, Hurd et al.[226] reported recently that after lasting self-administration of cocaine, less dopamine was released *in vivo* in the NA. The authors[226] attribute this surprising finding to depletion of dopamine, i.e., exhaustion of dopamine production in the A_{10} cells after repeated stimulation. Some other findings also indicate that an eventual depletion of dopamine stores in these cells must be calculated with. Imperato et al.[191] reported recently that daily cocaine administration results in gradual increase in the extracellular dopamine concentration in the NA, but only for 3 d. Thereafter, in spite of the continuing treatment the dopamine content drops below the level measured in untreated animals. Furthermore, withdrawal of both cocaine[193] and morphine[194] results in a profound decrease in the dopamine release in the ventral striatal/NA area. Obviously, decreased release of dopamine after repeated treatments might only enhance or at least explain the upregulation of the postsynaptic dopamine receptors, i.e., the enhanced sensitivity to exogenous dopamine receptor stimulants.

B. The Role of Excitatory Amino Acids

Some scattered data implicate the involvement of excitatory amino acids, more exactly the glutamate (*N*-methyl-DL-aspartate: NMDA) receptors in the process of sensitization. The above mentioned finding on inhibition of sensitization to metamphetamine by lesioning the fimbria fornix[205] points to the role of glutamate-releasing hippocampo-NA efferents in the process. In one laboratory, MK-801, a noncompetitive antagonist of NMDA receptors, abolished the sensitization to amphetamine and cocaine-induced stereotypy and convulsions[207] and the amphetamine-elicited sensitization to NMDA convulsions.[208] Simultaneously, Trujillo and Akil[227] reported that MK-801 treatment can abolish the tolerance development to the analgesic action of morphine, and it also attenuates the physical dependence induced by morphine. For the time being it is difficult to assess the significance of these findings. Anyhow, these data seem to open new vistas in this field.

C. The Role of Neurotensin

As discussed in Section II, neurotensin also activates the A_{10} neurons. Its eventual participation of sensitization is not proven (yet). Nevertheless, it is worth mentioning here that metamphetamine releases neurotensin in conjunction with dopamine liberation.[228] Furthermore, MK-801 blocks the neurotensin release in the NA induced by metamphetamine.[229]

IX. SUMMARY: SENSITIZATION AND SECONDARY LONG-TERM GENETIC ALTERATIONS: A SIMPLIFIED MODEL OF DRUG ADDICTION

Based on the data reviewed in this chapter and in Chapter 9 dealing with the alterations in enkephalin dynorphin synthesis during drug addiction, a simplified working hypothesis arises.

The gist of this proposal is outlined in the following two summaries and in Figure 1.

TABLE 5
The Biochemical Basis of Sensitization

Species examined	Experimental paradigm	Observed effect	Ref.
Rat	Effect of pimozide on sensitization to intra-VTA[a] applied morphine induced locomotor hyperactivity	Inhibition	209
Rat, mouse	Effect of haloperidol on sensitization to hyperlocomotion induced by cocaine, amphetamine, metamphetamine	Inhibition	206,207
Rat	Effect of i.p.-applied SCH-23390 on sensitization to		209
	Amphetamine (s.c.)	Inhibition	
	Intra-VTA morphine	No effect	
Rat	Effect of i.p.-applied sulpiride on sensitization induced by		209
	Systemic amphetamine	No effect	
	Intra-VTA morphine	No effect	
Rat	Modulation of sensitization to metamphetamine-induced locomotor hyperactivity by		211
	Sulpiride	Inhibition	
	YM-09151-2	Inhibition	
Rat	Alteration in the steady-state turnover of dopamine in the limbic system upon repeated amphetamine treatment	No change	205,212
Rat	Effect of repeated cocaine treatments on dopamine metabolism in the		
	NA[b]	Progressive enhancement	189 163
	VTA	Deceleration	188,201
Rat	Effect of repeated treatments with morphine, DALA,[c] DAGO[d] on dopamine metabolism in the		
	NA	Acceleration	156,199
	VTA	Deceleration	200
Rat tissue slice from VTA	Effect of repeated cocaine or morphine treatments on dopamine release induced by K^+ depolarization	Attenuation	201
Rat	Effect of repeated cocaine treatments	Attenuation of the inhibitory effect	216
	On the inhibitory effect of A_{10} cell firing seen with i.v. apomorphine or local phoretic cocaine		
	On the inhibition by phoretically applied dopamine of NA unites	Augmented inhibitory effect	

TABLE 5 (continued)
The Biochemical Basis of Sensitization

Species examined	Experimental paradigm	Observed effect	Ref.
	Effect of repeated metamphetamine treatments on the binding of spiroperidol to NA and frontal cortex	Increase in B_{max}	222
	Effect of repeated cocaine treatments on the dopamine$_2$ receptors in the NA	Increase in B_{max}	203,204
	Effect of lasting cocaine treatment on dopamine release from the NA	Initial increase followed by decrease	161,191
	Effect of morphine withdrawal on dopamine release from the NA	Decrease	191
	Effect of cocaine withdrawal on dopamine release from the NA	Decrease	193
	Effect of MK-801 on sensitization to amphetamine- and cocaine-induced motor excitation	Inhibition	207,208
	Effect of MK-801 on tolerance to morphine-induced analgesia and physical dependence	Inhibition	227

[a] VTA: Ventral tegmental area.
[b] NA: Nucleus accumbens.
[c] DALA: (D-Ala^2,Met^5)-enkephalinamide.
[d] DAGO: (D-Ala^2,NMePhe^4,Gly-ol^5)-enkephalin.

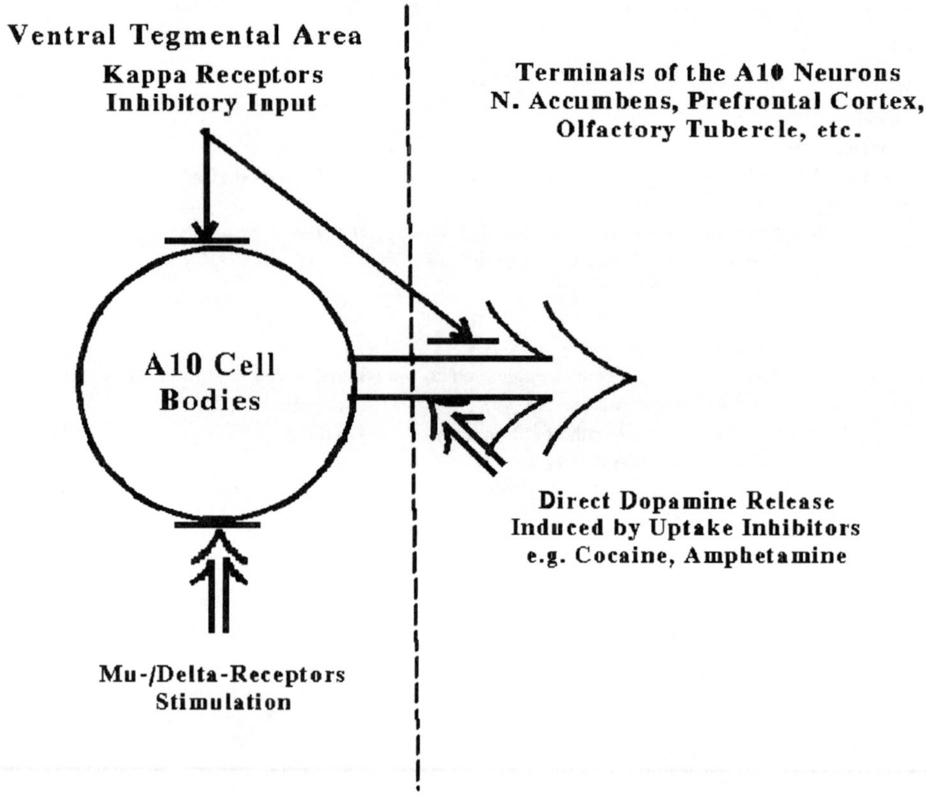

FIGURE 1. Facilitation by μ-opioids or psychostimulants and inhibition by κ-opioids of the A_{10} dopamine neurons.

COMMONALITIES IN SENSITIZATION BY μ-δ-OPIOIDS AND PSYCHOSTIMULANTS OF A_{10} NEURONS (THE MESOLIMBIC DOPAMINE SYSTEM)

(Based mostly on data reviewed in Chapters 4 and 7)

- Acute treatment with μ-/δ-opiates or psychostimulants, i.e., euphoria-inducing drugs, elicits locomotor stimulation and dopamine release in the mesolimbic dopamine system (n. accumbens, olfactory tubercle, prefrontal cortex, etc.).
- Repeated treatments with abused drugs induce enhanced locomotor stimulation and more intense release of dopamine.
- Eventually, the dopamine stores in the A_{10} neurons might be depleted.
- This process of sensitization is accompanied with subsensitivity (downregulation) of presynaptic and/or supersensitivity of the postsynaptic dopamine receptors.
- On the other hand, the dysphoric κ-agonists do not modify or rather decrease the mesolimbic dopamine release and do not induce sensitization.
- During withdrawal, dopamine release is decreased below normal from the mesolimbic dopamine terminals as seen with κ-agonists (see also Figure 1).

GENETIC ALTERATIONS IN THE TRANSCRIPTION OF ENDOGENOUS OPIOIDS: A SELF-PERPETUATING PROCESS?

(Based mostly on data reviewed in Chapter 9)

- In the nigrostriatal and, presumably, in the mesolimbic structures too, dopamine:
  tonically inhibits the expression of enkephalins;
  dopaminergic stimulants facilitate dynorphin synthesis.
- During substance abuse, dopamine release is enhanced (as discussed previously) and exogenous opioids exert a negative feedback effect on the synthesis of their endogenous counterparts. Therefore, during substance abuse both μ/δ-agonists and dopamine uptake blockers suppress the synthesis of enkephalins and facilitate the transcription of dynorphin.
- Due to decreased enkephalin/dynorphin ratio, the mood in the drug-free state might become more and more dysphoric, which can be temporarily overcome by further drug consumption.

Of course, in the summaries and in Figure 1 our present knowledge about drug abuse is grossly oversimplified. Their only merit, if any, is that they are consistent with the majority of data reviewed in Chapters 4, 7, and 9 and elsewhere.

REFERENCES

1. **Himmelsbach, C. K.,** Morphine with reference to physical dependence, *Fed. Proc. Fed. Am. Soc. Exp. Biol.*, 2, 201, 1943.
2. **Wikler, A.,** Dynamics of drug dependence: implications of a conditioning theory for research and treatment, in *Opiate Addiction: Origins and Treatment*, Fisher, S. and Friedman, A. M., Eds., Winston, V.H., Washington, D.C., 1973, 7.
3. **Hunt, G. H. and Oderoff, M. E.,** Follow-up study of narcotic drug addicts after hospitalization, *Public Health Rep.*, 77, 41, 1962.
4. **Canada, A. T.,** Methadone in a thirty-day detoxification program for narcotic addicts: a critical review, *Int. J. Addict.*, 7, 613, 1972.
5. **Wilson, B. K., Elms, R. R., and Thompson, C. P.,** Low-dosage use of methadone in expanded detoxification: an experimental comparison, *Arch. Gen. Psychiatr.*, 31, 233, 1974.
6. **Shiffman, A. S.,** The tobacco withdrawal syndrome, in *Cigarette Smoking as a Dependence Process*, Krasnegor, A. E., Ed., National Institute on Drug Abuse, Washington, D.C., 1979, 158.
7. **Jones, R. T.,** The pharmacology of cocaine, in *Cocaine: Pharmacology, Effects, and Treatment of Abuse*, Grabowski, J., Ed., National Institute on Drug Abuse, Washington, D.C., 1984, 34.
8. **Wise, R. A. and Bozarth, M. A.,** A psychomotor theory of addiction, *Psychol. Rev.*, 94, 469, 1987.
9. **Jaffe, J. H.,** Drug addiction and drug abuse, in *The Pharmacological Basis of Therapeutics*, Goodman Gilman, A., Rall, T. W., Nies, A. S., and Taylor, P., Eds., Pergamon Press, New York, 1990, 522.
10. **Werner, T., Smith, S. G., and Davis, W. M.,** An analysis of reinforcement during intragastric self-administration of morphine and methadone in rat, *Commun. Psychopharmacol.*, 1, 19, 1977.
11. **Pickens, R. and Thompson, R.,** Cocaine-reinforced behavior in rats: effects of reinforcement magnitude and fixed ratio size, *J. Pharmacol. Exp. Ther.*, 161, 122, 1968.
12. **Woods, J. H. and Schuster, C. R.,** Reinforcement properties of morphine, cocaine, and SPA as function of unit dose, *Int. J. Addict.*, 3, 231, 1968.
13. **Katz, J. L.,** Drugs as reinforcers: pharmacological and behavioral factors, in *The Neuropharmacological Basis of Reward*, Liebman, J. M. and Cooper, S. J., Eds., Oxford Unversity Press, Oxford, 1989, 164.
14. **Bozarth, M. A. and Wise, R. A.,** Anatomically distinct opiate receptor fields mediate reward and physical dependence, *Science*, 224, 516, 1984.

15. **Broekkamp, C. L. E., Van den Boggard, J. H., Heijnen, H. J., Rops, R. H., Cools, A. R., and Van Rossum, J. M.,** Separation of inhibiting and stimulating effects of morphine on self-stimulation behavior by intracerebral microinjections, *Eur. J. Pharmacol.*, 36, 443, 1976.
16. **Wikler, A. and Pescor, F. T.**, Classical conditioning of morphine abstinence phenomenon, reinforcement of opioid drinking behavior and "relapse" in morphine addicted rats, *Psychopharmacology*, 10, 255, 1967.
17. **Wikler, A., Pescor, F. T., Miller, D., and Norrel, H.**, Secondary reinforcement in morphine dependence, *Psychopharmacology*, 20, 103, 1971.
18. **Wikler, A.,** Conditioning factors in opiate addiction and relapse, in *Narcotics*, Wilner, D. I. and Kassebaum, D. D., Eds., McGraw-Hill, New York, 1965, 85.
19. **Scuster, C. R. and Woods, J. H.,** The conditioned reinforcing effects of stimuli associated with morphine reinforcement, *Int. J. Addict.*, 3, 223, 1968.
20. **Davis, W. M. and Smith, S. G.,** Naloxone use to eliminate opiate seeking behavior. Need for extinction of conditioned reinforcement, *Biol. Psychiatry*, 9, 181, 1974.
21. **Haertzen, C. A. and Hooks, N. T.,** Changes in personality and subjective experience associated with the chronic administration and withdrawal of opiates, *J. Nerv. Ment. Dis.*, 140, 606, 1969.
22. **McAuliffe, W. E. and Gordon, R. A.,** A test of Lindesmith's theory of addiction: the frequency of euphoria among long-term addicts, *Am. J. Sociol.*, 79, 795, 1974.
23. **Lett, B. T.,** Repeated exposures intensify rather than diminish the rewarding effects of amphetamine, morphine, and cocaine, *Psychopharmacology*, 98, 357, 1989.
24. **Glickman, S. E. and Schiff, B. B.,** A biological theory of reinforcement, *Psychol. Rev.*, 74, 81, 1967.
25. **Fog, R.,** Behavioral effects of amphetamine and morphine and of a combination of the two drugs, *Psychopharmacology*, 16, 305, 1970.
26. **Rethy, C. R., Smith, C. B., and Villareal, J. E.,** Effect of narcotic analgesics upon the locomotor activity and brain catecholamine content of the mouse, *J. Pharmacol. Exp. Ther.*, 176, 472, 1971.
27. **Babbini, M. and Davis, N. M.,** Time dose relationships for locomotor activity effects of morphine after acute and repeated treatments, *Br. J. Pharmacol.*, 46, 213, 1972.
28. **Ayhan, I. H. and Randrup, A.,** Behavioral and pharmacological studies on morphine-induced excitation of rats. Possible relation to brain catecholamines, *Psychopharmacology*, 29, 317, 1973.
29. **Ahtee, L.,** Catalepsy and stereotypies in rats treated with methadone: relation to striatal dopamine, *Eur. J. Pharmacol.*, 27, 221, 1974.
30. **Oliviero, A. and Castellano, C.,** Genotype-dependent sensitivity and tolerance to morphine and heroin: dissociation between opiate-induced running and analgesia in the mouse, *Psychopharmacology*, 39, 13, 1974.
31. **Van Rossum, J. M., van der Schoot, J. B., and Hurkmans, J. A. T. M.,** Mechanism of action of cocaine and amphetamine in the brain, *Experientia*, 18, 229, 1962.

32. **Ross, S. B. and Rényi, A. L.**, Inhibition of uptake of tritiated catecholamines into by antidepressants and related agents, *Eur. J. Pharmacol.*, 2, 181, 1967.
33. **Snyder, S. H. and Coyle, J. T.**, Regional differences in [^{3}H]-dopamine uptake into rat brain homogenates, *J. Pharmacol. Exp. Ther.*, 165, 78, 1969.
34. **Von Voigtlander, P. F. and Moore, K. E.**, Turning behavior of mice with unilateral 6-hydroxydopamine lesions in the striatum: effects of apomorphine, l-DOPA, amantadine, amphetamine and other psychostimulants, *Neuropharmacology*, 12, 451, 1973.
35. **Creese, I. and Iversen, S. D.**, The pharmacological and anatomical substrates of the amphetamine response in the rat, *Brain Res.*, 83, 419, 1975.
36. **Reith, M. E. A., Sershen, H., and Lajtha, A.**, Saturable [^{3}H]-cocaine binding in central nervous system of mouse, *Life Sci.*, 27, 1055, 1980.
37. **Hoffman, B. B. and Lefkowitz, R. J.**, Catecholamines and sympathomimetic drugs, in *The Pharmacological Basis of Therapeutics*, Goodman Gilman, A., Rall, T. W., Nies, A. S., and Taylor, P., Eds., Pergamon Press, New York, 1990, 189.
38. **Galloway, M. P.**, Regulation of dopamine and serotonin synthesis by acute administration of cocaine, *Synapse*, 6, 63, 1990.
39. **Izenwasser, S., Werling, L. L., and Cox, B. M.**, Comparison of the effects of cocaine and other inhibitors of dopamine uptake in rat striatum, nucleus accumbens, olfactory tubercle, and medial prefrontal cortex, *Brain Res.*, 520, 303, 1990.
40. **Wood, P. L. and Rao, T. S.**, Morphine stimulation of mesolimbic and mesocortical but not nigrostriatal dopamine release in the rat as reflected by changes in 3-methoxytyramine levels, *Neuropharmacology,* 30, 399, 1991.
41. **Kennedy, L. T. and Hanbauer, I.**, Sodium-sensitive cocaine binding to rat striatal membrane: possible relationship to dopamine uptake site, *J. Neurochem.*, 41, 172, 1983.
42. **Sershen, H., Reith, M. E. A., Hashim, A., and Lajtha, A.**, Reduction of dopamine uptake and cocaine binding in mouse striatum by N-methyl-4-phenyl-1,2,3,6-tetrahydropyridine, *Eur. J. Pharmacol.*, 102, 175, 1984.
43. **Grigoriadis, D. E., Wilson, A. A., Lew, R., Sharkey, S. J., and Kuhar, M. J.**, Dopamine transport sites selectively labeled by a novel photoaffinity probe: ^{125}I-DEEP, *J. Neurosci.*, 9, 2664, 1989.
44. **Salle, F. R., Fogel, E. L., Schwartz, E., Choi, S. M., Curran, D. P., and Niznik, H. B.**, Photoaffinity labeling of the mammalian transporter, *FEBS Lett.*, 256, 219, 1989.
45. **Berger, B., Martenson, R., Laing, P., Thurcauf, A., De Costa, B., Rice, K. C., and Paul, S. M.**, Photoaffinity labeling of the dopamine uptake carrier protein using a novel high affinity azidoderivative of GBR-12935, *Soc. Neurosci. Abstr.*, 16, 13, 1990.
46. **Missale, C., Castelletti, L., Govoni, S., Spano, P. F., Trabucchi, M., and Hanbauer, I.**, Dopamine uptake is differentially regulated in rat striatum and nucleus accumbens, *J. Neurochem.*, 45, 51, 1985.
47. **Boja, J. W. and Kuhar, M. J.**, [^{3}H]Cocaine binding and inhibition of [^{3}H]dopamine uptake is similar in both the rat striatum and nucleus accumbens, *Eur. J. Pharmacol.*, 173, 215, 1989.

48. **Marshall, J. F., O'Dell, S. J., Navarette, R., and Rosenstein, A. J.,** Dopamine high affinity transport site topography in rat brain: major differences between dorsal and ventral striatum, *Neuroscience*, 37, 11, 1990.
49. **Doucet, L., Descarries, L., and Garcia, S.,** Quantification of the dopamine innervation in adult rat neostriatum, *Neuroscience*, 9, 427, 1986.
50. **Stamford, J. A., Kruk, Z. L., Palij, P., and Millar, J.,** Diffusion and uptake of dopamine in rat caudate and nucleus accumbens compared using fast cyclic voltametry, *Brain Res.*, 448, 381, 1988.
51. **Lew, R., Vaughan, R., Simantov, R., Wilson, A., and Kuhar, M. J.,** Dopamine transporters in the nucleus accumbens and the striatum have different apparent molecular weights, *Synapse*, 8, 152, 1991.
52. **Yokel, R. A. and Pickens, R.,** Self-administration of optical isomers of amphetamine and methylamphetamine by rats, *J. Pharmacol. Exp. Ther.*, 187, 27, 1973.
53. **Ferris, R. M., Tang, F. L. M., and Maxwell, R. A.,** A comparison of capacities of isomers of amphetamine, deoxypipradol and methylphenidate to inhibit the uptake of tritiated catecholamines into rat cerebral cortex slices, synaptosomal preparations of rat cerebral cortex, hypothalamus and striatum and into adrenergic nerves of rabbit aorta, *J. Pharmacol. Exp. Ther.*, 181, 407, 1972.
54. **Ritz, M. C., Lamb, R. J., Goldberg, S. R., and Kuhar, M. J.,**Cocaine receptors on dopamine transporters are related to self-administration of cocaine, *Science*, 237, 1219, 1987.
55. **Cline, E. J., Scheffel, U., Boja, J. W., Carroll, F. I., Katz, J. L., and Kuhar, M. J.,** Behavioral effects of novel cocaine analogs: a comparison with *in vivo* receptor binding potency, *J. Pharmacol. Exp. Ther.*, 260, 1174, 1992.
56. **Nishino, H., Ono, T., Muramoto, K., Fukuda, M., and Sasaki, K.,** Neuronal activity in the ventral tegmental area (VTA) during motivated bar-press feeding in the monkey, *Brain Res.*, 413, 302, 1987.
57. **Randrup, A. and Munkvad, I.,** Role of catecholamines in the amphetamine response, *Nature*, 211, 540, 1966.
58. **Randrup, A. and Scheel-Krüger, J.,** Diethyl-dithiocarbamate and amphetamine stereotype behavior, *J. Pharm. Pharmacol.*, 18, 752, 1966.
59. **Weissman, A., Koe, B. K., and Tenen, S.,** Antiamphetamine effects following inhibition of tyrosine hydroxylase, *J. Pharmacol. Exp. Ther.*, 151, 339, 1966.
60. **Stolk, J. M. and Rech, R. H.,** Antagonism of d-amphetamine by α-methyl-p-tyrosine. Behavioral evidence for participation of catecholamine stores and synthesis in the amphetamine stimulant response, *Neuropharmacology*, 9, 249, 1970.
61. **Post, R. M. and Rose, H.,** Increasing effects of repetitive cocaine administration in the rat, *Nature*, 260, 731, 1976.
62. **Bhattacharyya, A. K. and Pradhan, S. N.,** Interaction between motor activity and stereotypy in cocaine treated rats, *Psychopharmacology*, 63, 311, 1979.
63. **Amalric, M. and Koob, G. F.,** Low doses of methylnaloxonium in the nucleus accumbens antagonize hyperactivity induced by heroin in the rat, *Pharmacol. Biochem. Behav.*, 23, 411, 1985.

64. **Székely, J. I.,** The most characteristic in vivo effects of opiates; changes in the spontaneous motor activity, in *Opioid Peptides,* Vol. I, Research Methods, Székely, J. I. and Rónai, A. Z., Eds., CRC Press, Boca Raton, FL, 1982, 46.
65. **Kelly, P. H. and Iversen, S. D.,** Selective-6-hydroxydopamine destruction of mesolimbic dopamine neurons: abolition of psychostimulant-induced locomotor activity in rats, *Eur. J. Pharmacol.*, 40, 45, 1976.
66. **Vaccarino, F. J., Amalric, M., Swerdlow, N. R., and Koob, G. F.,** Blockade of amphetamine but not opiate-induced locomotion following antagonism of dopamine function in the rat, *Pharmacol. Biochem. Behav.*, 24, 61, 1986.
67. **Pijnenburg, A. J. J., Honig, W. M. M., Van Rossum, J. M.,** Inhibition of d-amphetamine-induced locomotor activity by injection of haloperidol into the nucleus accumbens of the rat, *Psychopharmacology*, 41, 87, 1975.
68. **Kelly, P. H., Seviour, P. W., and Iversen, S. D.,** Amphetamine and apomorphine responses in the rat following 6-OHDA lesions of the nucleus accumbens, *Brain Res.*, 94, 507, 1975.
69. **Joyce, E. M. and Iversen, S. D.,** The effect of morphine applied locally to the mesencephalic dopamine cell bodies on spontaneous motor activity in the rat, *Neurosci. Lett.*, 14, 207, 1979.
70. **Broekkamp, C. L. E., Phillips, A. G., and Cools, A. R.,** Stimulant effect of enkephalin microinjection into the dopaminergic A10 area, *Nature*, 278, 560, 1979.
71. **Kelley, A. E., Stinus, L., and Iversen, S. D.,** Interactions between D-Ala^2-Met-enkephalin, A10 dopaminergic neurones and spontaneous behavior in the rat, *Behav. Brain Res.*, 1, 3, 1980.
72. **Joyce, E. M., Koob, G. F., Strecker, R., Iversen, S. D., and Bloom, F. B.,** The behavioral effects of enkephalin analogues injected into the ventral tegmental area and globus pallidus, *Brain Res.*, 221, 359, 1981.
73. **Stinus, L., Koob, F. G., Ling, N., Bloom, F. E., and Le Moal, M.,** Locomotor activation induced by infusion of endorphins into the ventral tegmental area: evidence for opiate-dopamine interactions, *Proc. Natl. Acad. Sci. U.S.A.*, 77, 2323, 1977.
74. **Latimer, L. G., Duffy, P. and Kalivas, P. W.,** Mu opioid receptor involvement in enkephalin activation of dopamine neurons in the ventral tegmental area, *J. Pharmacol. Exp. Ther.*, 241, 328, 1987.
75. **Holmes, L. J., Bozarth, M. A., and Wise, R. A.,** Circling from intracranial morphine applied to the ventral tegmental area in rats, *Brain Res. Bull.*, 11, 295, 1983.
76. **Jenck, F., Bozarth, M., and Wise, R. A.,** Contraversive circling induced by ventral tegmental microinjections of moderate doses of morphine and [D-Pen^2,D-Pen^5]enkephalin, *Brain Res.*, 450, 382, 1988.
77. **Holmes, L. J. and Wise, R. A.,** Contralateral circling induced by tegmental morphine: anatomical localization, pharmacological specificity, and phenomenology, *Brain Res.*, 326, 19, 1985.
78. **Delfs, J. M., Schreiber, L., and Kelley, A. E.,** Microinjection of cocaine into the nucleus accumbens elicits locomotor activation in the rat, *J. Neurosci.*, 10, 303, 1990.

79. **Clarke, P. B. S., Jakubovic, A., and Fibiger, H. C.,** Anatomical analysis of involvement of mesolimbocortical dopamine in the locomotor stimulant actions of d-amphetamine and apomorphine, *Psychopharmacology*, 96, 511, 1988.
80. **Swerdlow, N. R., Amalric, M., and Koob, G. F.,** Nucleus accumbens opiate-dopamine interactions and locomotor activation in the rat: evidence for a presynaptic locus, *Pharmacol. Biochem. Behav.*, 26, 765, 1987.
81. **Pert, A. and Sivit, C.,** Neuroanatomical locus for morphine and enkephalin-induced hyperactivity, *Nature*, 265, 645, 1977.
82. **Kalivas, P. W., Widerlöv, E., Stanley, D., Breese, G., and Prange, A. J., Jr.,** Enkephalin action on the mesolimbic system: a dopamine-dependent and a dopamine-independent increase in locomotor activity, *J. Pharmacol. Exp. Ther.*, 227, 229, 1983.
83. **Haveman, U. and Kuschinsky, K.,** Locomotor activity of rats after injection of various opioids into the nucleus accumbens and the septum mediale, *Naunyn-Schmiedeberg's Arch. Pharmacol.*, 331, 175, 1985.
84. **Atweh, S. F. and Kuhar, M. J.,** Autoradiographic localization of opiate receptors in rat brain. III. The telencephalon, *Brain Res.*, 134, 393, 1977.
85. **Pollard, H., Llorens, C., Bonnet, J. J., Costentin, J., and Schwartz, J. C.,** Opiate receptors on mesolimbic dopamine neurons, *Neurosci. Lett.*, 7, 295, 1977.
86. **Pijnenburg, A. J. J., Honig, W. M. M., and Van Rossum, J. M.,** Effects of antagonists upon locomotor stimulation induced by injection of dopamine and noradrenaline into the nucleus accumbens of nialamide pretreated rats, *Psychopharmacology*, 41, 175, 1975.
87. **Costall, B., Fortune, D. H., and Naylor, R. J.,** The induction of catalepsy and hyperactivity by morphine administered directly into the nucleus accumbens of rats, *Eur. J. Pharmacol.*, 49, 49, 1978.
88. **Polgár, K., Máté, I., Till, M., and Székely, J. I.,** Modulation by enkephalin analogues and neuroleptics of apomorphine-induced stereotypy and turning behavior in rats, *Neuropharmacology*, 26, 1309, 1987.
89. **Stinus, L., Herman, J. P., and Le Moal, M.,** GABAergic mechanisms within the ventral tegmental area: involvement of dopaminergic (A 10) and non-dopaminergic neurones, *Psychopharmacology*, 77, 186, 1982.
90. **Stinus, L., Kelley, A. E., and Iversen, S. D.,** Increased spontaneous activity following substance P infusion into A 10 dopaminergic area, *Nature*, 276, 616, 1978.
91. **Mogenson, G. J., Wu, M., and Manchanda, S. K.,** Locomotor activity initiated by microinfusions of picrotoxin into the ventral tegmental area, *Brain Res.*, 161, 311, 1979.
92. **Kalivas, P. W., Duffy, P., and Eberhardt, H.,** Modulation of A10 dopamine neurons by γ-aminobutyric acid agonists, *J. Pharmacol. Exp. Ther.*, 253, 858, 1990.
93. **Shi, W.-X. and Bunney, B. S.,** Neurotensin modulates autoreceptor mediated dopamine effects on midbrain dopamine cell activity, *Brain Res.*, 543, 315, 1991.
94. **Kalivas, P. W., Nemeroff, C. B., and Prange, A. J., Jr.,** Increase in spontaneous motor activity following infusion of neurotensin into the ventral tegmental area, *Brain Res.*, 229, 525, 1981.

95. **Cador, M., Kelley, A. E., Le Moal, M., and Stinus, L.,** Behavioral analysis of the effect of neurotensin injected into the ventral mesencephalon on investigatory and spontaneous motor behavior in the rat, *Psychopharmacology*, 85, 187, 1985.
96. **Quirion, R. P., Gaudreau, P., St-Pierre, S., Rioux, F., and Pert, C. B.,** Autoradiographic distribution of (^{3}H) neurotensin receptors in rat brain: visualization by tritium sensitive film, *Peptides*, 3, 757, 1982.
97. **Dilts, R. P. and Kalivas, P. W.,** Autoradiographic localization of μ-opioid and neurotensin receptors within the mesolimbic dopamine system, *Brain Res.*, 488, 311, 1989.
98. **Glimcher, P. W., Margolin, D., Giovino, A. A., and Hoebel, B. G.,** Neurotensin: a new "reward peptide", *Brain Res.*, 291, 119, 1984.
99. **Blaha, C. D., Coury, A., Fibiger, H. C., and Phillips, A. G.,** Effects of neurotensin on dopamine release and metabolism in the rat striatum and nucleus accumbens: cross-validation using *in vivo* voltametry and microdialysis, *Neuroscience*, 34, 699, 1990.
100. **Nemeroff, C. B.,** Neurotensin: per chance an endogenous neuroleptic?, *Biol. Psychiatry*, 15, 283, 1980.
101. **Ervin, G. N., Birkemo, L. S., Nemeroff, C. B., and Prange, A. J., Jr.,** Neurotensin blocks amphetamine-induced behaviors, *Nature*, 291, 73, 1981.
102. **Hasenöhrl, R. U., Gerhardt, P., and Huston, J. P.,** Positively reinforcing effects of the neurotachykinin substance P in the basal forebrain: mediation by its C-terminal sequence, *Exp. Neurol.*, 115, 282, 1982.
103. **Huston, J. P. and Oitzl, M. S.,** The relationship between reinforcement and memory: parallels in the rewarding and mnemonic effects of neuropeptide substance P., *Neurosci. Biobehav. Rev.*, 13, 171, 1989.
104. **Cador, M., Rivet, J.-M., Kelley, A. E., Le Moal, M., and Stinus, L.,** Substance P, neurotensin and enkephalin injections into the ventral tegmental area: comparative study on dopamine turnover in several forebrain structures, *Brain Res.*, 486, 357, 1989.
105. **Deutsch, A. Y., Maggio, J. E., Bannon, M. J., Kalivas, P. W., Tam, S. E., Goldstein, M., and Roth, R. H.,** Substance K and substance P differentially modulate mesolimbic and mesocortical systems, *Peptides*, 6, 113, 1985.
106. **Boix, F., Mattioli, M., Adams, F., Huston, J. P., and Schwarting, R. K. W.,** Effects of substance P on extracellular dopamine in neostriatum and nucleus accumbens, *Eur. J. Pharmacol.*, 216, 103, 1992.
107. **Pulvirenti, L., Swerdlow, N. R., and Koob, G. F.,** Microinjection of a glutamate antagonist into the nucleus accumbens reduces psychostimulant locomotion in rats, *Neurosci. Lett.*, 103, 213, 1989.
108. **Ostrowski, N. L., Hatfield, C. B., and Caggiula, A. R.,** The effect of low doses of morphine on the activity of dopamine-containing cells and on behavior, *Life Sci.*, 31, 2347, 1982.
109. **Gysling, K. and Wang, R. Y.,** Morphine induced activation of A10 dopamine neurons in the rat, *Brain Res.*, 277, 119, 1983.
110. **Matthews, R. T. and German, D. C.,** Electrophysiological evidence for excitation of rat ventral tegmental area dopamine neurons by morphine, *Neuroscience*, 11, 617, 1984.

111. **Trulson, M. E. and Araseth, K.,** Morphine increases the midbrain dopamine neurons in vitro, *Eur. J. Pharmacol.*, 114, 105, 1985.
112. **Lacey, M. G., Mercuri, N. B., and North, R. A.,** Two cell types in rat substantia nigra zona compacta distinguished by membrane properties and the actions of dopamine and opioids, *J. Neurosci.*, 9, 1233, 1989.
113. **Olpe, H.-R., Koella, W. P., Wolf, P., and Haas, H. L.,** The action of Baclofen on neurons of substantia nigra and of the ventral tegmental area, *Brain Res.*, 134, 577, 1977.
114. **Wolf, P., Olpe, H.-R., Avrith, D., and Haas, H. L.,** GABAergic inhibition of neurons in the ventral tegmental area, *Experientia*, 34, 73, 1978.
115. **German, D. C., Dalsass, M., and Kiser, R. S.,** Electrophysiological examination of the ventral tegmental (A10) area in the rat, *Brain Res.*, 181, 191, 1980.
116. **Yim, C. Y. and Mogenson, G. J.,** Effect of picrotoxin and nipecotic acid on inhibitory response of dopaminergic neurons in the ventral tegmental area to stimulation of the nucleus accumbens, *Brain Res.*, 199, 466, 1980.
117. **Nauta, W. J. H., Smith, G. P., Faull, R. L. M., and Domesick, V. B.,** Efferent connections and nigral afferents of nucleus accumbens septi in the rat, *Neuroscience*, 3, 385, 1978.
118. **Phillipson, O. T.,** Afferent projections to the ventral tegmental area of Tsai and interfascicular nucleus: a horseradish peroxidase study in the rat, *J. Comp. Neurol.*, 19, 117, 1979.
119. **Hakan, R. L. and Henriksen, S. J.,** Opiate influences on nucleus accumbens neuronal electrophysiology: dopamine and non-dopamine mechanisms, *J. Neurosci.*, 9, 3538, 1989.
120. **Einhorn, L. C., Johansen, P. A., and White, F. J.,** Electrophysiological effects of cocaine in the mesoaccumbens dopamine system: studies in the ventral tegmental area, *J. Neurosci.*, 8, 100, 1988.
121. **Brodie, M. S. and Dunwiddie, T. V.,** Cocaine effects in the ventral tegmental area: evidence for an indirect dopaminergic mechanism of action, *Naunyn-Schmiedeberg's Arch. Pharmacol.*, 342, 660, 1990.
122. **Clouet, D. H. and Ratner, M.,** Catecholamine biosynthesis in the brains of rats treated with morphine, *Science*, 168, 584, 1970.
123. **Freye, E. and Kuschinsky, K.,** Effects of fentanyl and droperidol on the dopamine metabolism of the rat striatum, *Pharmacology*, 14, 1, 1976.
124. **Moleman, P., Van Valkenburg, C. F. M., and v.d. Krogt, J. A.,** Effects of morphine on dopamine metabolism in rat striatum and limbic structures in relation to the activity of dopaminergic neurones, *Naunyn-Schmiedeberg's Arch. Pharmacol.*, 327, 208, 1984.
125. **Wood, P. L.,** Opioid regulation of CNS dopaminergic pathways: a review of methodology, receptor types, regional variations and species differences, *Peptides*, 4, 595, 1983.
126. **Van Loon, G. R. and Kim, C.,** β-Endorphin-induced increase in striatal dopamine turnover, *Life Sci.*, 23, 961, 1978.
127. **Deyo, S. N., Swift, R. M., and Miller, R. J.,** Morphine and endorphins modulate the dopamine turnover in rat median eminence, *Proc. Natl. Acad. Sci. U.S.A.*, 76, 3006, 1976.

128. **Gudelsky, G. A. and Porter, J. C.,** Morphine- and opioid peptide-induced inhibition of the release of dopamine from the tubero-infundibular neurons, *Life Sci.*, 25, 1697, 1979.
129. **Alper, R. H., Demarest, K. T., and Moore, K. E.,** Morphine differentially alters synthesis and turnover of dopamine in central neuronal system, *J. Neural. Trans.*, 48, 157, 1980.
130. **Lookingland, K. J. and Moore, K. E.,** Differential effects of morphine on the rates of dopamine turnover in the neural and intermediate lobes of the rat pituitary glands, *Life Sci.*, 37, 1225, 1985.
131. **Westerink, B. H. C.,** Effect of centrally acting drugs on regional dopamine metabolism, *Adv. Biochem. Psychopharmacol.*,19, 255, 1978.
132. **Algeri, S., Consolazione, A., and Plaznik, A.,** Increased metabolism of dopamine and serotonin induced in forebrain areas by etorphine microinjection in periaqueductal gray, *Eur. J. Pharmacol.*, 68, 383, 1980.
133. **Kim, H. S., Iyengar, S., and Wood, P. L.,** Opiate actions on mesocortical dopamine metabolism in the rat, *Life Sci.*, 39, 2033, 1986.
134. **Wood, P. L., Stotland, M., Richard, J. W., and Rackham, A. C.,** Actions of mu, kappa, sigma, delta and agonist/antagonists opiates on striatal dopaminergic function, *J. Pharmacol. Exp. Ther.*, 215, 697, 1980.
135. **Yonehara, N. and Clouet, D. H.,** Effect of delta and mu opioid peptides on the turnover and release of dopamine in rat striatum, *J. Pharmacol. Exp. Ther.*, 231, 38, 1984.
136. **Biggio, G., Casa, M., Corda, M., Di Bello, C., and Gessa, G. L.,** Stimulation of dopamine synthesis in caudate nucleus by intrastriatal enkephalins and antagonism by naloxone, *Science*, 200, 552, 1978.
137. **Chesselet, M. F., Chéramy, A., Reisine, T. D., and Glowinski, J.,** Morphine and delta opiate locally stimulate in vivo dopamine release in cat caudate nuclei, *Nature*, 291, 320, 1981.
138. **Zhang, A.-Z., Yang, H.-Y. T., and Costa, E.,** Nociception, enkephalin content and dipeptidyl carboxypeptidase activity in brain of mice treated with exopeptidase inhibitors, *Neuropharmacology*, 21, 625, 1982.
139. **Giorgi, O., Pibiri, M. G., Ongini, E., Trampus, M., and Biggio, G.,** The neutral endopeptidase-24.11 (enkephalinase inhibitor), SCH 32615, increases dopamine metabolism in the nucleus accumbens of the rat, *Eur. J. Pharmacol.*, 196, 137, 1991.
140. **Manzanares, J., Lookingland, K. J., and Moore, K. E.,** Kappa opioid receptor mediated regulation of dopaminergic neurons in the rat brain, *J. Pharmacol. Exp. Ther.*, 256, 500, 1991.
141. **Iwatsubo, K. and Clouet, D. H.,** Effects of morphine and haloperidol on the electrical activity of rat nigrostriatal neurons, *J. Pharmacol. Exp. Ther.*, 202, 429, 1977.
142. **Loh, H. H., Brase, D. A., Sampath-Khanna, S., Mar, J. B., and Way, W. L.,** β-Endorphin *in vitro* inhibition of striatal dopamine release, *Nature*, 264, 567, 1976
143. **Arbilla, S. and Langer, S. Z.,** Morphine and β-endorphin inhibit release of noradrenaline from cerebral cortex but not of dopamine from rats striatum, *Nature*, 271, 559, 1978.

144. **Lubetzki, C., Chesselet, M. F., and Glowinski, J.,** Modulation of dopamine release in rat striatal slices by delta opiate agonists, *J. Pharmacol. Exp. Ther.*, 222, 435, 1982.
145. **Petit, F., Hamon, M., Fournie-Zaluski, M.-C., Roques, B. P., and Glowinski, J.,** Further evidence for a role of δ-opiate receptors in the presynaptic regulation of newly synthesized dopamine release, *Eur. J. Pharmacol.*, 126, 1, 1986.
146. **Mulder, A. H., Wardeh, G., Hogenboom, F., and Frankhuyzen, A. L.,** κ- and δ-opioid receptor agonists differentially inhibit striatal dopamine and acetylcholine release, *Nature*, 308, 278, 1984.
147. **Werling, L. L., Frattali, A., Portoghese, P. S., Takemori, A. E., and Cox, B. M.,** Kappa receptor regulation of dopamine release from striatum and cortex of rats and guinea pigs, *J. Pharmacol. Exp. Ther.*, 246, 282, 1988.
148. **Heijna, M. H., Padt, M., Hogenboom, F., Portoghese, P. S., Mulder, A. H., and Scoeffelmeer, A. N. M.,** Opioid receptor mediated inhibition of dopamine and acetylcholine release from slices of rat nucleus accumbens, olfactory tubercle and frontal cortex, *Eur. J. Pharmacol.*, 18, 267, 1990.
149. **Di Chiara, G. and Imperato, A.,** Opposite effects of mu and kappa opiate agonists on dopamine release in the nucleus accumbens and in the dorsal caudate of freely moving rats, *J. Pharmacol. Exp. Ther.*, 244, 1067, 1988.
150. **Di Chiara, G. and Imperato, A.,** Drugs abused by humans preferentially increase synaptic dopamine concentrations in the mesolimbic system of freely moving rats, *Proc. Natl. Acad. Sci. U.S.A.*, 85, 5274, 1988.
151. **Wood, P. L. and Altar, A. C.,** Dopamine release in vivo from nigrostriatal, mesolimbic and mesocortical neurons: utility of 3-methoxytyramine measurements, *Pharmacol. Rev.*, 40, 163, 1989.
152. **Spanagel, R., Herz, A., and Shippenberg, T.,** The effects of opioid peptides on dopamine release in the nucleus accumbens: an in vivo microdialysis study, *J. Neurochem.*, 55, 1734, 1990.
153. **Lohse, M. and Wuttke, W.,** Release and synthesis rates of catecholamines in hypothalamic, limbic and midbrain structures following intraventricular injections of beta-endorphin in male rats, *Brain Res.*, 229, 389, 1981.
154. **Iyengar, K., Kim, H. S., Marien, M. R., McHugh, D., and Wood, P. L.,** Modulation of mesolimbic dopaminergic projections by beta-endorphin in the rat, *Neuropharmacology*, 28, 123, 1989.
155. **Spanagel, R., Herz, A., Bals-Kubik, R., and Shippenberg, T. S.,** β-Endorphin induced locomotor stimulation and reinforcement are associated with an increase in the dopamine release in the nucleus accumbens, *Psychopharmacology*, 104, 51, 1991.
156. **Kalivas, P. W. and Duffy, P.,** Effect of acute and daily neurotensin and enkephalin treatments on extracellular dopamine in the nucleus accumbens, *J. Neurosci.*, 10, 2940, 1990.
157. **Church, W. H., Justice, J. B., Jr., and Byrd, L. D.,** Extracellular dopamine in rat striatum following uptake inhibition by cocaine, nomifensine and benztropine, *Eur. J. Pharmacol.*, 139, 345, 1987.
158. **Hernandez, L. and Hoebel, B. G.,** Food reward and cocaine increase extracellular dopamine in the nucleus accumbens as measured by microdialysis, *Life Sci.*, 42, 1705, 1988.

159. **Bradberry, C. W. and Roth, R. H.,** Cocaine increases extracellular dopamine in rat nucleus accumbens and ventral tegmental area as shown by in vivo microdialysis, *Neurosci. Lett.*, 103, 97, 1989.
160. **Moghaddam, B. and Bunney, B. S.,** Differential effect of cocaine on extracellular dopamine levels in rat prefrontal cortex and nucleus accumbens: Comparison to amphetamine, *Synapse*, 4, 156, 1989.
161. **Hurd, Y. L., Weiss, F., Koob, G. F., Andén, N.-E., and Ungerstedt, U.,** Cocaine reinforcement and extracellular dopamine overflow in rat nucleus accumbens: an in vivo microdialysis study, *Brain Res.*, 498, 199, 1989.
162. **Galloway, M. P.,** Regulation of dopamine and serotonin synthesis by acute administration of cocaine, *Synapse*, 6, 63, 1990.
163. **Kalivas, P. W. and Duffy, P.,** Effect of acute and daily cocaine treatment on extracellular dopamine in the nucleus accumbens, *Synapse*, 5, 48, 1990.
164. **Hurd, Y. L. and Ungerstedt, U.,** Cocaine: an in vivo microdialysis evaluation of its acute action on dopamine transmission in rat striatum, *Synapse*, 3, 48, 1989.
165. **Pettit, H. O. and Justice, J. B., Jr.,** Dopamine in the nucleus accumbens during cocaine self-administration as studied by in vivo microdialysis, *Pharmacol. Biochem. Behav.*, 34, 899, 1989.
166. **Pettit, H. O. and Justice, J. B., Jr.,** Effect of dose on cocaine self-administration behavior and dopamine levels in the nucleus accumbens, *Brain Res.*, 539, 94, 1991.
167. **Pradhan, S., Roy, S. N., and Pradhan, S. N.,** Correlation of behavioral and neurochemical effects of acute administration of cocaine in rats, *Life Sci.*, 22, 1737, 1978.
168. **Phillips, A. G., Blaha, C. D., and Fibiger, H. C.,** Neurochemical correlates of brain stimulation measured by ex vivo and in vivo analyses, *Neurosci. Biobehav. Rev.*, 13, 99, 1989.
169. **Kaymakcalan, S. and Woods, L. A.,** Nalorphine-induced "abstinence-syndrome" in morphine tolerant albino rats, *J. Pharmacol. Exp. Ther.*, 117, 112, 1956.
170. **Eidelberg, E. and Schwartz, A. S.,** Possible mechanisms of action of morphine on brain, *Nature*, 225, 1152, 1970.
171. **Kumar, R., Mitchell, E., and Stolerman, I. P.,** Distributed patterns of behaviour in morphine tolerant and abstinent rats, *Br. J. Pharmacol.*, 42, 473, 1971.
172. **Vasko, M. R. and Domino, E. F.,** Tolerance development to the biphasic effect of morphine on locomotor activity and brain acetylcholine in the rat, *J. Pharmacol. Exp. Ther.*, 207, 848, 1978.
173. **Mucha, R. F., Volkovskis, C., and Kalant, H.,** Conditioned increases in locomotor activity produced by morphine as unconditioned stimulus, and the relation of conditioning to acute morphine effect and tolerance, *J. Comp. Physiol. Psychol.*, 95, 351, 1981.
174. **Hinson, R. E. and Siegel, S.,** Anticipatory hyperexcitability and tolerance to the narcotizing effect of morphine in the rat, *Behav. Neurosci.*, 97, 759, 1983.

175. **Vezina, P. and Stewart, J.,** Conditioning and place-specific sensitization of increase in activity induced by morphine in the VTA, *Pharmacol. Biochem. Behav.*, 20, 925, 1984.
176. **Kalivas, P. W., Taylor, S., and Miller, J. S.,** Sensitization to repeated enkephalin administration into the ventral tegmental area of the rat. I. Behavioral characterization, *J. Pharmacol. Exp. Ther.*, 235, 537, 1985.
177. **Downs, A. W. and Eddy, N. B.,** The effect of repeated doses of cocaine on the rat, *J. Pharmacol. Exp. Ther.*, 46, 199, 1932.
178. **Post, R. M. and Rose, H.,** Increasing effects of repetitive cocaine administration in the rat, *Nature*, 260, 731, 1976.
179. **Tilson, H. A. and Rech, R. H.,** Conditioned drug effects and absence of tolerance to d-amphetamine-induced motor activity, *Pharmacol. Biochem. Behav.*, 1, 149, 1973.
180. **Wallach, M. B. and Gerson, S.,** Sensitization to amphetamine, *Psychopharmacol. Bull.*, 7, 30, 1971.
181. **Segal, D. S. and Mandell, A. J.,** Long-term administration of amphetamine: progressive augmentation of motor activity and stereotypy, *Pharmacol. Biochem. Behav.*, 2, 249, 1974.
182. **Vezina, P., Giovino, A. A., Wise, R. A., and Stewart, J.,** Environment specific cross-sensitization between the locomotor activating effects of morphine and amphetamine, *Pharmacol. Biochem. Behav.*, 32, 581, 1989.
183. **Fujiwara, S., Kuribara, H., and Tadokoro, S.,** Effects of repeated administration of pentazocine on ambulatory activity in mice: comparison with the effects of amphetamine and metamphetamine, *Jpn. J. Pharmacol.*, 54, 61, 1990.
184. **Robinson, T. E. and Becker, J. B.,** Behavioral sensitization is accompanied by enhancement in amphetamine-stimulated dopamine release from striatal tissue, *in vitro*, *Eur. J. Pharmacol.*, 85, 253, 1982.
185. **Kolta, M. G., Shreve, P., De Souza, V., and Uretsky, N. J.,** Time course of the development of the enhanced behavioral and biochemical responses to amphetamine after pretreatment with amphetamine, *Neuropharmacology*, 24, 823, 1985.
186. **Castaneda, E., Becker, J., and Robinson, T. E.,** The long-term effect of repeated amphetamine treatment *in vivo* on amphetamine, KCl and electrical stimulation evoked striatal dopamine release *in vitro, Life Sci.*, 42, 2447, 1988.
187. **Kalivas, P. W., Duffy, P., DuMars, L. A., and Skinner, C.,** Behavioral and neurochemical effects of acute and daily cocaine administration in rats, *J. Pharmacol. Exp. Ther.*, 245, 485, 1988.
188. **Robinson, T. E., Jurson, P. A., Bennett, J. A., and Bentgen, K. M.,** Persistent sensitization of dopamine neurotransmission in ventral striatum (nucleus accumbens) produced by prior experience with (+)-amphetamine: microdialysis study in freely moving rats, *Brain Res.*, 462, 211, 1988.
189. **Kazahaya, Y., Akimoto, K., and Otsuki, S.,** Subchronic metamphetamine treatment enhances metamphetamine- or cocaine-induced dopamine efflux *in vivo*, *Biol. Psychiatry*, 25, 903, 1989.
190. **Brock, J. W., Ng, J. P., and Justice, J. B., Jr.,** Effect of cocaine on dopamine synthesis in the nucleus accumbens as determined by microdialysis perfusion with NSD-1015, *Neurosci. Lett.*, 117, 234, 1990.

191. **Imperato, A., Mele, A., Scrocco, M. G., and Puglisi-Allegra**, Chronic cocaine alters limbic extracellular dopamine. Neurochemical basis for addiction, *Eur. J. Pharmacol.*, 212, 299, 1992.
192. **Acquas, E., Carboni, E., and Di Chiara, G.,** Profound depression of mesolimbic dopamine release after morphine withdrawal in dependent rats, *Eur. J. Pharmacol.*, 193, 133, 1991.
193. **Parson, L. H., Smith, A. D., and Justice, J. B., Jr.,** Basal extracellular dopamine is decreased in the rat nucleus accumbens during abstinence from chronic cocaine, *Synapse*, 9, 60, 1991.
194. **Bunney, W. C., Massari, V. J., and Pert, A.,** Chronic morphine-induced hyperactivity in rats is altered by nucleus accumbens and ventral tegmental lesions, *Psychopharmacology*, 82, 318, 1984.
195. **Stinus, L., Winnock, M., and Kelley, A. E.,** Chronic neuroleptic treatment and mesolimbic dopamine denervation induce behavioral supersensitivity to opiates, *Neuropharmacology*, 85, 323, 1985.
196. **Schwartz, J. M., Ksir, C., Koob, G. F., and Bloom, F. E.,** Changes in locomotor response to beta-endorphin microinfusion during and after opiate abstinence syndrome — a proposal for a model of the onset of mania, *Psychiatr. Res.*, 7, 153, 1982.
197. **Stewart, J. and Vezina, P.,** Environment specific enhancement of the hyperactivity induced by systemic or intra-VTA morphine injections in rats preexposed to amphetamine, *Psychobiology*, 15, 144, 1987.
198. **DuMars, L. A., Rodger, L. D., and Kalivas, P. W.,** Behavioral cross-sensitization between cocaine and enkephalin in the A10 dopamine region, *Behav. Brain Res.*, 27, 87, 1988.
199. **Kalivas, P. W. and Duffy, P.,** Sensitization to repeated morphine injection in the rat: possible involvement of A10 dopamine neurons, *J. Pharmacol. Exp. Ther.*, 241, 204, 1987.
200. **Kalivas, P. W.,** Sensitization to repeated enkephalin administration into the ventral tegmental area. II. Involvement of the mesolimbic dopamine system, *J. Pharmacol. Exp. Ther.*, 235, 537, 1985.
201. **Kalivas, P. W. and Weber, B.,** Amphetamine injection into the ventral mesencephalon sensitizes rats to peripheral amphetamine and cocaine, *J. Pharmacol. Exp. Ther.*, 245, 1095, 1988.
202. **Vezina, P. and Stewart, J.,** Amphetamine administered to the ventral tegmental area but not to the nucleus accumbens sensitizes rats to systemic morphine: lack of conditional reflex, *Brain Res.*, 516, 99, 1990.
203. **Dougherty, G. G., Jr. and Ellinwood, E. H., Jr.,** Chronic d-amphetamine into the nucleus accumbens: lack of reverse tolerance of locomotor activity, *Life Sci.*, 28, 2295, 1981.
204. **Vezina, P., Kalivas, P. W., and Stewart, J.,** Sensitization occurs to the locomotor effect of morphine and the specific mu opioid receptor agonist, DAGO, administered repeatedly to the ventral tegmental area but not to the nucleus accumbens, *Brain Res.*, 417, 51, 1987.
205. **Yoshikawa, T., Shibuya, H., Kaneno, S., and Toru, M.,** Blockade of behavioral sensitization to metamphetamine by lesion of hippocampo-accumbal pathway, *Life Sci.*, 48, 1325, 1991.

206. **Kashiwabara, K., Akiyama, K., Sato, M., and Otsuki, S.,** Effect of haloperidol, sulpiride and coeruletide on metamphetamine-induced abnormal behaviors in animals, *Jpn. J. Neurpsychopharmacol.*, 6, 181, 1984 (in Japanese) cit. post: Hamamura et al., 1991.
207. **Karler, R., Calder, L. D., Chaudhry, I. A., and Turkanis, S. A.,** Blockade of ''reverse tolerance'' to cocaine and amphetamine by MK-801, *Life Sci.*, 45, 599, 1989.
208. **Karler, R., Chaudhry, I. A., Calder, L. D., and Turkanis, S. A.,** Amphetamine behavioral sensitization and the excitatory amino acids, *Brain Res.*, 537, 76, 1990.
209. **Vezina, P. and Stewart, J.,** The effect of dopamine receptor blockade on the development of sensitization to the locomotor activating effect of amphetamine and morphine, *Brain Res.*, 499, 108, 1989.
210. **Stewart, J. and Vezina, P.,** Microinjections of Sch-23390 into the ventral tegmental area and substantia nigra pars reticulata attenuate the development of sensitization to the locomotor activating effects of systemic amphetamine, *Brain Res.*, 495, 401, 1989.
211. **Hamamura, T., Akiyama, K., Akimoto, K., Kashihara, K., Okumura, K., Ujike, H., and Otsuki, S.,** Co-administration of either selective D_1 or D_2 dopamine antagonist with metamphetamine prevents metamphetamine-induced behavioral sensitization and neurochemical change, studied by in vivo intracerebral dialysis, *Brain Res.*, 546, 40, 1991.
212. **Nishikawa, T., Mataga, N., Takashima, M., and Toru, B.,** Behavioral sensitization and relative hyperresponsiveness of striatal and limbic dopaminergic neurons after repeated metamphetamine treatments, *Eur. J. Pharmacol.*, 88, 195, 1983.
213. **Steketee, J. D. and Kalivas, P. W.,** Effect of microinjections of apamine into the A10 dopamine region of rats: a behavioral and neurochemical analysis, *J. Pharmacol. Exp. Ther.*, 254, 711, 1990.
214. **Hughes, M., Romey, G., Duval, D., Vincent, J. P., and Lazdunski, M.,** Apamine as a selective blocker of calcium-dependent potassium channel in neuroblastoma cells: voltage-clamp and biochemical characterization of the toxin receptor, *Proc. Natl. Acad. Sci. U.S.A.*, 79, 1308, 1982.
215. **Muller, P. and Seeman, P.,** Presynaptic subsensitivity as a possible basis for sensitization by long-term dopamine mimetics, *Eur. J. Pharmacol.*, 55, 149, 1979.
216. **Henry, D. J., Greene, M. A., and White, F. J.,** Electrophysiological effects of cocaine in the mesoaccumbens system: repeated administration, *J. Pharmacol. Exp. Ther.*, 251, 833, 1989.
217. **Taylor, D. L., Ho, P. T., and Fagan, J. D.,** Increased dopamine receptor binding in rat brain by repeated cocaine injections, *Commun. Psychopharmacol.*, 3, 137, 1979.
218. **Lucchelli, A., Boselli, C., and Brunello, N.,** Influence of chronic morphine treatment on dopaminergic sensitivity, *Pharmacol. Res.*, 22 (Suppl. 3), 35, 1990.
219. **Kalivas, P. W. and Bronson, M.,** Mesolimbic dopamine lesions produce an augmented behavioral response to enkephalin, *Neuropharmacology*, 24, 931, 1985.

220. **Paulson, P. E. and Robinson, T. E.,** Sensitization to systemic amphetamine produces an enhanced locomotor response to a subsequent intra-accumbens amphetamine challenge in rats, *Psychopharmacology*, 104, 140, 1991.
221. **Abrous, D. N., Stinus, L., Le Moal, M., and Herman, J. P.,** Intra-accumbens implants of embryonic dopaminergic neurons reverse the behavioral supersensitivity to opiates evoked by lesion of the mesolimbic dopaminergic pathway, *Brain Res.*, 525, 155, 1990.
222. **Akiyama, K., Sato, M., and Otsuki, S.,** Increased ^{3}H-spiperone binding sites in mesolimbic area related to metamphetamine-induced behavioral hypersensitivity, *Biol. Psychiatry*, 17, 223, 1981.
223. **Goeders, N. E. and Kuhar, M. J.,** Chronic cocaine administration induces opposite changes in dopamine receptors in the striatum and nucleus accumbens, *Alcohol Drug Res.*, 7, 207, 1987.
224. **Peris, J., Boyson, S. J., Cass, W. A., Curella, P., Dwoskin, L. P., Larson, G., Lin, L.-H., Yasuda, R. P., and Zahniser, N. R.,** Persistence of neurochemical changes in dopamine systems after repeated cocaine administration, *J. Pharmacol. Exp. Ther.*, 253, 38, 1990.
225. **Volkow, N. D., Fowler, J. S., Wolf, A. P., Schyler, D., Shiue, C.-Y., Alpert, R., Dewey, S. L., Logan J., Bendriem, B., Christman, D., Hitzemann, R., and Henn, F.,** Effects of chronic cocaine abuse on postsynaptic dopamine receptors, *Am. J. Psychiatry*, 147, 719, 1990.
226. **Hurd, Y. L., Weiss, F., Koob, G., and Ungerstedt, U.,** The influence of cocaine self-administration on in vivo dopamine and acetylcholine neurotransmission in rat caudate-putamen, *Neurosci. Lett.*, 109, 227, 1990.
227. **Trujillo, K. A. and Akil, H.,** Inhibition of morphine tolerance and dependence by the NMDA receptor antagonist MK-801, *Science*, 251, 85, 1991.
228. **Wachi, M., Okuda, M., Togashi, S., Miyashita, O., and Wakahoi, T.,** Effects of metamphetamine administration on brain neurotensin-like immunoreactivity in rats, *Neurosci. Lett.*, 78, 222, 1987.
229. **Singh, N. A., Bush, L. G., Gibb, J. W., and Hanson, G. R.,** Dopamine-mediated changes in central nervous system neurotensin systems: a role for NMDA receptors, *Eur. J. Pharmacol.*, 187, 337, 1990.

8 THE ROLE OF STRESS IN ADDICTION

It is generally accepted without any scientific argument that humans are especially vulnerable to drug abuse in stressful situations. However obvious it is that people light up a cigarette, take a drink, or resort to illicit drugs to relieve anxiety, fear, or suffering caused by any kind of noxious physical or psychological stimuli, the corresponding human experimental observations provide mostly rather circumstantial evidence for this generally held belief. Thus, there are hardly any psychometric, experimental psychological data supporting the generally accepted view that substances are abused more frequently under stressful conditions. The only notable exception seems to be the abuse of nicotine. Smokers have been found to light up cigarettes if they were exposed to "naturalistic" stressors (simulated airplane noise, public speaking)[1] or to the kinds of stress used typically in experimental laboratories.[2–4]

Furthermore, as discussed in Chapter 11, drug addicts generally are dysphoric, depressed, they are less able to endure emotional or physical deprivations, and suffer from various kinds of personality disorders. This seems to apply to smokers as well, namely, smokers as a subgroup of the whole population show an especially high level of anxiety and depression and they are particularly sensitive to various kinds of stress.[4–6] Thus, using the word "stress" in its broadest sense, addicts can probably be considered stress-sensitive subjects. Nevertheless, discounting the smokers, in humans the eventual experimental confirmation of this kind of reasoning is still missing. (Probably for ethical reasons, namely, it might be improper to try to provoke consumption of abused substances [especially hard drugs] in order to clarify a scientific issue.)

However, just a few animal-experimental observations prove that stress might activate the endogenous opioid and mesolimbic dopamine systems and stressful manipulations might speed up the process of sensitization. Thus, in animals stress seems to be an additional factor in substance abuse.

I. STRESS AND ENDOGENOUS OPIOIDS

A. Stress and Opioid Analgesia

Several kinds of experimental manipulations which obviously are aversive or distinctly painful have been shown to produce "analgesia", i.e., elevation of the pain threshold in conventional antinociceptive assays. Thus, exposing rats to inescapable foot shocks,[7] forcing the rodents to swim in cold water,[8] food deprivation,[9] and restraining rats[10] induce "analgesia", i.e., elevation of the threshold of experimentally induced pain in various assays. Analgesia can also be elicited by "psychological stressing", e.g., by putting rats or mice into an environment where they are not exposed to foot shocks on the experimental days, but see others struggling, jumping, and vocalizing upon suffering from the painful stimuli.[11,12]

Certain kinds of experimental analgesia are dependent on mobilizing endogenous opioids, i.e., these types of antinociception can be abolished by opiate antagonists, and tolerance develops upon repeated application, whereas some other forms are not mediated by opioids (for review see Lewis et al.).[13] For example, the cold-water swim-induced analgesia is resistant to naloxone.[8] Or the analgesia induced by electric foot shocks may or may not be mediated by endogenous opioids, and the two forms of antinociception are differentially modulated by cimetidine.[14]

Actually, cold-water swimming, a manipulation resulting in a kind of nonopioid analgesia, might even inhibit the analgesic effect of morphine,[15] though it seems to be evident that antinociceptive effects of opiates and nonopioid forms of stress analgesia might add up.

It is more than probable that this divergence in experimental findings is due not only to the multiplicity of endogenous mechanisms of antinociception, but some apparent contradictions might be related to the fact that the μ- and κ-opiate receptor mediated forms of analgesia have different mechanisms (as discussed in Chapters 1 and 2). Truly, restraint stress has been reported to equally potentiate the analgesic effect of μ-selective morphine and (D-Ala^2,NMePhe^4,Gly^5-ol)-enkephalin (DAGO) and that of the δ-preferring (D-Ala^2,D-Leu^5)-enkephalin (DADLE).[10] However, in a subsequent, more careful study, the same group[16] found that restraint stress potentiated primarily the analgesic effect of μ-agonists, whereas the similar effects of (D-Pen^2,D-Pen^5)-enkephalin (DPDPE), a really δ-selective analog, and U-50488H, the prototypic κ-agonist, were hardly modified. Thus restraint stress might selectively potentiate the μ-opioid (analgesic) mechanisms. In another study the analgesic effect of "psychological stress" (in mice) was antagonized by nor-binaltorphimine (NBI), i.e., a selective antagonist of κ-receptors, whereas the analgesic effect of foot shock and forced swimming, i.e., other forms of stress analgesia, were not modified.[17] Shortly, various forms of experimental stress might be synergistic with either μ- or κ-receptor-mediated analgesia or eventually with neither of them.

Furthermore, in the same study, both repeated application of psychological stress and U-50488H inhibited the development of morphine tolerance, and this inhibition could also be reversed by NBI.[17]

In conjunction with the later data it is worth recalling that according to Yamamoto et al.,[18] coadministration of U-50488H inhibited the development of morphine tolerance in rats. In addition, though U-50488H failed to modify the development of morphine dependence in rats,[19] dynorphin was reported to suppress the morphine-withdrawal syndrome in monkeys[20] and rats.[21] In line with these data, Sofuoglu et al.[22] reported recently that the acute tolerance to the analgesic effect of morphine elicited by a single large dose of it was potentiated (extended in time) if the animals were also treated with NBI. This finding[22] also indicates that κ-receptor-mediated mechanisms might inhibit the development of tolerance to μ-agonists.

All these findings are pretty surprising and apparently not very frequently quoted in the literature. However, they are in line with the data on the aspecific character of analgesia induced by κ-agonists (see Chapter 2) and the antianalgesic effect of dynorphin (see Chapter 3). Summarily, these findings corroborate many others indicating that

- μ- and κ-receptor-mediated analgesic mechanisms are different.
- These different kinds of mechanisms might even be antagonistic.
- Activation of the κ-receptors might inhibit the development of tolerance to the analgesic action of morphine, at least in some experimental conditions.

B. Modulation by Stress of Synthesis of Endogenous Opioids

As discussed in Chapter 3, in rats suffering from chronic pain induced by Freund adjuvant, the dynorphin content and its synthesis (i.e., transcription of the corresponding gene) are significantly enhanced in the dorsal horn and along the ascending pain pathways in the spinal cord. Przewlocki et al.[23] detected similar changes upon repeated electrical foot shocks (30 min, twice daily for 7 d). However, no alterations in the enkephalin, endorphin, or dynorphin contents have been detected in the higher brain areas in these studies. Thus, whether it is chronic pain or stressful electrical stimulation, only the dynorphin system is affected in the spinal cord.

On the other hand, using i.p. injection of hypertonic saline or naloxone-precipitated morphine withdrawal as stressful stimuli, Lightman and Young[24] observed rapid accumulation of enkephalin mRNA in the parvicellular paraventricular nucleus of the hypothalamus. In the subsequent work of the same group these interventions (naloxone-precipitated opiate withdrawal or loading with hypertonic saline) a concomitant increase in the mRNA encoding for corticotrophin-releasing factor (CRF) was observed in the same brain region.[25]

Of course, CRF release upon stressful stimuli is just the response expected from this polypeptide. Its coexpression with enkephalin is, however, a new finding which might help to understand the apparently controversial data in the literature about regulation by opioids of the hypothalamo-pituitary-adrenal (HPA) stress system. Namely, this finding

TABLE 1
Modulation by Stress of the Synthesis of Endogenous Opioids

Species examined	Experimental paradigm	Observation	Ref.
Rat	Effect of chronic painful inflammation of the legs induced by Freund adjuvant on the level of endogenous opioids	Increased genetic expression and synthesis of dynorphin in the spinal cord along the pain pathways	See the data reviewed in detail in Chapter 3
	Effect of repeated foot shocks on the production of endogenous opioids	As seen after tretament with Freund adjuvant, i.e., enhancement	23
	Effect of stress induced by morphine withdrawal or by i.p. injection of hypertonic saline on the expression of enkephalin in the hypothalamus	Increase in its mRNA level, i.e., enhanced expression of enkephalin in the paraventricular nucleus	24
	As above	Simultaneous increase in expression of enkephalin and CRF[a] in the paraventricular nucleus	25

[a] CRF = Corticotropin-releasing hormone.

indicates that mobilization of the HPA system is accompanied by enhanced expression of the enkephalins in this area. On the other hand, the data are controversial as to whether or not the opposite of this finding also is valid, i.e., whether or not endogenous opioids modulate the activity of HPA (Table 1).

C. Modulation by Opioids of the Stress Reaction

Even today the eventual effect of μ-agonist on the activity of the HPA system is a rather controversial issue. Namely, in some relatively old experimental studies it was claimed that classical opiates induce ACTHcorticosterone release, i.e., they should activate the stress pathways (for review see Haracz et al.[26] and Bailey and Kitchen).[27] Natural enkephalins,[28,29] β-endorphin,[26,30] and even μ-receptor-preferring enkephalin analogs such as (D-Ala^2,Met^5)-enkephalinamide (DALA), FK 33–824, (D-Ala^2,NMePhe^4,Met(O)^5)-enkephalin, and (D-Met^2, Pro^5)-enkephalinamide[31–33] have also been reported to elevate the serum corticosterone level in rats and sheep.[34–36] In animal studies suppression was only seen in mice after Leu-enkephalin.[37,38]

However, based on many studies there is no doubt that in humans FK 33–824 reduces the serum cortisol level.[39–43] So do β-endorphin[44] and Leu-enkephalin[45] in healthy volunteers. Furthermore, it is well established that in opiate addicts the serum ACTH and cortisol levels are subnormal rather than elevated.[46,47]

To interpret these seemingly disparate data it is important to emphasize that in a recent study[48] chronic morphine treatment *per se* reduced the expression of the proenkephalin gene. This finding is in agreement with the findings on inhibition of unit activity by morphine and opioid peptides *in vitro* in the paraventricular nucleus.[49] Furthermore, Plotsky[50] reported on inhibition of CRF release into the hypophyseal-portal circulation in rats upon i.c.v. β-endorphin treatment.

As for the opiate antagonists, naloxone and naltrexone have been reported to elevate the serum corticosterone level in rats[27,51–54] and mice.[55] Naloxone elevated the serum ACTH and corticosterone levels in rats, with hypothalamic deafferentation as well.[56] In healthy humans, high doses of naloxone and naltrexone also increase the serum cortisol and ACTH levels.[42,43,57–59] Since opioids do not modify the liberation of ACTH from the pituitary,[60–62] it is logical to propose that under physiological conditions μ-opioids inhibit the release of stress hormones at the level of the hypothalamus by suppressing the liberation of CRF though the related data are not quite unequivocal. *In vitro,* μ- and δ-receptor-preferring opioids in most studies inhibited CRF release,[63–65] but in one work[66] increased liberation was seen. Naloxone enhances the release of CRF from isolated hypothalamic slice.[66] In view of these findings the earlier observations on facilitation by classical opiates and μ- and δ-receptor-preferring opioid peptides of ACTH/corticoid release in animals might be attributed to the stressful nature of dissipation of opiate euphoria. However, for a similar reason the elevation of serum cortisol/corticosterone levels upon treatment by opiate antagonist might also be due to the stressful character of displacement of endogenous opioids by their exogenous antagonists and not to an eventual tonic μ-receptor-mediated inhibition of CRF release.

A further confounding factor is the alleged ability of μ-opiates to release vasopressin,[67–69] which in turn might liberate ACTH from the anterior pituitary.

On the other hand, natural and synthetic ligands of κ-receptors such as dynorphin$_{1-13}$, ethylketocyclazocine, bremazocine, tifluadom, and U-50488H have been found to drastically elevate the serum ACTH and corticosterone levels in rats.[70–78] By all means, the effect of κ-receptors seems to be robust; they mobilize the stress hormones even in subanalgesic doses.[70] This effect is surely not related to eventual vasopressin release; namely, they induce water diuresis.[79,80] Furthermore, selective antagonists of the μ- and κ-receptors suppress the stress effects only of their corresponding agonists[75,77] and there is no cross tolerance in this respect between the two groups.[70,71] Nevertheless, it would be an oversimplification to conclude that μ-(δ)-opioids inhibit, whereas κ-ones facilitate CRF release. Namely, in one study,[50] i.c.v. dynorphin$_{1-13}$, similarly to β-endorphin, inhibited the liberation of CRF into the hypophyseal-portal blood. Furthermore, Nikolarikis et al.[81] reported that the ACTH-releasing effect of MR-2034, a κ-agonist, could be reversed by anti-CRF antiserum, but the similar action of morphine could not.

Thus the possibility arises that at the hypothalamic level, CRF liberation is tonically inhibited by μ- and κ-receptors as well. However,

- κ-Agonists activate the HPA system somewhere at higher level(s), whereupon CRF liberation is secondary to their stress effect.
- μ-Opioids induce ACTH release either by vasopressin release, and/or dissipation of their euphoriant action is stressful, per se.

Nevertheless, the data indicate that in humans their tonic effect is the suppression of the HPA system.

Summarily, the data reviewed in this section indicate that stressful events liberate endogenous opioids, at least in the hypothalamus, and mobilization of the endogenous opioid mechanisms, especially the κ-receptor-mediated ones, is stressful. Thus in view of the data reviewed in Chapter 7, the theoretical possibility arises that stress, similarly to consumption of abused substances, might take part in the process of sensitization. Furthermore, as reviewed in the next section, stress might interact with another link of sensitization as well (Table 2).

II. STRESS AND THE MESOLIMBIC DOPAMINERGIC SYSTEM

Facilitation, i.e., acceleration of the dopamine turnover, has been recorded in the ventral tegmental area (VTA) and various terminal areas of the A_{10} dopamine neurons exposing rats or mice to various forms of stress. The kind of stress applied included, primarily, mild foot shock,[82–87] but prolonged immobilization,[88] tail-pinch,[89,90] swim stress,[87] keeping the animals isolated from the others,[91] or social interaction with an aggressive male congener[89] all induced more or less the same effect. Of some various kinds of social interaction just the encounter with an aggressive male congener, i.e., that of aversive character, resulted in enhanced dopamine turnover.[89] Thus various kinds of stressful manipulations apparently induce the same type of reaction.

Interestingly, in the various studies not exactly the same structures showed enhancement of the dopamine turnover, but in all studies one or more of the structures known to contain terminals of the A_{10} neurons showed the response, including the nucleus accumbens (NA), the frontal cortex (FC), the cingular cortex (CC), the piriform cortex (PC), the olfactory tubercle (OT), and of course the VTA. In one study, elevation also

TABLE 2
Modulation by Opioid Agonists and Antagonists of the Stress Reaction

Species examined	Experimental paradigm	Observation	Ref.
Various	Effect of classical (μ) opiates on the serum ACTH and corticosterone levels	Increase	26
Rat	Effects of Met- and Leu-enkephalin, β-endorphin, enkephalin analogs as DALA,[a] FK 33-824,[b] DMPEA[c] on the corticosterone level	Elevation	26, 30–33
Mouse	Effect of Leu-enkephalin and FK 33-824 on serum corticosterone level	Decrease	37,38
Sheep	Effect of Met- and Leu-enkephalin and DAMGO[d] on serum corticosterone and ACTH levels	Elevation	34,36
Healthy humans	Effect of FK 33-824, Leu-enkephalin and β-endorphin on serum ACTH and cortisol levels	Reduction	39–44
Opiate addicts	Serum ACTH and cortisol levels	Reduction	45,46
Rat and mice	Effects of naloxone and naltrexone on serum corticosterone level	Elevation	27, 51–55
Rat with hypothalamic deafferentation	Effect of naloxone on ACTH release	Elevation	56
Healthy humans	Effects of naloxone and naltrexone on serum ACTH and cortisol levels	Elevation	42,43 57–59
Isolated rat hypothalamic slice preparation	Effect of natural enkephalins, DALA, DADLE,[e] FK 33-824, β-endorphin, dynorphin on CRF[f] release	Inhibition	63–65
Isolated rat hypothalamic slice preparation	Effect of Met- and Leu-enkephalin on CRF release	Enhancement	66
Isolated rat hypothalamic slice preparation	Effect of naloxone on CRF release	Enhancement	65
Isolated rat hypothalamic slice preparation	Modulation of CRF release by		
	β-Endorphin	Inhibition	50
	Dynorphin	Inhibition	50
Rat	Effect of U-50488H, a selective κ-agonist, on serum and corticosterone levels	Drastic elevation	70,71, 73,74, 76,78

[a] DALA = (D-Ala[2],Met[5])-enkephalinamide.
[b] FK 33–824 = (D-Ala[2],NMePhe[4],Met(*O*)(ol)[5])-enkephalin.
[c] DMPEA = (D-Met[2],Pro[5])-enkephalinamide.
[d] DAMGO = (D-Ala[2],NMePhe[4],Gly(ol)[5])-enkephalin.
[e] DADLE = (D-Ala[2],D-Leu[5])-enkephalin.
[f] CRF = Corticotropin releasing factor.

was seen in the amygdaloid complex,[85] but in these works the dopamine metabolism did not increase above the baseline level in the striatum and the substantia nigra[82,83,85,86,92] only in one study.[91] Of the "typical" mesolimbic terminal areas it was the septum where the response could not be demonstrated.[84,85,93]

For the time being, no explanation can be given for the fact that in the various laboratories the acceleration of dopamine turnover was restricted to various areas of the mesolimbic dopamine system, i.e., the response only could be seen in one or in some of the above listed structures. Difficulties in the exact delineation of the boundaries of these areas and differences in the methodology in measurement of dopamine metabolism might serve for at least a partial explanation. Namely, the dopamine turnover was characterized either by measuring the dopamine level after blockade of its synthesis by α-methyl-*p*-tyrosine[82] or by the relative concentration of its metabolite, i.e., by the DOPAC/dopamine ratio,[84,91] or by the DOPAC level in itself,[83,85,86,87,93] or by recording the latter via *in vivo* voltametry.[89,90] Of course, the multiplicity of the procedures applied might also explain the differences. Nevertheless, it cannot be excluded that various types of stress activate different subgroups of A_{10} neurons, which also might explain the contradictions. For example, Bannon et al.[92] proposed that mild foot shocks only enhance the dopamine turnover in the cortical areas, whereas higher intensities are needed to activate the pathways projecting to the NA. Anyhow, the main point is that in all these studies projection areas of the A_{10} and not those of the A_9 neurons showed enhanced dopamine metabolism.

Pairing the stressful manipulation with neutral stimuli, subsequently the conditional signals also elicited acceleration of the dopamine turnover.[85,87,93]

The facilitation by stress of A_{10} neurons could be demonstrated by other methodology as well. As described by Chiodo et al.[94] and Maeda and Mogenson,[95] there are dopaminergic cells in the VTA which respond to different kinds of peripheral sensory stimulation with increased or decreased firing. In cats, these multiresponsive units respond with enhanced activity to various kinds of stressful manipulations.[96] On the other hand the nigral (i.e., A_9) neurons did not react with activation to tail-pinch and arousing stimuli.[97]

Using another approach, Cabib et al.[98] observed enhancement of the apomorphine-elicited stereotypy in mice after repeated immobilization-induced stress, and this enhanced responsivity could be normalized by daily administration of naltrexone. Thus, upon repeated stress not only acceleration of dopamine turnover, but even the enhanced sensitivity to dopamine agonists can be detected just as seen when the sensitizing factor is an abused substance (see Chapter 7). Actually, the same group demonstrated subsequently that repeated exposure of mice to immobilization stress resulted in a decrease in the 3-methoxytyramine level in the NA, which is a sign of decreased dopamine release.[99] Thus enhanced turnover leads to exhaustion and secondary decrease in the amount of dopamine liberated from the A_{10} cells, just as seen in the case of sensitization to abused substances (see Chapter 7).

Bannon et al.[92] reported that injection of antisubstance P antibody into the VTA reversed the acceleration of dopamine turnover induced by foot shocks. As discussed in Chapter 7, substance P (in the VTA) contributes to the activation of A_{10} neurons. Nevertheless, further confirmation and data are needed to decide whether or not substance P is really involved in the opiate receptor-mediated activation of A_{10} neurons.

The process of stress-induced facilitation can be antagonized by diazepam.[83,84,86,87,90] As discussed in Chapter 7, A_{10} cells receive a tonically active inhibitory GABA-ergic input and the positive cooperative interaction (around chloride channels) of benzodiazepine and GABA (γ-amino butyric acid) receptors is well established (see the corresponding textbooks). Thus the inhibitory effect of diazepam also proves that stress activates the mesolimbic dopamine system via the cell bodies of A_{10} neurons. The antagonistic effect of diazepam could be abolished by CGS 8216, a benzodiazepine antagonist.[87]

Pentobarbital does not antagonize the stress-induced acceleration of dopamine,[84] indicating that the agonistic interaction with the benzodiazepine receptors and not the anxiolytic effect is the factor suppressing the activation of A_{10} neurons. On the other hand, in another laboratory not only diazepam, but zoldipem, zopiclone, and meprobamate also inhibited the acceleration of dopamine turnover.[87] Thus both potentiation of GABA-ergic inhibition of VTA (by positive cooperative interaction between the benzodiazepine and GABA receptors) and the anxiolytic effect *per se* might contribute to this effect (Table 3).

III. CROSS SENSITIZATION BETWEEN STRESS AND ABUSED SUBSTANCES

Several data indicate that stress might be a contributing factor in the process of sensitization. Antelman et al.[100] first demonstrated that repeated exposure of rats to mild tail pressure resulted in enhancement of amphetamine-induced stereotypy, just as seen in animals receiving repeated amphetamine treatments.

Similarly, repeated foot shock stress augmented the amphetamine-elicited locomotor hyperactivity.[101] Of course, these phenomena might not be related to the mesolimbic dopamine system. However, Kalivas et al.[102] reported that repeated mild foot shock stress further increased the facilitation of dopamine turnover in the NA induced by intra-VTA injection of DALA and vice-versa, i.e., repeated intra-VTA administration of this en-

TABLE 3
Stress and the Mesolimbic Dopamine System

Species examined	Experimental paradigm	Observation	Ref.
Rat	Foot shock-induced changes in the dopamine turnover in various terminal areas of A_{10} neurons	Acceleration	82–87
Mouse	Immobilization induced changes in dopamine turnover in the frontal cortex	Acceleration	88
Rat	Effect of tail pinch on the DOPAC release in the NA	Increase	89,90
Rat	Effect of swim stress on the dopamine turnover in the frontal cortex	Acceleration	87
Rat	Effect of long-term isolation on the dopamine turnover	Reduced in the frontal cortex, but enhanced in the accumbens and striatum	91
Rat (male)	Effect of encounter with an aggressive male congener	Increase in dopamine release from the NA[a]	89
Mouse	Effect of repeated exposure to immobilization stress	Reduced dopamine release in the NA[a]	98
Cat	Effect of stressful manipulations on the activity of multiresponsive A_{10} neurons	Enhanced firing	96
Rat	Effect of diazepam on the stress-induced facilitation of dopamine turnover	Inhibition	83,84, 86,87, 90

[a] NA = Nucleus accumbens.

kephalin analog augmented the metabolic facilitation seen after stressing the animals.

The cross sensitization between opioids and stressful manipulations does not necessarily mean that these abused substances are stressful. Mice show markedly reduced motility as a kind of conditioned stress[103] when put into an environment where they previously received electric foot shock. However, i.c.v. administration of DALA or DADLE, i.e., relatively μ- or δ-preferring analogs, attenuated this conditioned reaction whereas dynorphin$_{1-17}$ and E-2078, an enzymatically stabilized dynorphin analog, potentiated the behavioral response.[104] The μ-/δ- and κ-ligands mutually antagonized the effect of each other.[104] Thus opioids might attenuate the psychic effects of stressful manipulations. Nevertheless, they activate the same process of sensitization.

Actually, the sensitivity to stress is one of the factors determining the individual propensity to drug abuse. Piazza et al.[105] demonstrated that rats

TABLE 4
Cross-Sensitization Between Stress and Drugs of Abuse

Species examined	Experimental paradigm	Observation	Ref.
Rat	Effect of repeated stress (tail pinch, foot shock) on amphetamine-induced hypermotility	Enhancement	100,101
	Effect of repeated foot shock stress on the increase in dopamine turnover in the NA[a] induced by DALA[b] injected into the VTA[c]	Potentiation	102
	Effect of repeated intra-VTA DALA on the acceleration of dopamine turnover induced by foot shock	Potentiation	102
	Individual sensitivity to novelty stress and amphetamine-induced hyperactivity	Positive correlation	105
	Individual sensitivity to amphetamine-induced hyperactivity and its self-administration	Positive correlation	106

[a] NA = Nucleus accumbens.
[b] DALA = (D-Ala[2],Met[5])-enkephalinamide.
[c] VTA = Ventral tegmental area.

can be subgrouped according to their sensitivity to novelty stress. The sensitive ones responded to both a novel environment and a standard dose of amphetamine with more intense hyperlocomotion than the nonsensitive ones. Determining the sensitivity of individuals, the sensitive ones have been found to be more prone to drug abuse, i.e., in these animals the amphetamine self-administration behavior could be established more easily than in their less sensitive congeners.[105] Furthermore, in subjects previously exposed either to amphetamine injections or stressful tail-pinch sessions, the amphetamine self-administration behavior developed more easily than in their experimentally naive counterparts.[106]

(Summarily, these experimental data indicate that (see Table 4):

- Stress is an important factor in making subjects vulnerable to drug addiction.
- The process of sensitization can be initiated/aggravated by both repeated administration of abused substances and stressful manipulations.
- Though some drugs might be abused to alleviate stress, they actually speed up the process of sensitization.
- Enhanced sensitivity to stress might be a factor explaining the propensity of certain subjects to drug abuse.

REFERENCES

1. **Schachter, S.,** Pharmacological and psychological determinants of smoking, *Ann. Intern. Med.*, 88, 104, 1978.
2. **Dobbs, S. D., Strickler, D. P., and Maxwell, W. E.,** The effects of stress, and relaxation in the presence of stress on urinary pH and smoking behavior, *Addictive Behav.*, 6, 345, 1981.
3. **Rose, J. E., Ananda, S., and Jarvik, M. E.,** Cigarette smoking during anxiety provoking and monotonous tasks, *Addictive Behav.*, 8, 353, 1983.
4. **Pomerlau, O. F. and Pomerlau, C. S.,** Research on stress and smoking: progress and problems, *Br. J. Addict.*, 86, 599, 1991.
5. **Billings, A. G. and Moos, R. H.,** Social-environmental factors among light and heavy cigarette smokers. A controlled comparison with non-smokers, *Addictive Behav.*, 8, 381, 1983.
6. **Shiffman, S.,** Cigarette smoking, nicotine addiction, and relaxation, *J. Consult. Clin. Psychol.*, 50, 71, 1982.
7. **Hayes, R. L., Bennett, G. J., Newlon, P., and Mayer, D. J.,** Analgesic effects of certain noxious and stressful manipulations in the rat, *Soc. Neurosci. Abstr.*, 2, 1350, 1976.
8. **Bodnar, R. J., Kelly, D. D., Steiner, S. S., and Glusman, M.,** Stress-produced analgesia and morphine-produced analgesia: lack of cross-tolerance, *Pharmacol. Biochem. Behav.*, 8, 661, 1978.
9. **Spiaggia, A., Bodnar, R. J., Kelly, D. D., McManus, M. E., and Glusman, M.,** Biphasic alterations in nociceptive thresholds after food deprivation, *Soc. Neurosci. Abstr.*, 3, 492, 1977.
10. **Appelbaum, B. D. and Holtzman, S. G.,** Stress-induced changes in the analgesic and thermic effects of opioid peptides in the rat, *Brain Res.*, 377, 330, 1986.
11. **Jensen, T. S. and Smith, D. F.,** Effects of emotions on nociceptive threshold in rats, *Physiol. Behav.*, 28, 597, 1982.
12. **Takahashi, M., Tokuyama, S., and Kaneto, H.,** Implication of endogenous opioid mechanisms in the production of antinociceptive effect induced by psychological stress in mice, *Jpn. J. Pharmacol.*, 44, 283, 1987.
13. **Lewis, J. W., Cannon, J. T., and Liebeskind, J. C.,** Opioid and nonopioid mechanisms of stress analgesia, *Science*, 208, 623, 1980.
14. **Gogas, K. R., Hough, L. B., Glick, S. D., and Su, K. L.,** Opposing actions of cimetidine on naloxone-sensitive and naloxone-insensitive forms of footshock-induced analgesia, *Brain Res.*, 370, 370, 1986.
15. **Steinman, J. L., Faris, P. L., Mann, P. E., Olney, J. W., Komisaruk, B. R., Willis, W. D., and Bodnar, R. J.,** Antagonism of morphine analgesia by nonopioid cold-water swim analgesia: direct evidence for collateral inhibition, *Neurosci. Biobehav. Rev.*, 14, 1, 1990.
16. **Calcagnetti, D. J., Fleetwood, S. W., and Holtzman, S. G.,** Pharmacological profile of the potentiation of opioid analgesia by restraint stress, *Pharmacol. Biochem. Behav.*, 37, 193, 1990.

17. **Takahashi, M., Senda, T., Tokuyama, S., and Kaneto, H.,** Further evidence for the implication of a κ-opioid receptor mechanism in the production of psychological stress-induced analgesia, *Jpn. J. Pharmacol.*, 58, 487, 1990.
18. **Yamamoto, T., Ohno, M., and Ueki, S.,** A selective κ-opioid agonist, U-50,488H, blocks the development of tolerance to morphine analgesia in rats, *Eur. J. Pharmacol.*, 156, 173, 1988.
19. **Fukugawa, Y., Katz, J. L., and Suzuki, T.,** Effects of selective κ-opioid agonist, U-50,488H on morphine dependence in rats, *Eur. J. Pharmacol.*, 170, 47, 1989.
20. **Aceto, M. D., Dewey, W. L., Chang, J.-K., and Lee, N. M.,** Dynorphin-(1–13): effects in nontolerant and morphine-dependent rhesus monkeys, *Eur. J. Pharmacol.*, 83, 139, 1982.
21. **Green, P. G. and Lee, N. M.,** Dynorphin A-(1–13) attenuates withdrawal in morphine dependent rats: effects of route of administration, *Eur. J. Pharmacol.*, 145, 267, 1988.
22. **Sofuoglu, M., Portoghese, P. S., and Takemori, A. E.,** Maintenance of acute morphine tolerance in mice by selective blockage of κ opioid receptors with norbinaltorphimine, *Eur. J. Pharmacol.*, 210, 159, 1992.
23. **Przewlocki, R., Lasoń, W., Höllt, V., Silberring, J., and Herz, A.,** The influence of chronic stress on multiple opioid systems in the rat: pronounced effects upon dynorphin in spinal cord, *Brain Res.*, 413, 213, 1987.
24. **Lightman, S. L. and Young, W. S., III,** Changes in hypothalamic preproenkephalin A mRNA following stress and opiate withdrawal, *Nature*, 328, 643, 1987.
25. **Harbuz, M., Russell, J. A., Sumner, B. E. H., Kawata, M., and Lightman, S. L.,** Rapid changes in the content of proenkephalin A and corticotrophin releasing hormone mRNAs in the paraventricular nucleus during morphine withdrawal in urethane-anaesthetized rats, *Mol. Brain Res.*, 9, 285, 1991.
26. **Haracz, J. L., Bloom, A. S., Wang, R. I. H., and Tseng, L. F.,** Effect of intraventricular β-endorphin and morphine on hypothalamic-pituitary-adrenal activity and the release of pituitary β-endorphin, *Neuroendocrinology*, 33, 170, 1981.
27. **Bailey, C. and Kitchen, I.,** Developmental responses to opioids reveals a lack of effect of stress-induced corticosterone-levels in neonatal rats, *Br. J. Pharmacol.*, 91, 119, 1987.
28. **Lishmanov, Y. B., Lasukova, T. B., and Alekminskaya, L. A.,** Experimental study of enkephalins and hormonal-metabolic reactions in stress of different gravity, (in Russian), *Byull. Eksp. Biol. Med.*, 99, 269, 1985.
29. **Lishmanov, Y. B., Slepushkin, V. D., and Prum, I. A.,** Effect of enkephalins on the activity of some stress hormones in acute myocardial ischemia, (in Russian) *Patol. Fiziol. Eksp. Ter.*, 3, 22, 1985.
30. **Van Loon, G. R. and De Souza, E. B.,** Development of tolerance to the ACTH releasing effect of β-endorphin, *Res. Commun. Chem. Pathol. Pharmacol.*, 22, 203, 1978.
31. **Ferri, S., Arrigo-Reina, R., Scoto, G., Spadaro, C., and Spampinato, S.,** The role of opioid peptidergic system in body reactivity, in *Neural Peptides and Neural Communication*, Costa, E. and Trabucchi, M., Eds., Raven Press, New York, 1980, 347.

32. **De Souza, E. B. and Van Loon, G. R.,** D-Ala^2-Met-enkephalinamide, a potent opioid peptide, alters pituitary adrenocortical secretion in rats, *Endocrinology*, 111, 1483, 1982.
33. **Hashimoto, K., Suemaru, S., Hattori, T., Takao, T., Inoue, H., Sugawara, M., Kageyama, J., and Ota, Z.,** Effects of (D-Ala^2,Met^5)-enkephalinamide and naloxone on ACTH and corticosterone secretion, *Endocrinol. Jpn.*, 33, 813, 1986.
34. **Carson, G. D. and Challis, J. R. G.,** Changes in plasma cortisol and progesterone after administration of enkephalins to newborn lambs, *Can. J. Physiol. Pharmacol.*, 60, 1166, 1982.
35. **Bousquet, J., Lye, S. J., and Challis, J. R. G.,** Comparison of leucine-enkephalin and adrenocorticotrophin effects on adrenal function in fetal and adult sheep, *J. Reprod. Fertil.*, 70, 499, 1984.
36. **Wang, X. M., Coghlan, J. P., Congiu, M., Scoggins, B. A., and Wintour, E.,** Met-enkephalin stimulates secretion of ACTH in sheep, *J. Endocrinol.*, 11, 463, 1986.
37. **Gibson, A., Hart, S. L., and Shabib, A.,** Leucine-enkephalin and methionine-enkephalin produce opposing effects on plasma corticosterone levels in ether-stressed mice, *Br. J. Pharmacol.*, 70, 509, 1980.
38. **Kitchen, I. and McEwan, A. R.,** Plasma corticosterone levels after i.c.v. injection of opioids in normal and ether-stressed mice, *Br. J. Pharmacol.*, 79, 328P, 1983.
39. **Stubbs, W. A., Delitala, G., Jones, A., Jeffcoate, W. J., Edwards, C. R. W., Ratter, S. J., Besser, G. M., Bloom, S. R., and Alberti, K. G. M. M.,** Hormonal and metabolic responses to an enkephalin analogue in man, *Lancet*, 2, 1225, 1978.
40. **Del Pozo, E., Martin-Perez, J., Stadelman, A., and Girard, J.,** Inhibitory action of a Met-enkephalin on ACTH release in man, *J. Clin. Invest.*, 65, 1531, 1980.
41. **Del Pozo, E., von Graffenried, B., Brownell, J., Derrer, F., and Marbach, P.,** Endocrine effects of a methionine-enkephalin derivative (FK 33–824) in man, *Hormone Res.*, 13, 90, 1980.
42. **Delitala, G., Grossman, A., and Besser, G. M.,** The modulation of opiate-induced changes in anterior pituitary function by a dopamine agonist and antagonist, *Neuroendocrinol. Lett.*, 3, 147, 1981.
43. **Grossman, A., Gaillard, R. C., McCartney, P., Rees, L. H., and Besser, G. M.,** Opiate modulation of the pituitary-adrenal axis: effects of stress and circadian rhythm, *Clin. Endocrinol. (Oxford)*, 17, 279, 1982.
44. **Taylor, T., Dluhy, R. G., and Williams, G.,** β-Endorphin suppresses adrenocorticotropin and cortisol levels in normal human subjects, *J. Clin. Endocrinol. Metab.*, 57, 592, 1983.
45. **Smythe, G. A., McGinley, R., Chisholm, D. J., and Lazarus, L.,** Effect of leu-enkephalin on serum cortisol levels in man, *Clin. Exp. Pharmacol. Physiol.*, 7, 47, 1980.
46. **Volavka, J., Bauman, J., Pevnick, J., Reker, D., James, D., and Cho, D.,** Short-term hormonal effects of naloxone in man, *Psychoneuroendocrinology*, 5, 225, 1980.
47. **Gold, M. S., Pottash, A. L. C., Extein, I., and Kleber, H. D.,** Anti-endorphin effects of methadone, *Lancet*, 2, 972, 1980.

48. **Harbuz, M., Russell, J. A., Sumner, B. E. H., Kawata, M., and Lightman, S. L.,** Rapid changes in the content of proenkephalin A and corticotrophin releasing hormone mRNS in the paraventricular nucleus during morphine withdrawal in urethane-anaesthetized rats, *Mol. Brain Res.*, 9, 285, 1991.
49. **Pittman, Q. J., Hatton, J. D., and Bloom, F. E.,** Morphine and opioid peptide reduce paraventricular neuronal activity: studies on the rat hypothalamic slice preparation, *Proc. Natl. Acad. Sci. U.S.A.*, 77, 5527, 1980.
50. **Plotsky, P. M.,** Opioid inhibition of immunoreactive corticotropin-releasing factor secretion into the hypophyseal-portal circulation of rats, *Regul. Peptides*, 16, 235, 1986.
51. **Eisenberg, R. M.,** Effects of naloxone on plasma corticosterone in the opiate-naive rat, *Life Sci.*, 26, 935, 1980.
52. **Eisenberg, R. M.,** Effects of naltrexone on plasma corticosterone in opiate-naive rats: a central action, *Life Sci.*, 34, 1185, 1984.
53. **Ježova, D., Vigaš, M., and Jurcovičová, J.,** ACTH and corticosterone response to naloxone and morphine in normal, hypophysectomized and dexamethasone-treated rats, *Life Sci.*, 31, 307, 1982.
54. **Montilla, P., Toercillas, F., Varo, A., and Muñoz, M. C.,** Effects of various doses of naloxone on the plasma level of corticosterone in rats (in Spanish), *Rev. Esp. Fisiol.*, 40, 401, 1984.
55. **Gibson, A., Ginsburg, M., Hall, M., and Hart, S. L.,** The effects of opiate receptor agonists and antagonists on the stress-induced secretion of corticosterone in mice, *Br. J. Pharmacol.*, 65, 139, 1979.
56. **Siegel, R. A., Chowers, I., Conforti, N., Feldman, S., and Weidenfeld, J.,** Effects of naloxone on basal and stress-induced ACTH and corticosterone secretion in the male rat — site and mechanism of actions, *Brain Res.*, 249, 103, 1982.
57. **Blankstein, J., Reyes, F. I., Winter, J. S. D., and Faiman, C.,** Effects of naloxone upon prolactin and cortisol in normal women, *Proc. Soc. Exp. Biol. Med.*, 164, 363, 1980.
58. **Morley, J. E., Baranetsky, N. G., Wingert, T. D., Carlson, H. E., Hersham, J. M., Melmed, S., Levin, S. R., Jamison, K. R., Weitzman, R., Chang, R. J., and Varner, A. A.,** Endocrine effects of naloxone-induced opiate receptor blockade, *J. Clin. Endocrinol. Metab.*, 50, 251, 1980.
59. **Gaillard, R. C., Grossman, A., Smith, R., Rees, R. H., and Besser, G. M.,** The effects of a met-enkephalin analogue on ACTH, β-LPH, β-endorphin and met-enkephalin in patients with adrenocortical disease, *Clin. Endocrinol. (Oxford)*, 14, 471, 1981.
60. **Rivier, C., Vale, W., Ling, N., Brown, M., and Guillemin, R.,** Stimulation in vivo of the secretion of prolactin and growth hormone by β-endorphin, *Endocrinology*, 100, 238, 1977.
61. **Buckingham, J. C. and Hodges, J. R.,** The effects of opioid peptides on the secretion of corticotrophin releasing factor in the rat, *Br. J. Pharmacol.*, 70, 79P, 1980.
62. **Rittmaster, R. S., Cutler, G. B., Sobel, D. O., Goldstein, D.S., Koppelman, M. C. S., Loriaux, D. L., and Chrousos, G. P.,** Morphine inhibits the pituitary-adrenal response to ovine cortiocotrophin-releasing hormone in normal subjects, *J. Clin. Endocrinol. Metab.*, 60, 891, 1985.

63. **Hashimoto, K., Suemaru, S., Opo, N., Hattori, T., Inoue, H., Takao, T., Sugawara, M., Kageyama, J., and Ota, Z.,** Dual effects of (D-Ala^2,Met^5)-enkephalinamide on CRF and ACTH secretion, *Peptides,* 8, 113, 1987.
64. **Yajima, F., Suda, T., Tomori, N., Sumitomo, T., Nakagami, Y., Ushiyama, T., Demura, H., and Shizume, K.,** Effects of opioid peptides on immunoreactive corticotrophin-releasing factor release from the rat hypothalamus in vitro, *Life Sci.*, 39, 181, 1986.
65. **Buckingham, J. C.** Stimulation and inhibition of corticotrophin releasing factor secretion by beta-endorphin, *Neuroendocrinology*, 42, 148, 1986.
66. **Buckingham, J. C.,** Secretion of corticotrophin and its hypothalamic releasing factor in response to morphine and opioid peptides, *Neuroendocrinology*, 35, 111, 1982.
67. **Bodo de, R. D.,** The antidiuretic action of morphine and its mechanism, *J. Pharmacol. Exp. Ther.*, 82, 74, 1944.
68. **Rockhold, R. W., Crofton, J. T., Wang, B. C., and Share, L.,** Effect of intracarotid administration of morphine and naloxone on plasma vasopressin levels and blood pressure in the dog, *J. Pharmacol. Exp. Ther.*, 224, 386, 1983.
69. **Wu, W., Altszuler, N., Zbuzek, V., and Zbuzek, V. K.,** The effect of morphine on vasopressin release in vivo and in vitro, *Pharmacologist*, 25, 208, 1983.
70. **Iyengar, S., Kim, H. S., and Wood, P. L.,** Kappa opiate agonists modulate the hypothalamic-pituitary-adrenocortical axis in the rat, *J. Pharmacol. Exp. Ther.*, 238, 429, 1986.
71. **Iyengar, S., Kim, H. S., and Wood, P. L.,** μ-,δ-,κ- and ϵ-Opioid receptor modulation of the hypothalamo-pituitary-adrenocortical axis: subchronic tolerance studies of endogenous opioid peptides, *Brain Res.*, 435, 220, 1987.
72. **Eisenberg, R. M.,** Effect of ethylketocyclazocine and N-allyl-normetazocine on plasma corticosterone, *Pharmacologist*, 24, 115, 1982.
73. **Eisenberg, R. M.,** Plasma corticosterone changes in response to central and peripheral administration of kappa and sigma opiate agonists, *J. Pharmacol. Exp. Ther.*, 233, 683, 1985.
74. **Fuller, R. M. and Leander, J. D.,** Elevation of serum corticosterone in rats by bremazocine, a κ-opioid agonist, *J. Pharm. Pharmacol.*, 36, 345, 1983.
75. **Hayes, A. G. and Stewart, B. R.,** Effect of μ- and δ-opioid receptor agonists on rat plasma corticosterone level, *Eur. J. Pharmacol.*, 116, 75, 1985.
76. **Pechnick, R., George, R., and Poland, R. E.,** Identification of multiple opiate receptors through neuroendocrine responses. I. Effects of agonists, *J. Pharmacol. Exp. Ther.*, 232, 163, 1985.
77. **Pechnick, R., George, R., and Poland, R. E.,** Identification of multiple opiate receptors through neuroendocrine responses. II. Antagonists of mu, kappa and sigma agonists by naloxone and WIN 44,413–3, *J. Pharmacol. Exp. Ther.*, 232, 170, 1985.
78. **Pfeiffer, A., Herz, A., Loriaux, D. L., and Pfeiffer, D. S.,** Central kappa- and mu-receptors mediate ACTH release in rats, *Endocrinology*, 116, 2688, 1985.

79. **Slizgi, G. R. and Ludens, J. H.,** Studies on the nature and mechanism of diuretic activity of opioid analgesic ethylketocyclazocine, *J. Pharmacol. Exp. Ther.*, 220, 585, 1982.
80. **Leander, J. D.,** Further study of kappa opioids on urination, *J. Pharmacol. Exp. Ther.*, 227, 35, 1983.
81. **Nikolarakis, K., Pfeiffer, A., Stalla, G. K., and Herz, A.,** The role of CRF in the release of ACTH by opiate agonists and antagonists in rats, *Brain Res.*, 421, 373, 1987.
82. **Thierry, A. M., Tassin, J.-P., Blanc, G., and Glowinski, J.,** Selective activation of mesocortical DA system by stress, *Nature*, 263, 242, 1976.
83. **Fadda, F., Argiolas, A., Melis, M. R., Tissari, A. H., Onali, P. L., and Gessa, G. L.,** Stress-induced increase in 3,4-dihydroxyphenylacetic acid (DOPAC) levels in the cerebral cortex and in n. accumbens: reversal by diazepam, *Life Sci.*, 23, 2219, 1978.
84. **Lavielle, S., Tassin, J.-P., Thierry, A.-M., Blanc, G., Hervé, D., Barthelemy, C., and Glowinski, J.,** Blockade by benzodiazepines of the selective high increase in dopamine turnover induced by stress in mesocortical dopaminergic neurons of the rat, *Brain Res.*, 168, 585, 1978.
85. **Herman, J. P., Guillonneau, D., Dantzer, R., Scatton, B., Semerdjian-Rouquier, L., and Le Moal, W.,** Differential effects of inescapable footshocks and of stimuli previously paired with inescapable footshocks on dopamine turnover in cortical and limbic areas of the rat, *Life Sci.*, 30, 2207, 1982.
86. **Reinhard, J. R., Jr., Bannon, M. J., and Roth, R. H.,** Acceleration by stress of dopamine synthesis and metabolism in prefrontal cortex: antagonism by diazepam, *Naunyn-Schmiedeberg's Arch. Pharmacol.*, 318, 374, 1982.
87. **Claustre, Y., Rivy, J. P., Dennis, T., and Scatton, B.,** Pharmacological studies on stress-induced increase in frontal cortical dopamine metabolism in the rat, *J. Pharmacol. Exp. Ther.*, 238, 693, 1986.
88. **Tassin, J. P., Herve, D., Blanc, G., and Glowinski, J.,** Differential effects of a two minute open field session on dopamine utilization in the frontal cortices of BALBIC and C57BL/6 mice, *Neurosci. Lett.*, 17, 67, 1980.
89. **Louliot, A., Le Moal, M., and Simon, H.,** Differential reactivity of dopaminergic neurons in the nucleus accumbens in response to different behavioral situations. An *in vivo* voltametric study in free moving rats, *Brain Res.*, 397, 395, 1986.
90. **D'Angio, M., Serrano, A., Rivy, J. P., and Scatton, B.,** Tail-pinch stress increases extracellular DOPAC levels (as measured by in vivo voltametry) in the rat nucleus accumbens but not frontal cortex: antagonism by diazepam and zoldipem, *Brain Res.*, 409, 169, 1987.
91. **Blanc, G., Hervé, D., Simon, H., Lisoprawski, A., Glowinski, J., and Tassin, J. P.,** Response to stress of mesocortico-frontal dopaminergic neurones in rats after long-term isolation, *Nature*, 284, 265, 1980.
92. **Bannon, M. J., Elliott, P. J., Alpert, J. E., Goedert, M., Iversen, S. D., and Iversen, L. L.,** Role of endogenous substance P in stress-induced activation of mesocortical dopamine neurons, *Nature*, 306, 791, 1983.
93. **Deutsch, A. Y., Tam, S.-Y., and Roth, R. H.,** Footshock and conditioned stress increase 3,4-dihydroxyphenylacetic acid (DOPAC) in the ventral tegmental area but not substantia nigra, *Brain Res.*, 333, 143, 1985.

94. **Chiodo, L. A., Caggiula, A. R., Antelman, S. M., and Lineberry, C. G.,** Reciprocal influences of activating and immobilizing stimuli on the activity of nigro-striatal dopamine neurons, *Brain Res.*, 176, 385, 1979.
95. **Maeda, H. and Mogenson, G. J.,** Effects of peripheral stimulation on the activity of neurons in the ventral tegmental area, substantia nigra and midbrain reticular formation of rats, *Brain Res. Bull.*, 8, 7, 1982.
96. **Trulson, M. E. and Preussler, D. W.,** Dopamine containing ventral tegmental neurons in freely moving cats: activity during the sleep-waking cycle and effects of stress, *Exp. Neurol.*, 83, 367, 1984.
97. **Strecker, R. E. and Jacobs, B. L.,** Substantia nigra dopaminergic unit activity in behaving cats: effect of arousal on spontaneous discharges and sensory evoked activity, *Brain Res.*, 361, 339, 1985.
98. **Cabib, S., Puglisi-Allegra, S., and Oliverio, A.,** Chronic stress enhances apomorphine-induced stereotyped behavior in mice: involvement of endogenous opioids, *Brain Res.*, 298, 138, 1984.
99. **Cabib, S., Oliverio, A., and Puglisi-Allegra, S.,** Stress-induced decrease of 3-methoxytyramine in the nucleus accumbens of the mouse is prevented by naltrexone pretreatment, *Life Sci.*, 45, 1031, 1989.
100. **Antelman, S. M., Eichler, A. J., Black, C. A., and Kocan, D.,** Interchangeability of stress and amphetamine in sensitization, *Science*, 207, 329, 1980.
101. **Herman, J.-P., Stinus, L., and Le Moal, M.,** Repeated stress increases locomotor response to amphetamine, *Psychopharmacology*, 84, 431, 1984.
102. **Kalivas, P. W., Richardson-Carlson, R., and Van Orden, G.,** Cross-sensitization between foot-shock stress and enkephalin-induced motor activity, *Biol. Psychiatry*, 21, 939, 1986.
103. **Kameyama, T. and Nagasaka, M.,** Effects of apomorphine and diazepam is a quickly learned conditioned suppression in rats, *Pharmacol. Biochem. Behav.*, 17, 59, 1982.
104. **Katoh, A., Nabeshima, T., and Kameyama, T.,** Behavioral changes induced by stressful situations: effects of enkephalins, dynorphin, and their interactions, *J. Pharmacol. Exp. Ther.*, 253, 600, 1989.
105. **Piazza, P. V., Deminière, J.-M., Le Moal, M., and Simon, H.,** Factors that predict individual vulnerability to amphetamine self-administration, *Science*, 245, 1512, 1989.
106. **Piazza, P. V., Deminière, J.-M., Le Moal, M., and Simon,H.,** Stress- and pharmacologically-induced behavioral sensitization increases vulnerability to acquisition of amphetamine self-administration, *Brain Res.*, 514, 22, 1990.

9 Altered Genetic Expression of Endogenous Opioids During Addiction

I. INTRODUCTION

It is well established that ex-addicts are more prone to get addicted again than drug-naive subjects. It is also generally accepted that patients with addiction to any drug in their history have greater propensity to be "hooked on" another abused substance than subjects with no drug abuse in their history. Thus, once somebody has been addicted, the danger of relapse seems to be perpetuated "indefinitely". The same conclusion is dictated by the common clinical practice of keeping ex-addicts on some kind of maintenance therapy as long as possible.[1] Permanent, quasi-lifetime changes in human behavior indicate that the process of addiction might induce some kind of alterations at the genetic level.

Several scattered old clinical and experimental studies also indicate that the process of addiction leaves behind very long-lasting functional changes. In a fundamental study, Himmelsbach[2] already reported in 1942 that in ex-opiate addicts a period of 6 to 9 months is needed to complete recovery. Parameters such as body weight, basal metabolic rate, blood pressure, specific gravity of the blood and hematocrit reading did not normalize within the first months of withdrawal.[2] The opiate addiction-induced long-term behavioral changes are well known: irritability, excitability. It is worth mentioning that in ex-opiate addicts morphine induces insomnia: it reduces the duration of both REM and non-REM phases of sleep, i.e., it shortens its total duration.[3] Thus morphine derivatives, i.e., drugs named after a Roman deity supposed to generate dreams (Morpheus), induce insomnia in postaddicts. These findings indicate that in ex-addicts the behavioral changes brought about by opiates are permanently altered even long after cessation of their ingestion.

Similar findings have been obtained in animal experiments as well. Martin et al.[4] observed that in postaddict rats, at least a 3- to 6-month-long drug-free period was needed for normalization of body weight, fluid intake, body temperature, and spontaneous activity. It is also an old observation that rats addicted earlier to classical opiates, i.e., the ones later termed μ-agonists, remain more active[4] and react with intense hyperactivity to a repeated dose even after many months of withdrawal.[5,6] In good harmony with these findings, Cochin and Kornetsky[7] reported that after morphine withdrawal or just after administration of a single high dose of morphine, its analgesic effect, in hot plate test, was attenuated over a 9- to 12-month-long period. (The average life span of rats is 1 to 3 years.) Thus a single course of treatments or just a single high dose of μ-opiates can induce very long-lasting changes in the organism. Thus it is probably reasonable to suppose that such changes are not simply functional, i.e., opiates might induce some kind of alteration at the genetic level.

Such experiments have not been conducted with κ-agonists. Namely, as discussed in Chapter 4, they do not induce dependence, and the above mentioned alterations are characteristic of postdependent subjects.[5–7] Nevertheless, an interesting series of EEG studies provide some information regarding the differences in the long-term effects of μ- and κ-opiates, respectively. In drug-naive animals both morphine and ethylketocyclazocine induce biphasic changes in the spontaneous electrographic activity, and an initial phase of synchronization is followed by desynchronization.[8,9] Upon repeated opiate treatments the phase of initial synchronization gradually shortens and is replaced by "secondary" desynchronization. In morphine-treated animals, only the phase of desynchronization, i.e., the electrographic equivalent of motor hyperactivity, can be elicited even a month after cessation of morphine treatment.[8,9] However, in ethylketocyclazocine post-tolerant subjects, this κ-agonist elicits the same electrographic alterations as seen in drug-naive animals, i.e., it does not induce post-treatment hyperexcitability.[8,9]

These observations show that in postaddict subjects the mixed behavioral effects of opiates rather the excitatory ones prevail, indicating an eventual dopaminergic supersensitivity as discussed in Chapter 7. This phenomenon apparently only develops after μ-, but not after κ-agonists. On the other hand, the long-lasting elevation of hematocrit reading and blood specific gravity in ex-addict humans[2] and polydipsia in rats treated previously with morphine[4] point to lasting changes in water intake/diuresis, i.e., to alterations, which might probably be due to hyperactivity of the κ-opioid system.[10] As reviewed in the subsequent sections, there are more direct data indicating that chronic morphine treatment really induces hyperactivity of the κ-system, most probably by enhancement of the genetic expression of dynorphin.

II. DOPAMINERGIC REGULATION OF GENETIC EXPRESSION OF ENKEPHALINS

As discussed in Chapter 7, acute treatment with μ-opiates results in dopamine release from its mesolimbic terminals, and upon repeated treatments the dopamine-mediated behavioral hyperactivity becomes more and more intense. However, as discussed below, dopamine shifts the enkephalin production just in the opposite direction. As demonstrated first by Hong et al.,[11] upon repeated systemic treatment with various dopamine receptor blockers (haloperidol, pimozide, chlorpromazine) the enkephalin content increases in the nucleus accumbens and the striatum, but not in the hypothalamus, the septum, or the medulla oblongata. Single injection with neuroleptics has no effect.[11] This action of haloperidol is due to increased biosynthesis of enkephalins and to an eventual inhibition of their release.[12] The role of genetic alterations in increased synthesis is proven by enhancement by neuroleptics of the (pre)proenkephalin mRNA production in the striatum.[13–18] In most of these studies, only the striatum was examined, or it was not separated from the n. accumbens for obvious technical reasons. Nevertheless, in the ones where n. accumbens and the striatum were separated[11,16] the changes were the same. Thus probably it might be supposed that the effect of these drugs on A_9 neurons in the substantia nigra and A_{10} neurons in the ventral tegmental area is similar.

Administration of reserpine, i.e., depletion of the brain monoamines[15,19] or destruction of the dopamine neurons by 6-hydroxydopamine injected i.c.v.[20] or directly into the substantia nigra[21] results in a similar increase in the mRNA content coding for enkephalin. Thus the expression of enkephalins (transcription of the preproenkephalin gene) seems to be under a tonic dopaminergic inhibition in the nigrostriatal and mesolimbic dopamine systems. In agreement with these findings, bromocriptine has been found to decrease the striatal mRNA content.[15] Apomorphine, *per se,* had no effect in one study over a 9-d period of administration,[15] but it did reduce the levels of both Met- and Leu-enkephalin if given for 1 or 3 months[18] and it reversed the effects of haloperidol and other dopamine receptor blockers.[18] Scopolamine also reverses the effect of haloperidol, hence the possibility arises that the effect might be mediated by cholinergic interneurons.[22] On the other hand, the data available on the role of two subtypes of dopamine receptors are contradictory. Namely, in some studies[23,24] SCH 23390, a selective $dopamine_1$ receptor blocker, induced an increase in the enkephalin content as seen above with the nonselective antagonists, whereas sulpiride, a selective antagonist of $dopamine_2$ receptor, decreased the Met-enkephalin content[23] or it had no effect at all.[24] However, in another study SCH 23390 just decreased the striatal Met-enkephalin content.[16] Truly, in the latter study[16] the dose of SCH 23390 was much higher than in the other studies,[23,24] thus its effect might not

TABLE 1
Dopaminergic Regulation of Genetic Expression of Enkephalins

Species	Experimental paradigm	Effect observed	Ref.
Rat	Effect of long-term treatment with various dopamine receptor blockers such as haloperidol, chlorpromazine, or pimozide on enkephalin content of striatum and n. accumbens	Elevation	11,12
	Effect of long-term treatment with various dopamine receptor blockers on the production of (pre)proenkephalin mRNA in the striatum (eventually in n. accumbens)	Increased level, i.e., enhanced transcription	13–18
	Effect of depletion of brain dopamine by reserpine or 6-hydroxydopamine on the striatal mRNA encoding for enkephalins	Increased level	15, 19–21
	Effect of a selective D_1 receptor antagonist, SCH-23390, on the level of enkephalin or mRNA coding for it in the		
	Striatum	Increase	23,24
	N. accumbens	Decrease	16
	Effect of selective D_2 receptor antagonist, sulpiride, on enkephalin production	Decrease No effect	23,24
	Effect of apomorphine on enkephalin production in the striatum alone and in combination with SCH-23390	No effect per se, but reversal of the increase induced by SCH-23390	15,18, 24

have been selective enough. Furthermore, in the majority of the related studies (see Chapters 5 and 7) the role of D_1 (dopamine) receptors seemed to predominate.

Unfortunately, for the time being no data are available on the eventual effects of abused dopamine-releasing drugs (amphetamine, cocaine, etc.) on the biosynthesis/genetic expression of natural enkephalins.

Nevertheless, as a whole these observations indicate that dopamine in the nigrostriatal and mesolimbic dopamine systems exerts a tonic inhibitory effect on enkephalin synthesis at the level of transcription, and the effect is mediated primarily by the D_1 subtype of dopamine receptors (Table 1).

III. DOPAMINERGIC REGULATION OF GENETIC EXPRESSION OF DYNORPHIN

Another line of data indicates that in the same circuitry dopamine rather facilitates the synthesis of dynorphin. With the exception of one group,[25] all authors found that chronic haloperidol does not modify the dynorphin level (or prodynorphin mRNA) in the striatum.[16,26–29] In addition, both SCH 23390 and 6-hydroxydopamine lesion of the mesostriatal pathway failed to modify the dynorphin synthesis.[16] However, in another study[24] nigral injection of 6-hydroxydopamine resulted in simultaneous depletion of striatal dopamine and a decrease in its dynorphin content. On the other hand, apomorphine enhanced the dynorphin synthesis (the amount of the mRNA of preprodynorphin) in the striatum.[24,27,28] Examining various brain regions, the elevation was the most conspicuous in the striatum and substantia nigra, and a less robust elevation was seen in the nucleus accumbens and amygdala, but no change was detected in the hypothalamus, the hippocampus, and the frontal cortex.[28] The apomorphine-induced elevation of dynorphin could be antagonized by simultaneous administration of haloperidol[24,27] and by SCH 23390, a dopamine$_1$ receptor antagonist, but not by sulpiride, a selective dopamine$_2$ receptor blocker.[24] Examining the effects of specific agonists of the dopamine receptors, SKF 38393, a dopamine$_1$ receptor agonist, induced similar increase in the striatal dynorphin content, but LY-171555, a selective dopamine$_2$ receptor agonist, had no effect.[24]

The facilitation by dopamine agonist apomorphine of prodynorphin synthesis and the ineffectiveness of haloperidol to induce opposite changes, per se, indicate that

- In the nigrostriatal and mesolimbic dopamine system prodynorphin synthesis can be facilitated by enhanced production of dopamine.
- However, contrary to the synthesis of enkephalins, it is not proven that prodynorphin production is also tonically regulated by dopamine.
- Dopamine$_1$ and not dopamine$_2$ receptors seem to mediate this action of dopamine as well.

Based on these findings it is not surprising that repeated (but not acute) administration of dopamine-releasing (reuptake blocking) psychostimulants such as amphetamine[28,30] and cocaine[31,32] have also been shown to elevate the transcription of (pre)prodynorphin. In these studies, too, mostly the striatum was examined, but a similar increase was detected in the n. accumbens and olfactory tubercle in the studies where these minuscule brain regions were separated from the striatum.[30] Acute administration of amphetamine results in a decrease in the striatonigral dopamine content,[29] indicating that dopamine agonists induce an immediate release of dynorphin and the enhanced synthesis is secondary to its more intense liberation.

In a recent study, MK-801, a noncompetitive NMDA receptor antagonist, blocked the cocaine-induced elevation of striatal dynorphin,[32] indicating that glutamate might mediate the effect of dopamine.

These experiments prompted Trujillo and co-workers[29,30] to propose that the increased concentration of dynorphin might contribute to the aver-

TABLE 2
Dopaminergic Regulation of Genetic Expression of Dynorphin

Species	Experimental paradigm	Effect observed	Ref.
Rat	Effect of haloperidol on the dynorphin level or prodynorphin mRNA content in the striatum (eventually n. accumbens)	No effect	15, 26–29
	Effect of apomorphine on prodynorphin mRNA in various brain regions	Increase in striatum, n. accumbens, amygdala	24,28
	Modulation of prodynorphin mRNA level by		24
	SKF 38393, a D_1 agonist	Increase	
	LY-171555, a D_2 agonist	No effect	
	Effect of repeated amphetamine treatment on the level of prodynorphin mRNA in the striatum and n. accumbens	Increase	28,30
	Effect of repeated cocaine treatment on the level of prodynorphin mRNA in the striatum	Increase	31.32

sive effect of withdrawal of psychostimulants. While the reviewer shares this view, based on the data reviewed in Chapters 4 and 5, caution is warranted since:

- In the majority of these studies the striatum was examined, though there is reason to believe that in many cases it was not separated from the nucleus accumbens, located just along its caudal border.
- The aversive character of opiate withdrawal cannot be interpreted based solely on these data. Nevertheless, as discussed below, some data indicate that similar changes are elicited by classical μ-opiates (Table 2).

IV. MODULATION BY CLASSICAL OPIATES OF THE SYNTHESIS OF ENKEPHALINS, ENDORPHINS AND DYNORPHIN

The eventual modulation by exogenous opiates of the production of their endogenous counterparts is one the oldest issues in enkephalin/endorphin research. Kosterlitz and Hughes[33] first proposed in 1975 that during tolerance development the release of enkephalins is reduced via a negative feedback mechanism exerted by exogenous opiates. In the subsequent year,

Simantov and Snyder[34] reported that acute morphine treatment did not modify the whole brain enkephalin content, but its chronic administration resulted in significant elevation of its level. The unexpected finding was explained by supposing that this increase was due to a decrease in enkephalin release brought about by activation via morphine of feedback mechanisms.[34] Subsequently, three other groups[35–37] failed to replicate this initial finding. Examining various brain regions separately, the level of Met-enkephalin was found decreased in the striatum upon repeated morphine treatment in some studies[38,39] and naloxone induced just an elevation.[40] However, these early studies are difficult to interpret today, since either the total opioid content was determined by receptor displacement techniques or the "enkephalin immunoreactivity" may have included several other peptides which contain Met- or Leu-enkephalin sequences. Actually, most of these studies were carried out before clarification of the exact structure of proopiomelanocortin, proenkephalin, and the discovery of dynorphin. Obviously, reliable answers can only be given either by measuring in various brain regions the release of endogenous opioids directly and/or by determining separately the levels of different kinds of mRNAs, the ones coding for enkephalins, dynorphin-related peptides, and proopiomelanocortin, respectively.

A further confounding factor is the eventual up- and downregulation of opiate receptors upon chronic treatment with opiate antagonists and agonists, respectively. Whereas the eventual downregulation by agonists of various subclasses of opiate receptors is a highly disputed issue (for review see Székely),[41] it is well established that chronic treatment with naltrexone results in an increase in the B_{max} of μ- and δ-receptors.[42–45] It must be added that naltrexone failed to modify the B_{max} of κ-receptors,[44] though it is a nonspecific antagonist of all three classes of opiate receptors.

In the latter studies[42–43] the most robust increase was seen in the ventral tegmental area, other regions of the mesolimbic circuitry, the neocortex, the striatum, the substantia nigra, and the hypothalamus, i.e., mostly in structures involved in the rewarding circuitry of the brain (see Chapter 6).

Returning to the modulation by exogenous opioids of the production of their endogenous counterparts, the data indicate that the biosynthesis and release of enkephalins, endorphins, and dynorphin-related peptides are regulated differentially.

In experiments recording the release of endogenous opioids, mostly Met-enkephalin was examined. *In vitro,* in longitudinal muscle myenteric plexus of guineapig ileum the electrical stimulation-induced release of enkephalin is suppressed by morphine and enhanced by naloxone as expected on the basis of negative feedback via autoreceptors.[46] Similar observations have been made by measuring the spontaneous or electrical stimulation-induced release of immunoreactive Met-enkephalin using the technique of ventriculocisternal and spinal superfusion of anesthetized cats.[47,48] The K^+-induced release of Met-enkephalin from rat brain stem is also suppressed by morphine and enhanced by naloxone.[49] Using the same preparation to compare the effects of various receptor-selective agonists, dynorphin and U-50488H have been found the most effective,[50] which indicates that primarily κ-receptors might mediate the presynaptic feedback inhibition of Met-enkephalin release. Naloxone also enhanced the Met-enkephalin release from rat hypothalamic slice.[51] However, both morphine and naloxone failed to modify the K^+-induced release of Met- and Leu-enkephalins from the rat striatum.[52–54] Thus the autoreceptor-mediated feedback inhibition might not be present everywhere in the brain. Unfortunately, there are no data on the eventual autofeedback in the mesolimbic dopamine system. But the data to be reviewed below are easier to interpret by supposing that the negative feedback-induced inhibition of enkephalin release also is missing in the mesolimbic circuitry.

As for the other opioid peptides, morphine decreases the release of β-endorphin into the hypothalamo-hypophyseal blood of rats.[55] In hypothalamic slice preparation naloxone enhances both the spontaneous and corticotropin-releasing factor-induced release of β-endorphin and dynorphin.[56] Examining the effects of receptor subtype-selective antagonists in the same preparation (hypothalamic slice), the same group found that Met-enkephalin release was only enhanced by δ-specific ICI 174864, but the liberation of β-endorphin was potentiated by both this δ-ligand and (D-Tic[1])CTAP, a μ-antagonist.[57] Furthermore, the liberation of dynorphin could be augmented by this μ-ligand and nor-binaltorphimine, a selective κ-ligand as well.[57]

Of course, the most reliable measure of biosynthesis is the corresponding mRNA content. As for enkephalins, the subsequent studies, based on measurement of their mRNA level ultimately confirm the original postulate of Kosterlitz and Hughes.[33] The preproenkephalin gene expression was examined first by Uhl et al.,[58] who found it decreased in the striatum of morphine tolerant rats. (In this and most subsequent studies reviewed below, preproenkephalin mRNA or the other kinds of mRNA content was determined by hybridization of mRNA with the corresponding cDNA and subsequent Northern blot analysis.) In another related study, the morphine induced a decrease in Met-enkephalin content in the striatum, but the effect was not significant.[59] Finally, 9-d continuous s.c. infusion of FK 33–824, a highly potent enkephalin analog, resulted in a 20% reduction of the striatal Met-enkephalin concentration.[15]

On the other hand, 8-d continuous naltrexone infusion (implantation of two 30-mg pellets s.c.) resulted in a dramatic, 12-fold increase in the striatal preproenkephalin mRNA content and small increases in the hippocampus and hypothalamus, but no change was seen in the frontal cortex.[60] All these studies confirm that the expression of enkephalin might be under a tonic opioidergic control, primarily in the striatum. The receptor subtype involved in this feedback effect is not clarified yet. In an interesting study the enkephalin coding mRNA synthesis was also enhanced upon intrastriatal injection of β-chlornaltrexamine, but basically unaffected upon β-funaltrexamine.[61] Since β-chlornaltrexamine is a nonselective, irreversible alkilator of opiate receptors, whereas β-funaltrexamine selectively alkilates the μ-receptors,[62] the conclusion is simple. Non-μ-receptors might mediate the negative feedback effect of opiates. Since few data point to the δ-receptors, it is obviously the κ-ones which might exert a direct inhibitory feedback action on the biosynthesis of enkephalins.

Thus dynorphins might not be solely functional antagonists of enkephalins; they even seem to inhibit the genetic expression of the latter ones. However tempting this hypothesis is, obviously no such far-reaching conclusion can be drawn from a single study.

A further way of interaction might be modulation of the dopamine release. μ-Opiates enhance the release of dopamine in the striatum which might be mediated by opiate receptors located on the terminals of nigrostriatal dopaminergic terminals[63–65] and, as discussed in Chapters 5 and 7, the same might apply to the mesolimbic dopamine terminals as well. As discussed in Section II, the expression of enkephalins is under tonic dopaminergic control. Thus the negative feedback effect of enkephalins/classical opiates might be brought about indirectly due to their dopamine-releasing action.

Contrary to enkephalins, dynorphin production is enhanced by opiates. Weissman and Zamir[66] first reported that 3-d heroin treatment led to an increase in the dynorphin-B content of the globus pallidus and the ventral tegmental area. A significant increase in the striatal concentration of dynorphin-A and -B has been reported by another group as well.[30,59] In good analogy with these findings, levorphanol and fentanyl were found to elevate the prodynorphin-related peptide levels in guineapig myenteric plexus.[67] Thus, excitation of the μ-receptors by classical opiates seems to facilitate the expression of dynorphin, the endogenous ligand of κ-receptors. Finally, chronic ethylketocyclazocine treatment was reported to reduce the prodynorphin mRNA level in the hippocampus and hypothalamus.[68] Being an agonist of κ-receptors, its effect in the opposite direction probably does not contradict the other findings, though the mesolimbic-striatal structures have not been examined in this work.

As for the proopiomelanocortin-derived peptides, i.e., β-endorphin, its transcription was not significantly modified by morphine in the hypothalamus.[59] However, the same group reported subsequently that chronic morphine apparently inhibits the release of β-endorphin from the hypothalamus, and its resulting accumulation leads to a secondary decline of proopiomelanocortin mRNA production.[69] Thus exogenous opiates seem to modulate, inhibit the transcription of β-endorphin indirectly, via inhibition of its release (Table 3).

TABLE 3
Modulation by Classical Opiates of the Synthesis and Release of Enkephalins, Endorphins and Dynorphin

Species	Experimental paradigm[a]	Effect observed	Ref.
Guinea pig	Modulation of electrical stimulation-induced release of Met-enkephalin by		46
	Morphine	Reduction	
	Naloxone	Enhancement	
	(acute drug treatment)		
Cat (anesthetized)	Modulation by morphine of Met-enkephalin release into superfusate of spinal cord and perfusate of the aqueduct (acute drug treatment);	Reduction	48
	Effect of naloxone after long-term morphine treatment	Enhancement	47
Rat	Modulation of the release of Met-enkephalin from brain stem slice preparation by		49,50
	Morphine, U-50488H; DAGO,[b] DPDPE[c]	Reduction	
	Naloxone, MR2266,[d] ICI 154129[e]	Enhancement	
	(acute drug treatment)		
	Modulation by naloxone Met-enkephalin, β-endorphin and dynorphin release from hypothalamic slice preparation (acute drug treatment)	Enhancement	51
	Modulation of K^+-induced release of Met- and Leu-enkephalin from striatal slice by	No effect by any of them	52–54
	Morphine		
	Naloxone		
	(acute drug treatment)		
	Effect of chronic morphine treatment on striatal preproenkephalin mRNA level	Decrease	58,59
	Effect of chronic treatment with FK 33-824[f] on striatal preproenkephalin mRNA level	Decrease	15
	Modulation of the striatal preproenkephalin mRNA level by		
	Systemic naltrexone	Elevation	42,43
	Intrastriatal β-CNA[g]	Elevation	61
	Intrastriatal β-FNA[h]	No effect	61
	Effect of cronic treatment with morphine, heroin on striatal mRNA level coding for dynorphin	Elevation	30,69, 66

TABLE 3 (continued)
Modulation by Classical Opiates of the Synthesis and Release of Enkephalins, Endorphins and Dynorphin

Species	Experimental paradigm[a]	Effect observed	Ref.
	Effect of cronic morphine treatment on β-endorphin level in the hypothalamus	Transient increase due to supposed reduction in release	59
Guinea pig	Effect of single *in vivo* drug treatment on immunoreactive dynorphin level in the myenteric plexus After levorphanol: After naloxone:	Time-dependent increase after both kinds of treatment; combination results in additive effect	67

[a] Drug treatment applied is chronic, unless indicated otherwise.
[b] DAGO = (D-Ala[2],NMePhe[4],Gly[5])-enkephalin-ol, a selective μ-agonist.
[c] DPDPE = (D-Pen[2],D-Pen[5])-enkephalin, a selective δ-agonist.
[d] MR2266 = a selective κ-antagonist.
[e] ICI 154129 = a selective δ-antagonist.
[f] FK-33824 = (D-Ala[2],NMePhe[4],Met(O)[5])-enkephalin-ol, a selective μ-agonist.
[g] β-CNA = β-chlornaltrexamine, a nonselective irreversible opiate antagonist.
[h] β-FNA = β-funaltrexamine, a μ-selective irreversible opiate antagonist.

V. SUMMARY: A POSSIBLE GENETIC MECHANISM OF PERPETUATION OF DRUG ADDICTION

Summarily, these data indicate that exogenous opiates exert a complex action on endogenous opioid production which involves facilitation of dynorphin production, probably an inhibition of enkephalin synthesis and suppression of β-endorphin release with a secondary reduction of its synthesis. Namely, as summarized in Figure 1, nearly all the data reviewed above indicate that upon chronic opiate treatment the balance of enkephalins (+endorphins) vs. dynorphin is shifted in favor of the latter one. As discussed in Sections I and II, actually similar genetic changes are induced by direct release of the mesolimbic and striatal dopamine upon amphetamine and cocaine administration. Thus, whether opiates or psychostimulants are administered, the final outcome seems to be similar: a hypertrophy of the dynorphin-producing system and downregulation of enkephalin/β-endorphin biosynthesis mediated by enhancement of dopamine release during consumption of drugs of abuse. Upon interruption of drug consumption the enhanced dynorphin release + reduction of enke-

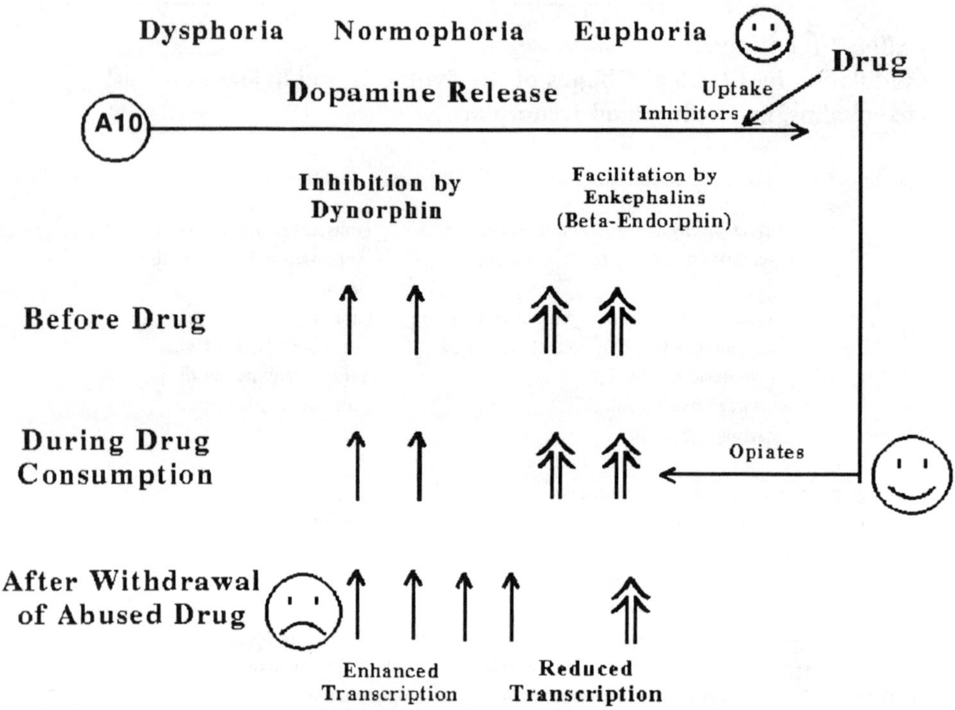

FIGURE 1. A proposed model of genetic alterations during drug addiction. As illustrated in the figure, μ-opiates are supposed to activate the dopamine-releasing A_{10} neurons via interneurons either at their cell bodies in the ventral tegmental area or at the level of the axon terminals in the nucleus accumbens. On the other hand, amphetamines, such as cocaine, are known to act directly on the terminals of these neurons, inducing dopamine release or inhibiting its reuptake. Whichever is their way of activation, the dopamine released from these neurons inhibits the transcription of enkephalins (and β-endorphin) and enhances the dynorphin production at genetic level. The relative overproduction of dynorphin and reduced release of enkephalins shift the mood towards dysphoria. In turn, the resulting dysphoria prompts the addicts to keep consuming substances of abuse in order to overcome the ''withdrawal dysphoria''. Hence the genetic expression of enkephalins is permanently suppressed and that of dynorphin will be enhanced. Thus the abuse of drugs becomes ''genetically'' perpetuated.

phalin production induces dysphoria. Then the resulting dysphoria prompts the addicts to resort to drugs again to restore their mood. Thus the dopamine-mediated shift in genetic expression of dynorphin and enkephalin might perpetuate the process of drug addiction (see also Figure 1 for visual illustration of the process).

REFERENCES

1. **Martin, W. R., Jasinski, D. R., Haertzen, C. A., Kay, D. C., Jones, B. E., Mansky, P. A., and Carpenter, R. W.,** Methadone — a reevaluation, *Arch. Gen. Psychiatr.*, 28, 286, 1973.
2. **Himmelsbach, C. K.,** Clinical studies of drug addiction. Physical dependence, withdrawal and recovery, *Arch. Int. Med.*, 69, 766, 1942.
3. **Kay, D. C., Eisenstein, R. B., and Jasinski, D. R.,** Morphine effects on human REM state, waking state and NREM sleep, *Psychopharmacology*, 14, 404, 1969.
4. **Martin, W. R., Wikler, A., Eades, C. G., and Pescor, F. T.,** Tolerance to and dependence on morphine in rats, *Psychopharmacology*, 4, 247, 1963.
5. **Babbini, M., Gaiardi, M., and Bartoletti, M.,** Persistence of chronic morphine effects upon activity in rats 8 months after ceasing treatment, *Neuropharmacology*, 14, 611, 1975.
6. **Brady, L. S. and Holtzman, S. G.,** Locomotor activity in morphine-dependent and post-dependent rats, *Pharmacol. Biochem. Behav.*, 14, 361, 1981.
7. **Cochin, J. and Kornetsky, C.,** Development and loss of tolerance to morphine in single and multiple injections, *J. Pharmacol. Exp. Ther.*, 145, 1, 1964.
8. **Young, G. A. and Khazan, N.,** Differential tolerance and cross-tolerance to repeated daily injections of mu and kappa opioid agonists in the rat, *Neuropharmacology*, 23, 505, 1984.
9. **Young, G. A. and Khazan, N.,** Differential protracted effects of morphine and ethylketocyclazocine challenges on EEG and behavior in the rat, *Eur. J. Pharmacol.*, 125, 265, 1986.
10. **Leander, J. D., Zerbe, R. L., and Hart, J. C.,** Diuresis and suppression of vasopressin by kappa opioids: comparison with mu and delta opioids and clonidine, *J. Pharmacol. Exp. Ther.*, 234, 463, 1985.
11. **Hong, J. S., Yang, H.-Y. T., Fratta, W., and Costa, E.,** Rat striatal methionine-enkephalin content after chronic treatment with cataleptogenic and noncataleptogenic antischizophrenic drugs, *J. Pharmacol. Exp. Ther.*, 205, 141, 1978.
12. **Hong, J. S., Yang, H.-Y. T., Gillin, J. C., Di Giulio, A. M., Fratta, W., and Costa, E.,** Chronic treatment with haloperidol accelerates the biosynthesis of enkephalins in rat striatum, *Brain Res.*, 160, 192, 1979.
13. **Sabol, S. L., Yoshikawa, K., and Hong, J.-S.,** Regulation of methionine-enkephalin precursor messenger RNA in rat striatum by haloperidol and lithium, *Biochem. Biophys. Res. Commun.*, 113, 391, 1983.
14. **Tang, F., Costa, E., and Schwartz, J. P.,** Increase of proenkephalin mRNA and enkephalin content of rat striatum after daily injection of haloperidol for 2 to 3 weeks, *Proc. Natl. Acad. Sci. U.S.A.*, 80, 3841, 1983.
15. **George, S. R. and Kertesz, M.,** Met-enkephalin concentrations in striatum respond reciprocally to alterations in dopamine neurotransmission, *Peptides*, 8, 487, 1987.

16. **Morris, B. J., Höllt, V., and Herz, A.,** Dopaminergic regulation of striatal proenkephalin mRNA and prodynorphin mRNA: contrasting effects of D_1 and D_2 antagonists, *Neuroscience*, 25, 525, 1988.
17. **Petrack, B., Emmett, M. R., Rao, T. S., Kim, H. S., and Wood, P. L.,** Increases in rat striatal preproenkephalin mRNA levels following chronic treatment with the depot neuroleptic, haloperidol decanoate, *Life Sci.*, 46, 687, 1990.
18. **Herman, Z. S., Huzarska, M., Kmieciak-Kolada, K., and Kowalski, J.,** Chronic treatment with chlorpromazine, thioridazine or haloperidol increases striatal enkephalins and their release from rat brain, *Psychopharmacology*, 104, 106, 1991.
19. **Mocchetti, I., Guidotti, A., Schwartz, J., and Costa, E.,** Reserpine changes the dynamic state of enkephalin stores in rat striatum and adrenal medulla by different mechanisms, *J. Neurosci.*, 5, 3379, 1985.
20. **Sivam, S. P., Breese, G. R., Napier, T. C., Müller, R. A., and Hong, J.-S.,** Dopaminergic regulation of proenkephalin-A gene expression in the basal ganglia, *NIDA Res. Monogr.*, 75, 389, 1986.
21. **Angulo, J. A., Davis, L. G., Burkhart, B. A., and Christoph, G. R.,** Reduction of striatal dopaminergic neurotransmission elevates striatal proenkephalin mRNA, *Eur. J. Pharmacol.*, 130, 341, 1986.
22. **Hong, J. S., Yang, H.-Y. T., Gillin, J. C., and Costa, E.,** Effect of long-term administration of antipsychotic drugs on enkephalinergic neurons, *Adv. Biochem. Psychopharmacol.*, 24, 223, 1980.
23. **Mocchetti, I., Naranjo, J. R., and Costa, E.,** Regulation of striatal enkephalin turnover in rats receiving antagonists of specific dopamine receptor subtypes, *J. Pharmacol. Exp. Ther.*, 241, 1120, 1987.
24. **Jiang, H.-K., McGinty, J. F., and Hong, J. S.,** Differential modulation of striatonigral dynorphin and enkephalin by dopamine receptor subtypes, *Brain Res.*, 507, 57, 1990.
25. **Quirion, R., Gaudreau, P., Martel, J.-C., St-Pierre, S., and Zamir, N.,** Possible interactions between dynorphin and dopaminergic systems in rat basal ganglia and substantia nigra, *Brain Res.*, 331, 358, 1985.
26. **Peterson, M. R. and Robertson, H. A.,** Effect of dopaminergic agents on levels of dynorphin 1–8 in rat striatum, *Prog. Neuropsychopharmacol. Biol. Psychiatr.*, 8, 725, 1984.
27. **Li, S. J., Sivam, S. P., and Hong, J. S.,** Regulation of the concentration of dynorphin A_{1-8} in the striato-nigral pathway by the dopaminergic system, *Brain Res.*, 398, 390, 1986.
28. **Li, S. J., Sivam, P., McGinty, J. F., Jiang, H. K., Douglass, J., Calavetta, L., and Hong, J. S.,** Regulation of the metabolism of striatal dynorphin by the dopaminergic system, *J. Pharmacol. Exp. Ther.*, 246, 403, 1988.
29. **Trujillo, K. A., Day, R., and Akil, H.,** Regulation of the striato-nigral prodynorphin peptides by dopaminergic agents, *Brain Res.*, 518, 244, 1990.
30. **Trujillo, K. A. and Akil, H.,** Changes in prodynorphin peptide content following treatment with morphine and amphetamine: possible role in mechanisms of action of drug of abuse, *NIDA Res. Monogr.*, 95, 550, 1990.
31. **Sivam, S. P.,** Cocaine selectively increases striatonigral dynorphin levels by a dopaminergic mechanism, *J. Pharmacol. Exp. Ther.*, 250, 818, 1989.

32. **Johnson, M., Bush, L. G., Gibb, J. W., and Hanson, G. R.,** Role of N-methyl-D-aspartate (NMDA) receptors in the response of extrapyramidal neurotensin and dynorphin A systems to cocaine and GBR 12909, *Biochem. Pharmacol.*, 41, 649, 1991.
33. **Kosterlitz, H. W. and Hughes, J.,** Some thoughts on the significance of enkephalin, the endogenous ligand, *Life Sci.*, 17, 91, 1975.
34. **Simantov, R. and Snyder, S. H.,** Elevated levels of enkephalin in morphine-dependent rats, *Nature*, 262, 505, 1976.
35. **Fratta, W., Yang, H.-Y. T., Hong, J., and Costa, E.,** Stability of Met-enkephalin content in brain structures of morphine-dependent or foot shock-stressed rats, *Nature*, 268, 452, 1977.
36. **Childers, S., Simantov, R., and Snyder, S. H.,** Enkephalin: radio-immunoassay and radioreceptor assay in morphine dependent rats, *Eur. J. Pharmacol.*, 46, 289, 1977.
37. **Wesche, D., Höllt, V., and Herz, A.,** Radioimmunoassay of enkephalins. Regional distribution in rat brain after morphine treatment and hypophysectomy, *Naunyn-Schmiedeberg's Arch. Pharmacol.*, 301, 79, 1977.
38. **Bergström, L. and Terenius, L.,** Enkephalin levels decrease in rat striatum during morphine abstinence, *Eur. J. Pharmacol.*, 60, 349, 1979.
39. **Przewlocki, R., Höllt, V., Duka, T., Kleber, G., Gramsch, C., Haarmann, I., and Herz, A.,** Long-term morphine treatment decreases endorphin levels in rat brain and pituitary, *Brain Res.*, 174, 357, 1979.
40. **Wüster, M., Schulz, R., and Herz, A.,** Inquiry into enkephalinergic feedback mechanisms during the development of opiate tolerance/dependence, *Brain Res.*, 189, 403, 1980.
41. **Székely, J. I.,** Induction of tolerance to and dependence on opioid peptides, in *Opioid Peptides*, Vol. IV, Biochemistry and Applied Physiology, Székely, J. I. and Ramabadran, K., Eds., CRC Press, Boca Raton, FL, 1990, 253.
42. **Zukin, R. S., Sugarman, J. R., Fitz-Syage, M. L., Gardner, E. L., Zukin, S. R., and Gintzler, A. R.,** Naltrexone-induced opiate receptor supersensitivity, *Brain Res.*, 245, 285, 1982.
43. **Tempel, A., Gardner, E. L., and Zukin, R. S.,** Visualization of opiate receptor upregulation by light microscopy autoradiography, *Proc. Natl. Acad. Sci. U.S.A.*, 81, 3893, 1984.
44. **Tempel, A., Garner, E. L., and Zukin, R. S.,** Neurochemical and functional correlates of naltrexone-induced opiate receptor up-regulation, *J. Pharmacol. Exp. Ther.*, 232, 439, 1985.
45. **Yoburn, B. C., Goodman, R. R., Cohen, A. H., Pasternak, G. W., and Inturrisi, C. E.,** Increased analgesic potency of morphine and increased brain opioid binding sites in the rat following chronic naltrexone treatment, *Life Sci.*, 36, 2325, 1985.
46. **Glass, J., Chan, W. C., and Gintzler, A. R.,** Direct analysis of the release of methionine-enkephalin from guinea pig myenteric plexus: modulation by endogenous opioids and exogenous morphine, *J. Pharmacol. Exp. Ther.*, 239, 742, 1986.
47. **Jhamandas, K., Yaksh, T. L., and Go, V. L. W.,** Acute and chronic morphine modifies the in vivo release of methionine-enkephalin-like immunoreactivity from the cat spinal cord and brain, *Brain Res.*, 297, 91, 1984.

48. **Yaksh, T. L. and Elde, R. P.,** Factors governing release of methionine enkephalin-like immunoreactivity from mesencephalon and spinal cord of the cat in vivo, *J. Neurochem.*, 46, 1056, 1981.
49. **Ueda, H., Fukushima, N., Kitao, T., Ge, M., and Takagi, H.,** Low doses of naloxone produce analgesia in the mouse brain by blocking presynaptic autoinhibition of the enkephalin release, *Neurosci. Lett.*, 65, 247, 1986.
50. **Ueda, H., Fukushima, N., Ge, M., Takagi, H., and Satoh, M.,** Presynaptic opioid κ-receptor and the regulation of the release of Met-enkephalin in the rat brainstem, *Neurosci. Lett.*, 81, 309, 1987.
51. **Nikolarakis, K. E., Almeida, O. F. X., and Herz, A.,** Multiple factors influencing the in vitro release of [Met^5]-enkephalin from rat hypothalamic slices, *J. Neurochem.*, 52, 428, 1989.
52. **Richter, J. A., Wesche, D. L., and Frederickson, R. C. A.,** K^+-Stimulated release of Met-and Leu-enkephalin from rat striatal slices: lack of effect of morphine and naloxone, *Eur. J. Pharmacol.*, 56, 105, 1979.
53. **Osborn, H. O. and Herz, A.,** K^+-Evoked release of Met-enkephalin from rat striatum in vitro: effect of putative neurotransmitters and morphine, *Naunyn-Scmiedeberg's Arch. Pharmacol.*, 310, 203, 1980.
54. **Sawynok, J., Labella, F. S., and Pinsky, C.,** Effects of morphine and naloxone on K^+-stimulated release of methionine-enkephalin from slices of rat striatum, *Brain Res.*, 189, 483, 1980.
55. **Koenig, J., Meltzer, H., and Gudelsky, G.,** Morphine and capsaicin administration alters the secretion of β-endorphin into the hypophyseal vasculature of the rat, *Neuroendocrinology*, 43, 611, 1986.
56. **Nikolarakis, K. E., Almeida, O. F. X., and Herz, A.,** Feedback inhibition of opioid peptide release in the hypothalamus of the rat, *Neuroscience*, 23, 143, 1987.
57. **Nikolarakis, K. E., Almeida, O. F. X., Yassouridis, A., and Herz, A.,** Presynaptic auto- and allelo-receptor regulation of hypothalamic opioid peptide release, *Neuroscience*, 31, 269, 1989.
58. **Uhl, G. R., Ryan, J. P., and Schwartz, J. P.,** Morphine alters preproenkephalin gene expression, *Brain Res.*, 459, 391, 1988.
59. **Bronstein, D. M., Trujillo, K., and Akil, H.,** Effects of morphine treatment on endogenous opioid biosynthesis, *NIDA Res. Monogr.*, 90, 256, 1988.
60. **Tempel, A., Kessler, J. A., and Zukin, R. S.,** Chronic naltrexone treatment increases expression of preproenkephalin and prepro-tachykinin mRNA in discrete brain regions, *J. Neurosci.*, 10, 741, 1990.
61. **Morris, B. J., Höllt, V., and Herz, A.,** Opioid gene expression in rat striatum, *Neurosci. Lett.*, 89, 80, 1988.
62. **Takemori, E. A., Ideda, M., and Porthogese, P. S.,** The μ, δ and κ properties of various opioid agonists, *Eur. J. Pharmacol.*, 123, 357, 1986.
63. **Gardner, E. L., Zukin, R. S., and Makman, M. H.,** Modulation of opiate receptor binding in striatum and amygdala by selective mesencephalic lesions, *Brain Res.*, 194, 232, 1980.
64. **Murrin, L. C., Coyle, J. T., and Kuhar, M. J.,** Striatal opiate receptors: pre- and postsynaptic localization, *Life Sci.*, 27, 1175, 1980.

65. **Eghbali, M., Santoro, C., Paredes, W., Gardner, E. L., and Zukin, R. S.,** Visualization of multiple opioid-receptor types in rat striatum after specific mesencephalic lesions, *Proc. Natl. Acad. Sci. U.S.A.*, 84, 6582, 1987.
66. **Weissman, B. A. and Zamir, N.,** Differential effects of heroin on opioid levels in the rat brain, *Eur. J. Pharmacol.*, 139, 121, 1987.
67. **Schulz, R., Metzner, K., Dandekar, T., and Gramsch, C.,** Opiates induce long-term increases in prodynorphin-derived peptide levels in the guinea-pig myenteric plexus, *Naunyn-Schmiedeberg's Arch. Pharmacol.*, 333, 381, 1986.
68. **Romuladi, P., Lesa, G., and Ferri, S.,** An opiate chronic treatment affects prodynorphin gene expression, *Pharmacol. Res.*, 21, 477, 1989.
69. **Bronstein, D. M., Przewlocki, R., and Akil, H.,** Effects of morphine treatment on pro-opiomelanocortin systems in rat brain, *J. Pharmacol. Exp. Ther.*, 519, 102, 1990.

10 The Mode of Action of Other Abused Substances

As reviewed below, the other classes of abused substances also seem to stimulate the rewarding circuitry described in Chapter 6 and in other sections of this monograph either by releasing endogenous opioids or by activating the A_{10} dopamine (DA) neurons via yet unspecified inputs. The related data are scanty, but summarily indicate that there are common elements in the mode of actions of various types of drugs of abuse.

I. ETHANOL

Most behavioral studies have been made with ethanol. Namely, as discussed below, in all frequently applied experimental paradigms alcohol has been shown to have rewarding potency just as seen with opiates and psychostimulants. Drug-naive rats avoid it if free choice is given between water and ethanol. However, strong preference develops if the animals receive ethanol solution as the only source of fluid for a certain while (for review see Froehlich et al.).[1] Ethanol is readily self-administered i.v. by monkeys[2,3] and rats as well.[4–6] In rats, similarly to other abused substances, moderate doses of ethanol apparently potentiate the rate of bar pressing rewarded by electric stimulation of the lateral hypothalamus/medial forebrain bundle,[7–10] but the threshold of self-stimulation (the minimal intensity of current needed for reward) remains the same.[9] In addition, ethanol induces conditioned place preference in rats.[11]

Its rewarding potency seems to be mediated by opioid mechanisms. In rats, voluntary ethanol consumption is enhanced upon pretreatment with fentanyl or morphine whether the latter is given systemically[12–17] or intracerebroventricularly.[18] Tiorphan, an agent known to elevate the brain enkephalin content by inhibiting its catabolism,[19] also enhanced the voluntary ethanol consumption.[1] However, suppression was seen with extremely high (60 mg/kg) systemic doses of morphine,[20,21] or intracerebroventricular administration of an enormous amount (200 μg) of Met-enkephalin.[21] Subcutaneous injection of Leu-enkephalin and D-Ala[2],Met[5]-enkephalinamide (DAME) also suppressed the voluntary ethanol intake.[22] The apparent suppressant effect of these high, nearly toxic doses of morphine is most probably due to its general depressant action. Actually, examining the dose-response relationship between the intracerebroventricular dose of morphine and voluntary ethanol intake, an inverted U-shaped curve was obtained.[23] The descending limb of this curve corresponded to the morphine doses which also decreased the water consumption.[23] The nonspecific character of the inhibitory effect of high morphine doses was evidenced by the inability of naltrexone to antagonize the suppressant effect of the 60-mg/kg dose in one of these studies.[21] However, the suppression by peripherally injected opioid peptides of ethanol intake could be reversed by naloxone.[22] Nevertheless, as discussed elsewhere (Székely),[24] opioid peptides hardly penetrate the blood-brain barrier, thus the effect should be of peripheral origin. Actually, as reviewed in Chapter 4, peripherally injected opioid peptides elicit aversion by activation of κ-receptors located in the periphery. Therefore, it is not surprising that Leu-enkephalin suppresses the ethanol consumption if injected peripherally (s.c.).[22] Namely, as also discussed in Chapter 4, opioid peptides hardly pass through the blood-brain barrier. For this reason, administering them systemically, their effects on the peripheral κ-receptors will prevail over the activation of centrally located μ-receptors, mediating among others their rewarding/euphoriant action. The inhibition by s.c.-injected β-endorphin of ethanol preference[25] can obviously be explained in the same way. (All the more, since the 31 amino acid-containing polypeptide has even "less chance" to cross into the brain than pentapeptides as enkephalins.)

On the other hand, naloxone dose dependently decreases the voluntary ethanol intake in rats,[1,12,14,26–28] indicating that opioid mechanisms are involved in the reinforcing effect of ethanol. Furthermore, naloxone inhibited the enhancement of spontaneous ethanol intake in rats exposed to inescapable foot shocks.[29] In addition, naloxone blocks the excitatory effect of ethanol on lateral hypothalamic self-stimulation,[10] the self-administration of ethanol in monkeys,[3] and the ethanol-induced increase in locomotion in mice.[30] Thus the rewarding effect of ethanol must be related to activation of opiate receptors. Since naloxone blocks the rewarding effect of ethanol, it is not surprising that it also attenuates the development of ethanol dependence.[31]

Naloxone also blocks the acute ethanol intoxication in mice[32] and humans[33] as well. Nevertheless, these findings, obtained upon applying naloxone in relatively high doses,[32,33] must be interpreted with caution. Namely, this opiate antagonist also shows GABA antagonist activity in high doses.[34] Thus the above mentioned reversal by naloxone of ethanol intoxication and the similar antagonism of barbiturate anesthesia[35] might be due to its effect on the GABA receptor complex, consequently unrelated to opioid mechanisms.

Since, of the various subtypes of opiate receptors so far, the μ-ones have been found to mediate euphoria (see Chapter 4), it is not surprising that ethylketocyclazocine, contrary to morphine, did not enhance ethanol consumption.[13] On the other hand, it is difficult to explain that in a recent study[1] ICI 174864 also attenuated the ethanol consumption, though this compound is a highly selective agonist of the δ-receptors.[36] In line with these data, the locomotor facilitation seen after morphine and buprenorphine was enhanced by ethanol.[37]

The way of activation of opioid reward mechanisms upon ethanol intake is not clarified yet. One theoretical possibility is a direct modulation of the opiate receptor/ligand interaction upon ethanol treatment (consumption). In one *in vitro* study,[38] ethanol enhanced the binding of μ-receptor-preferring dihydromorphine to its receptors in homogenized caudate tissue, but reduced the binding of (D-Ala^2,D-Leu^5)-enkephalin in higher concentrations. In another laboratory, chronic exposure of neuroblastoma-glioma hybrid cells to ethanol increased the B_{max} of (D-Ala^2)-Met-enkephalinamide.[39] Pfeiffer et al.[40] also observed an increase in the B_{max} of (D-Ala^2,D-Leu^5)-enkephalin, but no change in the binding of dihydromorphine in rat brain after 3-week treatment with ethanol. Based on these data[38–40] it is tempting to speculate that ethanol exerts its rewarding effect by upregulating the opiate receptors and/or increasing their affinity to their ligands. Unfortunately, these intriguing initial findings have not been confirmed or extended during the last decade. Thus the eventual potentiation by ethanol of opioid binding remains an unexplored possibility.

The other theoretical possibility is enhanced production of endogenous opioids in the rewarding circuitry. The related *in vivo* data are scarce and contradictory. In a frequently quoted human study,[41] the CSF β-endorphin content was significantly lower in alcoholic patients than in healthy controls. Of course, it is difficult to find out whether the decrease in CSF β-endorphin is due to the excessive release of this neurohormone upon ethanol consumption or if its "deficient" production prompts the patients to substitute the endogenous rewarding opioids with drugs of abuse. In rats, upon acute ethanol treatment the Met-enkephalin content increased in the striatum, but no changes were seen in other brain areas.[42] Chronic treatment led to a decrease in the striatum, pons/medulla, and midbrain in rats and guineapig as well,[42] but no change was observed in the hypothalamus,[42] whereas elevation was reported by others.[43,44] Examining the whole brain opioid content in mice, the enkephalin level was decreased and β-endorphin concentration increased in an alcohol-preferring strain as compared to an insensitive one.[45]

In vitro in primary cultures of rat fetal hypothalamic neurons, ethanol enhanced the spontaneous release of β-endorphin.[46] The effect of ethanol could be reversed by calcium channel blockers such as verapamil, but not by the sodium channel blocking tetrodotoxin.[46] Thus the β-endorphin-releasing effect of ethanol seems to be direct, i.e., no interneurons might be involved. In line with these observations, ethanol enhanced both the basal and depolarization-induced β-endorphin liberation from rat hypothalamic slice.[47] In both studies the dose-effect curves were of an inverted U-shape, i.e., in higher concentrations ethanol was ineffective or it just induced inhibition.[46,47] However, as discussed in Chapter 6, the hypothalamus is not the only main site of the rewarding effect of abused substances, and the eventual changes in the hypothalamic levels of opioids might also be secondary to decreased water intake seen upon chronic ethanol treatment.[43] Thus the alleged alterations may be irrelevant to substance abuse. Furthermore, measurement of the "stable" levels of neurohormones does not provide reliable information about their metabolism/utilization. Therefore, more informative is a recent study of Przewlocka and Lason[48] reporting on decreased release of Met-enkephalin-related and -enhanced production of dynorphin-related peptides upon chronic ethanol treatment from the hypothalamus. Thus ethanol might somehow downregulate the enkephalin system and upregulate dynorphin production, at least in the hypothalamus, as seen in the cases of classical opiates and psychostimulants (Chapters 7 and 9).

In principle, ethanol might act directly on the mesolimbic dopamine system. In fact, ethanol consumption is accompanied by enhanced release of dopamine in the nucleus accumbens,[49] the medial prefrontal cortex, and the olfactory tubercle.[50] As demonstrated by Imperato and Di Chiara,[49] ethanol only enhanced the dopamine release in doses inducing simultaneous motor activation, whereas higher doses eliciting behavioral depression reduced the dopamine production as well.[49] In good harmony with these data, others found the locomotor facilitation induced by dopamine injection into the nucleus accumbens enhanced in rats chronically pretreated with ethanol.[51] In rats selectively bred for alcohol preference the acceleration of dopamine release was accentuated.[50] All these findings indicate that chronic ethanol treatment (or consumption) might speed up the dopamine production in mesolimbic system or induce dopamine receptor supersensitivity. However, this effect is not direct. Namely, ethanol *in vitro* failed to enhance the basal or K^+-induced dopamine release.[52] Thus ethanol most probably indirectly activates the mesolimbic system, more exactly, the A_{10} neurons.[53]

The path of this stimulation is yet unknown. Nevertheless, it is worth mentioning that alcohol has been repeatedly shown to enhance the GABA release (for review see Hwang et al.)[54] and in rats selectively bred for ethanol preference the number of GABA-ergic terminals is increased in the nucleus accumbens.[54] Since the A_{10} neurons receive an inhibitory GABA-ergic input (see Chapter 7), based on these data ethanol should be aversive, which is actually true in acute experiments. Thus the answer might be related to some change in the activity of the GABA-releasing inhibitory cells upon chronic ethanol administration. Of course, this explanation is highly hypothetical and the way of activation by ethanol of the rewarding circuitry remains to be clarified. Anyhow, as a whole, all these data indicate that in (and beyond) the mesolimbic dopaminergic system the reinforcing mechanism of ethanol is similar to those of classical opiates and psychostimulants (Table 1).

II. NICOTINE

The behavioral effects of nicotine apparently do not resemble those evoked by ''hard drugs'' or ethanol. Nevertheless, a more careful analysis of the literature reveals that in some aspects its mechanism of action is similar to those of other abused substances. Nicotine also is readily self-administered by monkeys,[56] dogs,[57] and rats.[58,59] Nicotine also induces conditioned place preference in rats[60] and mice.[61] In higher doses, just as in the case of ethanol, its rewarding efficacy is reduced.[60] Inhibition by mecamylamine, but not by hexamethonium of the rewarding effect of nicotine[56,60] indicates that the effect is mediated by central nicotinergic receptors. The involvement of opiate receptors in the reinforcing effect of nicotine is indicated primarily by human studies where naloxone reduced cigarette smoking.[62,63] However, one group failed to replicate these data,[64] and in a recent study naltrexone also failed to suppress the self-administration of nicotine in rats.[65]

In rats, nicotine partially antagonized the morphine analgesia blocking effect of β-funaltrexamine,[66] an agent known to alkylate the μ-opiate receptors. Thus nicotine might act by mobilization of endogenous opioids which protect their receptors from irreversible binding.[66] In a recent study[67] nicotine enhanced the production of enkephalin metabolite Tyr-Gly-Gly in several brain regions, primarily in the nucleus accumbens. Unfortunately, in this study of Houdi et al.[67] the ventral tegmental area was not examined.

Other findings point more directly to the eventual participation of the mesolimbic dopaminergic system. In a recent work,[68] self-administration of nicotine by rats was antagonized by dopamine receptor blockers such as haloperidol (a nonselective antagonist), SCH-23390, and spiperone, i.e., selective antagonists of $dopamine_1$ and $dopamine_2$ receptors, respectively. However, in this study all these agents dose dependently decreased the operant responding for drugs,[68] whereas in the case of psychostimulants (see Chapter 5) a ''compensatory'' increase in responding was seen if the dopamine receptors were blocked by antagonists. Anyhow, this observation confirms the participation of the mesolimbic dopamine system in the rewarding effect of nicotine. Actually, some data indicate that the mesolimbic dopamine system might be activated by nicotine even without participation of the opioid system. Namely, nicotine receptors have been found both on the perikarya of A_{10} neurons and their mesolimbic terminals in the nucleus accumbens and olfactory tubercle.[69] The midbrain dopamine neurons are activated by i.p. nicotine and inhibited by mecamylamine.[70,71] Therefore,

TABLE 1
Opioids and Ethanol (E) Abuse

Species examined	Experimental paradigm	Observed effect	Ref.
Rat	Modulation of E consumption		12–17
	By small systemic doses of morphine	Enhancement	23
	By high systemic doses of morphine	Suppression	20,21,23
	By low to moderate doses of fentanyl	Enhancement	16
	By Met-enkephalin given i.c.v.	Suppression	21
	By Leu-enkephalin or (D-Ala^2, Met^5)-enkephalinamide given s.c.	Suppression	22
	By β-endorphin given s.c.	Suppression	25
	By ethylketocyclazocine	No effect	13
	Effect of naloxone given systemically on E consumption	Suppression	1,12,14, 26,27
Monkey	Effect of naltrexone given systemically on E consumption	Suppression	55
Rat	Effect of naltrexone on foot shock-induced facilitation of E consumption	Suppression	29
Rat	Effect of naloxone on facilitation of hypothalamic self-stimulation caused by E	Inhibition	10
Mouse	Effect of naloxone on development of dependence on E	Inhibition	31
Monkey	Effect of naloxone on self-administration of E	Inhibition	3
Mouse	Effect of naloxone on ethanol-induced general anesthesia	Inhibition	32
Human	Effect of naloxone on acute E intoxication	Inhibition	33
Mouse	Effect of naloxone on E-induced locomotor stimulation	Reversal	30
Rat	Effect on a δ-selective antagonist (ICI 174864) on E consumption	Inhibition	1
Mouse	Effect of E on morphine-induced locomotor facilitation	Potentiation	37
Rat and guineapig	Effect of chronic E treatment on Met-enkephalin content in various brain areas	Decrease in striatum, pons, and medulla, but no change in hypothalamus	42,43
Rat	Effect of chronic E treatment on β-endorphin content of the hypothalamus	Increase	43,44
		No change	42
Rat	Effect of E on β-endorphin release from the hypothalamus *in vitro*	Inhibition of spontaneous and depolarization-induced release	46,47

TABLE 1 (continued)
Opioids and Ethanol (E) Abuse

Species examined	Experimental paradigm	Observed effect	Ref.
Rat	Effect of chronic E treatment on hypothalammic release of		48
	Met-enkephalin and	Decrease	
	Dynorphin	Increase	
Human alcohol addicts	CSF β-endorphin content	Decrease	41
Rat	Effect of chronic E treatment on binding of		40
	DADLE[a]	Decrease in B_{max}	
	DHM[b]	No alteration	
	to whole-brain homogenate		
Mouse	Effect of E *in vitro* on the binding of		38
	DADLE	Inhibition	
	DHM	Biphasic effect	
	to homogenate of caudate nucleus		
Neuroblastoma-glioma cell culture	Effect of E on binding of Met-enkephalinamide	Acutely: inhibition Chronically: increase in B_{max}	39
Rat	Effect of chronic E treatment on dopamine release from nucleus accumbens, olfactory tubercle, prefrontal cortex	Increase	1,5,49
	Effect of E on the activity of A_{10} neurons	Facilitation	53

[a] DADLE = (D-Ala^2,D-Leu^5)-enkephalin.
[b] DHM = Dihydromorphine.

these cells might have a tonic nicotinergic excitatory input. According to more recent studies the excitation by nicotine of ventral tegmental neurons cannot be blocked by tetrodotoxin, which indicates that the effect is direct, i.e., no interneurons are involved.[72] The release by nicotine of endogenous opioids[66,67] and the direct excitation by nicotine of A_{10} neurons[72] indicates that smoking might activate the mesolimbic dopamine system in opiate-dependent and -independent ways as well. Not surprisingly, nicotine has been found to enhance the dopamine release in the mesolimbic system, primarily in the nucleus accumbens both *in vitro*[73] and *in vivo,* whether given systemically[74,75] or directly into the nucleus accumbens.[76] Thus it is

TABLE 2
Opioids and Nicotine (N) Abuse

Species examined	Experimental paradigm	Observed effect	Ref.
Rat	Effect of naltrexone on self-administration of N	Weak inhibition	65
Human volunteers	Effect of N on smoking	Reduction	62,63
		No effect	64
Rat	Effect of N on enkephalin release in various brain areas	Enhancement	67
Rat	Effect of N on μ-opiate receptor antagonism	Protection indicating release of endogenous opioids	66

easy to explain that in the above mentioned studies the "antinicotine" effect of naloxone was questionable, but dopamine antagonists and 6-hydroxydopamine lesions of the nucleus accumbens unequivocally inhibited the self-administration in rats.[77]

Nicotine also induces sensitization. In drug-naive animals it enhances spontaneous locomotion, whereas higher doses induce behavioral depression.[59,78] Nicotine also elicits behavioral excitation if injected directly into the ventral tegmental area, an effect reversible by mecamylamine, i.e., mediated by nicotinergic receptors.[79] However, nicotine does not elicit hyperactivity if injected into the nucleus accumbens or other terminal areas of the A_{10} neurons.[79] Upon repeated administration, just as seen with opiates and psychostimulants, its depressant effect gradually tapers off and the locomotor stimulation becomes more and more conspicuous.[80–83] The nicotine-induced locomotor hyperactivity is also accompanied with enhanced release of dopamine from the nucleus accumbens as demonstrated by Imperato et al.[74] However, one group observed just the opposite phenomenon, i.e., a gradual decrease in the nicotine-elicited locomotor excitation upon repeated administration[84] accompanied with an increase in the B_{max} of domperidone binding, i.e., the density of dopamine receptors.[85] Summarily, the above data indicate that nicotine also activates the mesolimbic circuitry as seen with ethanol. Nevertheless, unlike ethanol, its effect might be mediated not only by liberation of endogenous opioids, but it is also capable of activating the A_{10} cells directly (Table 2).

III. MISCELLANEOUS

A. Phencyclidine

Phencyclidine ("angel dust") is also readily self-administered by monkeys,[86] dogs,[87] and rats[88] as well, similarly to its congener ketamine.[88] As seen with other abused substances, conditioned taste aversion can be established on its stimulus effect.[89] As with other abused substances, phencyclidine also decreases the threshold of rewarding brain stimulation.[90] There are no data on liberation by phencyclidine of endogenous opioids. However, phencyclidine itself has been reported to interfere with the binding of some opioid ligands.[91–93] However, it is not simply an opiate receptor agonist. It apparently binds to many other types of receptors, including the muscarinic one[93] and "cocaine receptors", i.e., to the dopamine transporter.[94–96] Examining its structural analogs, close correlation has been found between the inhibition of dopamine uptake and binding to the phencyclidine receptors.[95] Thus it inhibits dopamine uptake and, similarly to psychostimulants, it enhances the extracellular dopamine concentration.[97,98] These biochemical actions explain the extremely intense stereotypy in rodents and other species[86,99] and its peculiar effects in humans. Namely, in men, upon its chronic administration, a characteristic "psychotic state" develops which reminds us of amphetamine psychosis.[100,101] The composite way of its action might explain the contradictory reports on the effects of naloxone. In some studies this opiate antagonist reversed the phencyclidine-elicited hyperactivity[102] and its autonomic effects;[103] however, in others it failed to affect the excitatory action[104] or to induce withdrawal hypothermia,[105] a characteristic sign of opiate dependence.

However complex its action is, the most important point is that as with other abused substances, it activates the mesolimbic dopamine system. The A_{10} dopamine neurons are activated after its systemic administration,[106,107] but not upon its iontophoresis into these cells.[106] Consequently, it might activate the mesolimbic system indirectly. Systemic administration of phencyclidine increases the dopamine metabolism and release in the nucleus accumbens,[108,109] and other agonists of the phencyclidine receptors such as ketamine[109] and MK-801,[110] also facilitate the dopamine turnover in some terminal regions of the mesolimbic system. Similarly to nicotine, phencyclidine also enhances the dopamine release when infused directly into the nucleus accumbens.[111] Thus phencyclidine seems to activate the A_{10} dopamine neurons both indirectly via opioid mechanisms and directly, as psychostimulants do.

B. Benzodiazepines

Several lines of evidence indicate that the abuse of benzodiazepines might also be related to activation of the endogenous opioids and eventually the mesolimbic dopamine system. Nevertheless, even if such a conclusion is really warranted, the way of activation of the rewarding circuitry must be rather indirect. Namely, contrary to other abused substances, self-administration of benzodiazepines and their preference over placebo have not been demonstrated in experimental animals. Truly, neither aversive nor reinforcing properties have been reported. Thus minor tranquilizers or specifically benzodiazepines might be inactive in the classical experimental paradigms of addiction (self-administration, conditioned place preference, etc.), indicating an eventual lack of both positive and negative reinforcing potencies. On the other hand, self-administration of minor tranquilizers and their preference over placebo was described in humans.[112] However, as discussed in Chapter 11, not everybody prefers tranquilizers to placebo. Addicts with a history of abuse of benzodiazepines, barbiturates, or opioids showed preference for diazepam and triazolam over placebo under blind conditions.[113] On the other hand, healthy subjects with no history of substance abuse preferred placebo to benzodiazepines when free choice was offered.[114] The only paradigm in which "positive" findings were reported in experimental animals is facilitation of the rate of (hypothalamic) self-stimulation.[115–117] This facilitation (seen after chlordiazepoxide) was found naloxone reversible in one study,[10] but not in another one.[117] Neither data are conclusive, since in this paradigm it is the lowering of threshold for self-stimulation and not the rate of responding which is the most specific indicator of rewarding potency (see Chapters 4 and 5), and chlordiazepoxide was found inactive in this regard.[116] Furthermore, chlordiazepoxide, contrary to morphine, failed to increase the spontaneous intake of ethanol sucrose solution.[12]

Nevertheless, in some other behavioral paradigms the effect of benzodiazepines is reversible with opiate antagonists. The most characteristic behavioral effect of benzodiazepines is the release of food-rewarded operant responding behavior suppressed by concomitant painful foot shocks. This "anticonflict" effect of benzodiazepines such as chlordiazepoxide, diazepam, and meprobamate, i.e., that of a nonbenzodiazepine anxiolytic agent, were abolished by naloxone.[118–120] Therefore, the antianxiety effect of benzodiazepines (and functionally related drugs) might be related to activation of endogenous opioid mechanisms. Furthermore, it is well known that anxiolytic drugs increase the locomotor activity which is generally explained by reducing the "neophobia" induced by the unfamiliar experimental environment. Naloxone has been found to suppress the locomotor facilitation in mice seen after various benzodiazepines.[118,121–123] Furthermore, in a recent study of Nowakowska and Chodera,[123] repeated administration of nitrazepam resulted in enhancement of the locomotor facilitation, i.e., behavioral sensitization could be induced.

In the few related biochemical examinations diazepam was found to increase the Met-enkephalin release in the striatum and hypothalamus.[124,125] Unfortunately, these experiments have not been repeated and it is questionable whether or not they have bearing on the abuse potential of benzodiazepines. From this point of view, more important might be the observations of O'Brien and White,[126] who recorded facilitation of the A_{10} cells accompanied with inhibition of the nondopamine neurons in the ventral tegmental area upon i.v. administration of diazepam. In the same study,[126] direct iontophoretic application into the A_{10} neurons had no such effect. Thus the facilitation is indirect. It is tempting to speculate that benzodiazepines, known to enhance the GABA-ergic functions,[127] act on the descending GABA-ergic fibers controlling the nondopamine neurons in the ventral tegmental area.[128] However, the nondopamine neurons inhibit the firing of A_{10} cells. Therefore, just the opposite effect is expected on the basis of such a proposal. Thus the way of facilitation remains unclear.

C. Cannabinol

Some data indicate that $\nabla.^9$-tetrahydrocannabinol (the active ingredient of marijuana) might activate the above mentioned circuitry. It has been demonstrated to facilitate the self-stimulation behavior[129] and to enhance the dopamine release from the nucleus accumbens in a naloxone-reversible way.[130] $\nabla.^9$-tetrahydrocannabinol does not bind to the opiate receptors, but it does allosterically modulate the δ- and μ-receptors.[131] Thus it also might act via the opioid input of the mesolimbic dopamine system.

D. Caffeine

As for caffeine, the data relating to this probably most broadly abused substance are scarce and contradictory. It was already examined in the pioneering work of Denau et al.,[2] but the data were inconclusive. Caffeine induced self-administration in monkeys, but only in some of the subjects in a narrow dose range.[2] In rats it also maintains self-administration in a narrow dose range with moderate efficacy.[6] In one study caffeine[132] induced conditioned place preference if low doses were used, whereas higher doses resulted in place aversion. In the related human studies[133,134] only a subgroup of the healthy normal subjects preferred caffeine to placebo at lower dose levels, but never at high ones. Comparing the two groups of healthy subjects, the choosers reported positive mood effects, whereas the non-choosers experienced dysphoria and increased anxiety.[134] Thus though the data are scanty, those obtained in animals and humans are in good harmony. There are no data on the eventual modulation by caffeine of the mesolimbic system. Nevertheless, some clue is provided by a study[135] reporting on increase and decrease in the striatal dopamine release in rats upon acute administration of caffeine in low and high doses, respectively.

These experimental findings are in good agreement with the everyday experience that the slight "euphoria" elicited by coffee is superseded by dysphoria and overstimulation if the consumption is not limited. It is probably the aversive effect of higher doses which explains that caffeine "addiction" and "physical dependence" are relatively mild and they only develop in a subgroup of subjects[133] (Table 3).

TABLE 3
Opioids and Other Abused Substances

Species examined	Experimental paradigm	Observed effect	Ref.
Rat	Effect of naloxone on phencyclidine-induced stereotypy	No effect Inhibition	104 102
Rat	Effect of naloxone on phencyclidine-induced hyperthermia	Inhibition	103
Rat	Effect of naloxone on body temperature after chronic phencyclidine treatment	No effect	105
Rat	Effect of naloxone on chlordiazepoxide-induced facilitation of hypothalamic self-stimulation	Reversal No specific effect	10 117
Rat	Effect of naloxone on the anticonflict, i.e., anxiolytic effect of various benzodiazepines	Reversal	118–120
Mouse	Effect of naloxone on the locomotor facilitation induced by benzodiazepines	Reversal	118, 121–123

E. Conclusion

Summarily, all these data indicate that there is something common in the mechanism of action of abused substances. Namely, all seem to activate the mesolimbic dopamine system in opiate-dependent and/or -independent ways. Of course, activation of the A_{10} neuron-mediated rewarding circuitry does not represent the only site of the brain eventually affected by various abused substances. Therefore, aside from their "common" rewarding properties, various drugs of abuse might induce very different subjective states via different receptor populations and brain areas. Nevertheless, the commonality in their mode of action, activation of the rewarding mechanisms, and sensitization have broad clinical significance.

At least two basic phenomena of substance abuse can be explained on the basis of these observations:

- The polyabuse of drugs, e.g., the frequent combined consumption of "hard drugs", with alcohol, nicotine, etc.
- The common experience that abusers readily switch from one drug to the other one, especially if some of them are not available.

REFERENCES

1. **Froehlich, J. C., Zweifel, M., Harts, J., Lumeng, L., and Li, T.-K.**, Importance of delta opioid receptors in maintaining high alcohol drinking, *Psychopharmacology*, 103, 467, 1991.
2. **Denau, G., Yanagita, T., and Seevers, M. H.**, Self-administration of psychoactive substances by the monkey, *Psychopharmacology,* 16, 30, 1969.
3. **Altshuler, H. L., Phillips, P. E., and Feinhandler, D. A.**, Alteration of ethanol self-administration by naltrexone, *Life Sci.*, 26, 679, 1980.
4. **Smith, S. G., Werner, T. E., and Davis, W. M.**, Comparison between intravenous and intragastric alcohol self-administration, *Physiol. Psychol.*, 4, 91, 1976.
5. **Sinden, J. D. and LeMagnen, J.**, Parameters of low-dose ethanol intravenous self-administration in the rat, *Pharmacol. Biochem. Behav.*, 16, 181, 1982.
6. **Collins, R. J., Weeks, J. R., Cooper, M. M., and Russel, R. R.**, Prediction of abuse liability of drugs using IV self-administration in rats, *Psychopharmacology*, 82, 6, 1984.
7. **Vrtunski, P., Murray, R., Wolin, L. R.**, The effect of alcohol on intracranially reinforced response, *Q. J. Stud. Alcohol*, 34, 718, 1973.
8. **Wayner, M. J., Greenberg, I., Fraley, S., and Fisher, S.**, Effect of tetrahydrocannabinol and ethyl alcohol on adjunctive behavior and the lateral hypothalamus, *Physiol. Behav.*, 10, 109, 1973.
9. **Carlson, R. H. and Lydic, R.**, The effect of ethanol upon threshold and response rate for self-stimulation, *Psychopharmacology*, 50, 61, 1976.
10. **Lorens, S. A. and Sainati, S. M.**, Naloxone blocks the excitatory effect of ethanol and chlordiazepoxide on lateral hypothalamic self-stimulation behavior, *Life Sci.* 23, 1359, 1978.
11. **Reid, L. D., Hunter, G. A., Beaman, C. M., and Hubbel, C. L.**, Toward understanding ethanol's capacity to be reinforcing: a conditioned place preference following injections of ethanol, *Pharmacol. Biochem. Behav.*, 22, 483, 1985.
12. **Beaman, C. M., Hunter, G. A., Dunn, L. L., and Reid, L. D.**, Opioids, benzodiazepines and intake of ethanol, *Alcohol*, 1, 39, 1984.
13. **Hunter, G. A., Beaman, C. M., Dunn, L. L., and Reid, L. D.**, Selected opioids, ethanol and intake of ethanol, *Alcohol*, 1, 43, 46.
14. **Hubbel, C. L., Czirr, S. A., Hunter, G. A., Beaman, C. M., LeCann, N. C., and Reid, L. D.**, Consumption of ethanol solution is potentiated by morphine and attenuated by naloxone persistently across repeated administrations, *Alcohol*, 3, 39, 1986.
15. **Hubbell, C. L., Abelson, M. L., Wild, K. D., Neuman, R., and Reid, L. D.**, Further studies of opioids and intake of sweetened alcoholic beverage, *Alcohol*, 5, 141, 1988.
16. **Czirr, S. A., Hubbell, C. L., Milano, W. C., Frank, J. M., and Reid, L. D.**, Selected opioids modify intake of sweetened ethanol solution among female rats, *Alcohol*, 4, 157, 1987.

17. **Choi, H. W., Wild, K. D., Hubbell, C. L., and Reid, L. D.,** Chronically administered morphine, circadian cyclicity, and intake of an alcoholic beverage, *Alcohol*, 7, 7, 1990.
18. **Linseman, M. A. and Harding, S.,** Intracerebroventricular morphine enhances alcohol consumption by rats, *Pharmacol. Biochem. Behav.*, 36, 405, 1990.
19. **Roques, B. P., Fournie-Zaluski, M. C., Soroca, E., Lecomte, J. M., Malfroy, B., Llorens, C., and Schwartz, J. C.,** The enkephalinase inhibitor tiorphan shows antinociceptive activity in mice, *Nature*, 288, 286, 1980.
20. **Sinclair, J. D., Adkins, J., and Walker, S.,** Morphine-induced suppression of voluntary alcohol drinking in rats, *Nature*, 246, 425, 1973.
21. **Ho, A. K. S. and Rossi, N.,** Suppression of ethanol consumption by Met-enkephalin in rats, *J. Pharm. Pharmacol.*, 34, 118, 1982.
22. **Sandi, C., Borrell, J., Guaza, C.,** D-Ala^2-Met^5-enkephalinamide impairs the acquisition of ethanol preference without influencing sucrose preference, *Physiol. Behav.*, 48, 435, 1990.
23. **Wild, K. D. and Reid, L. D.,** Modulation of ethanol intake by morphine: evidence for a central site of action, *Life Sci.*, 47, PL-49, 1990.
24. **Székely, J. I.,** Regional distribution of enkephalins and endorphins in the brain and peripheral tissues, in *Opioid Peptides*, Vol. III, Opiate Receptors and Their Ligands, Székely, J. I. and Rónai, A. Z., Eds., CRC Press, Boca Raton, FL, 1983, 100.
25. **Sandi, C., Borrell, J., and Guaza, C.,** β-Endorphin administration interferes with the acquisition and initial maintenance of ethanol preference in the rat, *Physiol. Behav.*, 45, 87, 1989.
26. **Marfaing-Jallat, P. and LeMagnen, J.,** Decrease in ethanol consumption by naloxone in naive and dependent rats, *Pharmacol. Biochem. Behav.*, 18, 537, 1983.
27. **Samson, II. II. and Doyle, T. F.,** Oral ethanol self administration in the rat: effect of naloxone, *Pharmacol. Biochem. Behav.*, 22, 91, 1985.
28. **Froehlich, J. C., Harts, J., Lumeng, L., Li, T.-K.,** Naloxone attenuates ethanol intake in rats selectively bred for high ethanol preference, *Pharmacol. Biochem. Behav.*, 35, 385, 1990.
29. **Volpicelli, J. R., Davis, M. A., and Olgin, J. E.,** Naltrexone blocks the post-shock increase of ethanol consumption, *Life Sci.*, 38, 841, 1986.
30. **Kiianmaa, K., Hoffman, P. L., and Tabakoff, P.,** Antagonism of the behavioral effects of ethanol by naltrexone in BALB/c, C57BL/6, and DBA/2 mice, *Psychopharmacology*, 79, 291, 1983.
31. **Blum, K., Futterman, S., Wallace, J. E., and Schwertner, H. A.,** Naloxone-dependent inhibition of ethanol dependence in mice, *Nature*, 265, 49, 1977.
32. **Ho, A. K. S. and Ho, C. C.,** Toxic interactions of ethanol with other central depressants: antagonism by naloxone to narcosis and lethality, *Pharmacol. Biochem. Behav.*, 11, 111, 1979.
33. **Sorensen, S. C. and Mattison, K.,** Naloxone as an antagonist in severe alcohol intoxication, *Lancet*, 2, 688, 1978.

34. **Dingledine, R., Iversen, L. L., and Breuker, E.,** Naloxone as GABA antagonist: evidence from iontophoretic, receptor binding and convulsant studies, *Eur. J. Pharmacol.*, 47, 19, 1978.
35. **Fürst, Z., Földes, F. F., and Knoll, J.,** The influence of naloxone on barbiturate anesthesia and toxicity, *Life Sci.*, 20, 921, 1977.
36. **Cotton, R., Giles, M. G., Miller, L., Shaw, J. S., and Timms, D.,** ICI-174864: a highly selective antagonist for the opioid delta receptor, *Eur. J. Pharmacol.*, 97, 331, 1984.
37. **Kuribara, H., Asashi, T., and Tadokoro, S.,** Enhancement of the ambulation-increasing effect of opioid analgesics by ethanol in mice, *Jpn. J. Pharmacol.*, 56, 457, 1991.
38. **Tabakoff, B. and Hoffman, P. L.,** Alcohol interactions with brain opiate receptors, *Life Sci.*, 32, 197, 1983.
39. **Charness, M. E., Gordon, A. S., and Diamond, I.,** Ethanol modulation of opiate receptors in cultured neural cells, *Science*, 222, 1246, 1983.
40. **Pfeiffer, A., Seizinger, B. R., and Herz, A.,** Chronic ethanol inhibition interferes with δ-, but not μ-opiate receptors, *Neuropharmacology*, 20, 1229, 1981.
41. **Genazzani, A. R., Nappi, G., Facchinetti, F., Mazzella, G. L., Parrini, D., Sinforiani, E., Petraglia, F., and Savoldi, F.,** Central deficiency of β-endorphin in alcohol addicts, *J. Clin. Endocrinol. Metab.*, 55, 583, 1982.
42. **Schulz, R., Wüster, M., Duka, T., and Herz, A.,** Acute and chronic ethanol treatment changes endorphin levels in brain and pituitary, *Psychopharmacology*, 68, 221, 1980.
43. **Cheng, S. S. and Tseng, L.-F.,** Chronic administration of ethanol on pituitary and hypothalamic β-endorphin in rats and golden hamsters, *Pharmacol. Res. Commun.*, 14, 1001, 1982.
44. **Patel, V. A. and Pohorecky, L. A.,** Acute and chronic ethanol treatment on beta-endorphin and catecholamine levels, *Alcohol*, 6, 59, 1989.
45. **Blum, K. and Briggs, A. H.,** Opioid peptides and genotypic responses to ethanol, *Biogenic Amines*, 5, 527, 1988.
46. **Sarkar, D. K. and Minami, S.,** Effect of acute ethanol on beta-endorphin secretion from rat hypothalamic neurons in primary cultures, *Life Sci.*, 47, PL-31, 1990.
47. **Gianoulakis, C.,** Characterization of the effects of acute ethanol administration on the release of β-endorphin peptides by the rat hypothalamus, *Eur. J. Pharmacol.*, 180, 21, 1990.
48. **Przewlocka, B. and Lason, W.,** The effect of single and repeated ethanol administration on hypothalamic systems activity — an *in vitro* release study, *Drug Alcohol Dependence*, 27, 63, 1991.
49. **Imperato, A. and Di Chiara, G.,** Preferential stimulation of dopamine release in the nucleus accumbens of freely moving rats by ethanol, *J. Pharmacol. Exp. Ther.*, 239, 219, 1986.
50. **Fadda, F., Mosca, E., Colombo, G., and Gessa, G. L.,** Effect of spontaneous ingestion of ethanol on brain dopamine release, *Life Sci.*, 44, 281, 1989.
51. **Engel, J. and Liljequist, S.,** The effect of long-term ethanol treatment on the sensitivity of dopamine receptors in the nucleus accumbens, *Psychopharmacology*, 49, 253, 1976.

52. **Seeman, P. and Lee, T.,** The dopamine releasing action of neuroleptics and ethanol, *J. Pharmacol. Exp. Ther.*, 190, 131, 1974.
53. **Gessa, G. L., Muntoni, F., Collu, M., Vargiu, L., and Mereu, G.,** Low doses of ethanol activate dopaminergic neurons in the ventral tegmental area, *Brain Res.*, 348, 201, 1985.
54. **Hwang, B. H., Lumeng, L., Wu, J.-Y., and Li, T.-K.,** Increased number of GABAergic terminals in the nucleus accumbens is associated with alcohol preference in rats, *Alcoholism: Clin. Exp. Res.*, 14, 503, 1990.
55. **Kornet, M., Goosen, C., and Van Ree, J. M.,** Effect of naltrexone on alcohol consumption during chronic alcohol drinking and after a period of imposed abstinence in free-choice drinking in rhesus monkeys, *Psychopharmacology*, 104, 367, 1991.
56. **Goldberg, S. R., Spealman, R. D., and Goldberg, D. M.,** Persistent behavior at high rates maintained by intravenous self-administration of nicotine, *Science*, 214, 573, 1981.
57. **Risner, M. E. and Goldberg, S. R.,** A comparison of nicotine and cocaine self-administration in the dog. Fixed-ratio and progressive ratio schedules of intravenous drug infusion, *J. Pharmacol. Exp. Ther.*, 224, 319, 1983.
58. **Singer, G., Simpson, F., and Lang, W. L.,** Schedule induced self injections of nicotine with recovered body weight, *Pharmacol. Biochem. Behav.*, 9, 387, 1978.
59. **Cox, B. M., Goldstein, A., and Nelson, W. T.,** Nicotine self-administration in rats, *Br. J. Pharmacol.*, 83, 49, 1984.
60. **Fudala, P. J., Teoh, K. W., and Iwamoto, E. I.,** Pharmacologic characterization of nicotine-induced conditioned place preference, *Pharmacol. Biochem. Behav.*, 22, 237, 1985.
61. **Costall, B., Kelly, M. E., Naylor, R. J., and Onaivi, E. S.,** The actions of nicotine and cocaine in a mouse model of anxiety, *Pharmacol. Biochem. Behav.*, 33, 197, 1989.
62. **Karras, A. and Kane, J. M.,** Naloxone reduces cigarette smoking, *Life Sci.*, 27, 1541, 1980.
63. **Gorelick, D. A., Rose, J., and Jarvik, M. E.,** Effect of naloxone on cigarette smoking, *J. Subs. Abuse*, 1, 153, 1989.
64. **Nemeth-Coslett, R. and Griffiths, R. R.,** Naloxone does not affect cigarette smoking, *Psychopharmacology,* 89, 261, 1986.
65. **Corrigall, W. A. and Coen, K. M.,** Opiate antagonists reduce cocaine but not nicotine self-administration, *Psychopharmacology*, 104, 167, 1991.
66. **Davenport, K. E., Houdi, A. A., and Van Loon, G. R.,** Nicotine protects against μ-opioid receptor antagonism by β-funaltrexamine: evidence for nicotine-induced release of endogenous opioids in brain, *Neurosci. Lett.*, 113, 40, 1990.
67. **Houdi, A. A., Pierzchala, K., Marson, L., Palkovits, M., and Van Loon, G. R.,** Nicotine-induced alteration in Tyr-Gly-Gly and Met-enkephalin in discrete brain nuclei reflects altered enkephalin neuron activity, *Peptides*, 12, 161, 1991.
68. **Corrigal, W. A. and Coen, K. M.,** Selective dopamine agonists reduce nicotine self-administration, *Psychopharmacology*, 104, 171, 1991.

69. **Clarke, P. B. S. and Pert, A.,** Autoradiographic evidence for nicotine receptors on nigrostriatal and mesolimbic dopaminergic neurons, *Brain Res.*, 348, 355, 1985.
70. **Greenhoff, J., Aston-Jones, G., and Svensson, T. H.,** Nicotinic effects on the firing pattern of midbrain dopamine neurons, *Acta Physiol. Scand.*, 128, 351, 1986.
71. **Mereu, G., Yoon, K.-W., P., Boi, V., Gessa, G. L., Naes, L., and Westfall, T. C.,** Preferential stimulation of ventral tegmental area dopaminergic neurons by nicotine, *Eur. J. Pharmacol.*, 141, 395, 1987.
72. **Calabresi, P., Lacey, M. G., and North, R. A.,** Nicotinic excitation of rat ventral tegmental neurones in vitro studied by intracellular recording, *Br. J. Pharmacol.*, 98, 135, 1989.
73. **Rowell, P. P., Carr, L. A., and Garner, A. C.,** Stimulation of [^{3}H]dopamine release by nicotine in rat nucleus accumbens, *J. Neurochem.*, 49, 1449, 1987.
74. **Imperato, A., Mulas, A., and Di Chiara, G.,** Nicotine preferentially stimulates dopamine release in the limbic system of freely moving rats, *Eur. J. Pharmacol.*, 132, 337, 1986.
75. **Clarke, P. B. S., Fu, D. S., Jakubovic, A., and Fibiger, H. C.,** Evidence that mesolimbic dopaminergic activation underlies the locomotor stimulant action of nicotine in rats, *J. Pharmacol. Exp. Ther.*, 246, 701, 1988.
76. **Mifsud, J.-C., Hernandez, L., and Hoebel, B. G.,** Nicotine infused into the nucleus accumbens increases synaptic dopamine as measured by in vivo microdialysis, *Brain Res.*, 478, 365, 1989.
77. **Corrigall, W. A.,** Understanding brain mechanisms in nicotine reinforcement, *Br. J. Addict.*, 85, 507, 1991.
78. **Bättig, K., Driscoll, P., Schlatter, J., and Ulster, H. J.,** Effects of nicotine on the exploratory locomotion pattern of female Roman high- and low-avoidance, rats, *Pharmacol. Biochem. Behav.*, 4, 435, 1976.
79. **Reavill, C. and Stolerman, I. P.,** Locomotor activity in rats after administration of nicotinic agonists intracerebrally, *Br. J. Pharmacol.*, 99, 273, 1990.
80. **Morrison, C. F. and Stephenson, J. A.,** The occurrence of tolerance to a central depressant effect of nicotine, *Br. J. Pharmacol.*, 45, 151, 1972.
81. **Stolerman, I. P., Fink, R., and Jarvik, M. E.,** Acute and chronic tolerance to nicotine measured by activity in rats, *Psychopharmacology*, 30, 329, 1973.
82. **Ksir, C., Hakan, R., Hall, D. P., Jr., and Kellar, K. J.,** Exposure to nicotine enhances the behavioral stimulant effect of nicotine and increases binding of [^{3}H]acetylcholine to nicotinic receptors, *Neuropharmacology*, 24, 527, 1985.
83. **London, E. D., Fanelli, R. J., Kimes, A. S., and Moses, R. L.,** Effects of nicotine on glucose utilization in the rat, *Brain Res.*, 520, 208, 1990.
84. **Lapin, E. P., Maker, H. S., Sershen, H., and Lajtha, A.,** Action of nicotine on accumbens dopamine and attenuation with repeated administration, *Eur. J. Pharmacol.*, 160, 53, 1989.
85. **Reilly, M. A., Lapin, E. P., Maker, H. S., and Lajtha, A.,** Increased ^{3}H-domperidone binding in the nucleus accumbens induced by repeated nicotine administration in rats, *J. Neurosci. Res.*, 18, 621, 1986.

86. **Balster, R. L. and Chait, L. D.,** The effects of phencyclidine on amphetamine stereotypy in rats, *Eur. J. Pharmacol.*, 48, 445, 1978.
87. **Risner, M. E.,** Intravenous self-administration of phencyclidine and related compounds in the dog, *J. Pharmacol. Exp. Ther.*, 221, 637, 1982.
88. **Marquis, K. L., Gussio, R., Webb, M. G., and Moreton, J. E.,** Cortical EEG changes during the self-administration of phencyclidine, *Neuropharmacology,* 28, 1193, 1989.
89. **Jackson, A. and Sanger, D. J.,** Conditioned taste aversions induced by phencyclidine and other antagonists of N-methyl-D-aspartate, *Neuropharmacology*, 28, 459, 1989.
90. **Kornetsky, C., Markowitz, R. A., and Esposito, R. U.,** Phencyclidine and naloxone: effects on sensitivity to aversive and rewarding stimulation in the rat, in *PCP (Phencyclidine): Historical and Current Perspectives*, Domino, E. F., Ed., NPP Books, Ann Arbor, MI, 1981, 321.
91. **Su, T. P., Cone, E. J., Shannon, H., and Vaupel, D. B.,** Relative potencies of phencyclidine and analogs in the opiate receptor binding assay: correlation with relative potencies determined in vivo in mouse and rat, *Res. Commun. Subs. Abuse*, 1, 85, 1980.
92. **Itzhak, Y., Kalir, A., Weissman, B. A., and Cohen, S.,** Receptor binding and antinociceptive properties of phencyclidine opiate-like derivatives, *Eur. J. Pharmacol.*, 72, 305, 1981.
93. **Vincent, J. P., Cavey, D., Kamenka, J. M., Geneste, P., and Lazdunski, M.,** Interaction of phencyclidine with muscarinic and opiate receptors in the central nervous system, *Brain Res.*, 152, 176, 1978.
94. **Vignon, J. and Lazdunski, M.,** Structure-function relationships in the inhibition of synaptosomal dopamine uptake by phencyclidine and analogues: potential correlation with binding site identified with [^{3}H]phencyclidine, *Biochem. Pharmacol.*, 34, 700, 1984.
95. **Vignon, J., Pinet, V., Cerruti, C., Kamenka, J.-M., and Chicheportiche, R.,** [^{3}H]N-{1-(2-benzo(b)thiophenyl)cyclohexyl} piperidine ([^{3}H]BTCP): a new phencyclidine analog selective for the dopamine uptake complex, *Eur. J. Pharmacol.*, 148, 427, 1988.
96. **Kuhar, M. J., Boja, J. W., and Cone, E. J.,** Phencyclidine binding to striatal cocaine receptors, *Neuropharmacology,* 29, 295, 1990.
97. **Deutsch, A. Y., Tam, S.-Y., Freeman, A. S., Bowers, M. B., Jr., and Roth, R. H.,** Mesolimbic and mesocortical dopamine activation by phencyclidine: contrasting pattern to striatal response, *Eur. J. Pharmacol.*, 134, 257, 1987.
98. **Chapman, C. D., Gazzara, R. A., and Howard, S. G.,** Effects of phencyclidine on extracellular levels of dopamine, dihydroxy-phenylacetic acid and homovanillic acid in conscious and anesthetized rats, *Neuropharmacology*, 29, 319, 1990.
99. **Martin, J. R., Berman, M. H., Krewsun, I., and Small, S. F.,** Phencyclidine-induced stereotyped behavior and serotonergic syndrome in rat, *Life Sci.*, 24, 1699, 1979.
100. **Fauman, B., Aldinger, G., Fauman, M., and Rosen, P.,** Psychiatric sequelae of phencyclidine abuse, *Clin. Toxicol.*, 9, 529, 1976.
101. **Young, T., Lawson, G. W., and Gacono, C. B.,** Clinical aspects of phencyclidine (PCP), *Int. J. Addict.*, 22, 1, 1987.

102. **Castellani, S., Giannini, A. J., and Adams, P. M.,** Effects of naloxone, Met-enkephalin, and morphine on phencyclidine-induced behavior in the rat, *Psychopharmacology,* 78, 76, 1982.
103. **Glick, S. D. and Guido, R. A.,** Naloxone antagonism of the thermoregulatory effects of phencyclidine, *Science,* 217, 1273, 1982.
104. **Verebey, K., Kogan, M. J., and Mulè, J.,** Phencyclidine-induced stereotypy in rats: effects of methadone, apomorphine, and naloxone, *Psychopharmacology,* 75, 44, 1981.
105. **Pechnik, R. N. and George, R.,** Naloxone does not produce withdrawal hypothermia in chronically phencyclidine-treated rats, *Eur. J. Pharmacol.,* 166, 295, 1989.
106. **Freeman, A. and Bunney, B. S.,** The effects of phencyclidine and N-allylnormetazocine on midbrain dopamine neuronal activity, *Eur. J. Pharmacol.,* 104, 287, 1984.
107. **Pawlowski, L., Mathé, J. M., and Svensson, T. H.,** Phencyclidine activates rat A10 dopamine neurons but reduces burst activity and causes regularization of firing, *Acta Physiol. Scand.*, 139, 529, 1990.
108. **Carboni, E., Imperato, A., Perezzani, L., and Di Chiara, G.,** Amphetamine, cocaine, phencyclidine and nomifensine increase extracellular dopamine concentrations preferentially in the nucleus accumbens of freely moving rats, *Neuroscience,* 28, 653, 1989.
109. **Rao, T. S., Kim, H. S., Lehman, J., Martin, L. L., and Wood, P. L.,** Differential effects of phencyclidine (PCP) and ketamine on mesocortical and mesostriatal dopamine release in vivo, *Life Sci.,* 45, 1065, 1989.
110. **Rao, T. S., Kim, H. S., Lehman, J., Martin, L. L., and Wood, P. L.,** Interactions of phencyclidine receptor agonist MK-801 with dopaminergic system: regional studies in the rat, *J. Neurochem.,* 54, 1157, 1990.
111. **Hernandez, L., Auerbach, S., and Hoebel, B. G.,** Phencyclidine (PCP) injected in the nucleus accumbens increases extracellular dopamine and serotonin as measured by microdialysis, *Life Sci.,* 42, 1713, 1988.
112. **Bigelow, G. E., Griffiths, R. R., and Liebson, I. A.,** Effects of response requirement upon human sedative self-administration and drug seeking behavior, *Pharmacol. Biochem. Behav.,* 5, 681, 1976.
113. **Roache, J. D. and Griffiths, R. R.,** Diazepam and triazolam self-administration in sedative abusers: concordance of subject ratings, performance and drug self-administration, *Psychopharmacology,* 99, 309, 1989.
114. **De Wit, H., Uhlenhuth, E. H., and Johanson, C. E.,** Lack of preference for flurazepam in normal volunteers, *Pharmacol. Biochem. Behav.,* 21, 865, 1984.
115. **Olds, M. E.,** Facilitatory action of diazepam and chlordiazepoxide on hypothalamic reward behavior, *J. Comp. Physiol. Psychol.,* 62, 136, 1966.
116. **Domino, E. F. and Olds, M. E.,** Effects of d-amphetamine, scopolamine, chlordiazepoxide and diphenylhydantoin on self-stimulation behavior and brain acetylcholine, *Psychopharmacology,* 23, 1, 1972.
117. **Ichitani, Y., Iwasaki, T., and Satoh, T.,** Effects of naloxone and chlordiazepoxide on lateral hypothalamic self-stimulation in rats, *Physiol. Behav.,* 34, 779, 1985.

118. **Duka, T., Cumin, R., Haefely, W., and Herz, A.,** Naloxone blocks the effect of diazepam and meprobamate on conflict behavior in rats, *Pharmacol. Biochem. Behav.*, 15, 115, 1981.
119. **Soubrié, P. A., Jobert, A., and Thiebot, M. H.,** Differential effects of naloxone against diazepam-induced release of behavior in rats in three aversive situations, *Psychopharmacology*, 69, 101, 1980.
120. **Billingsley, M. L. and Kubena, R. K.,** The effects of naloxone and picrotoxin on the sedative and anticonflict effects of benzodiazepines, *Life Sci.*, 22, 897, 1978.
121. **Castellano, C., Filibeck, U., and Pavone, F.,** Naltrexone-reversible effects of flunitrazepam on locomotor activity and passive avoidance behavior in mice, *Eur. J. Pharmacol.*, 104, 111, 1984.
122. **Sansone, M. and Vetulani, J.,** Effect of naloxone on the locomotor stimulatory action chlordiazepoxide in mice, *Pharmacol. Biochem. Behav.*, 31, 371, 1988.
123. **Nowakowska, E. and Chodera, A.,** Studies on the involvement of opioid mechanisms in the locomotor effects of benzodiazepines in rats, *Pharmacol. Biochem. Behav.*, 38, 265, 1991.
124. **Duka, T., Wüster, M., and Herz, A.,** Rapid changes in enkephalin levels in rat striatum and hypothalamus induced by diazepam, *Naunyn-Schmiedeberg's Arch. Pharmacol.*, 309, 1, 1979.
125. **Hársing, L. G., Jr., Yang, H.-Y., and Costa, E.,** Evidence for a γ-aminobutyric acid (GABA) mediation in the benzodiazepine inhibition of release of Met^5-enkephalin elicited by depolarization, *J. Pharmacol. Exp. Ther.*, 220, 616, 1982.
126. **O'Brien, D. B. and White, F. J.,** Inhibition of non-dopamine cells in the ventral tegmental area by benzodiazepines: relationship to A10 dopamine cell activity, *Eur. J. Pharmacol.*, 142, 343, 1987.
127. **Costa, E., Guidotti, A., Mao, C. C., and Suria, A.,** New conceptions on the mechanism of action of benzodiazepines, *Life Sci.*, 17, 167, 1975.
128. **Fuxe, K., Hökfelt, T., Ljungdahl, A., and Agnati, L., Johansson, O., and Perez de la Mora, M.,** Evidence for an inhibitory GABAergic control of the mesolimbic dopamine neurons, possibility of improving treatment of schizophrenia by combined treatment with neuroleptics and GABAergic drugs, *Med. Biol.*, 53, 177, 1975.
129. **Gardner, E. L., Paredes, W., Smith, D., Donner, A., Milling, C., Cohen, D., and Morrison, D.,** Facilitation of brain stimulation reward by Δ^9-tetrahydrocannabinol, *Psychopharmacology*, 96, 142, 1988.
130. **Chen, J., Paredes, W., Li, J., Smith, D., Lowinson, J., and Gardner, E. L.,** Δ^9-Tetrahydrocannabinol produces naloxone-blockable enhancement of presynaptic basal dopamine efflux in nucleus accumbens of conscious, freely-moving rats as measured by intracerebral microdialysis, *Psychopharmacology*, 102, 156, 1990.
131. **Vaysse, P. J.-J., Gardner, E. L., and Zukin, R. S.,** Modulation of rat brain opioid receptors by cannabinoids, *J. Pharmacol. Exp. Ther.*, 241, 534, 1987.
132. **Brockwell, N. T., Eikelboom, R., and Beninger, R. J.,** Caffeine-induced place and taste conditioning: production of dose-dependent preference and aversion, *Pharmacol. Biochem. Behav.*, 38, 513, 1991.

133. **Griffiths, R. G. and Woodson, P. P.,** Reinforcing effects of caffeine in humans, *J. Pharmacol. Exp. Ther.*, 246, 21, 1988.
134. **Stern, K. N., Chait, L. D., and Johanson, C. E.,** Reinforcing effect of caffeine in normal human volunteers, *Psychopharmacology*, 98, 81, 1989.
135. **Morgan, M. E., Dunn, D., and Vestal, R. E.,** Low dose caffeine modulation of caudate dopamine release as measured by in vivo electrochemistry in freely moving rat, *Soc. Neurosci. Abstr.*, 13, 914, 1987.

11 Individual Differences in Propensity to Drug Addiction

I. INTRODUCTION

One of the most intriguing questions related to substance abuse is that some individuals easily become addicted to drugs, while others do not. Even if our present knowledge on the process of sensitization is correct, the data reviewed in Chapters 7 and 9 do not explain the individual differences in propensity to drug addiction.

For example, if activation of the mesolimbic dopamine circuitry is equally rewarding, then dopamine stimulants should be pleasurable for everybody. This is, however, not the case. Johanson and Uhlenhuth[1] reported that 9 of 31 healthy volunteers did not prefer *d*-amphetamine over placebo when free choice was given them under laboratory conditions. In other words, this highly potent and widely abused substance failed to induce ''drug-seeking behavior'' in about one third of subjects examined, though, according to the simultaneously performed psychometric examinations (Profile of Mood States), amphetamine induced similar mood changes in all the volunteers.[1] Furthermore, as an apparent defiance of the generally accepted theory on sensitization (see Chapter 7), repeated exposure to amphetamine did not increase, but rather decreased the proportion of subjects who selected amphetamine vs. placebo.[1] In view of these experimental data it is not surprising that only a small fraction of people become drug addicts after an initial or ''accidental'' familiarization with them. Based on a careful review of the related literature, Erickson and Alexander[2] concluded that only 5 to 10% of those who ever try cocaine will use it regularly, i.e., at least weekly or more frequently. Thus, apparently, the majority of subjects do not become drug addicts after an initial ''experimentation'' with the drug. On the other hand, a vast literature indicates that subjects liable to substance abuse easily switch from one drug to another or they simultaneously consume more than one. For example, according to a recent report, in a cohort of young people tobacco use was the ''...gateway to drug, alcohol abuse''.[3]

As for the sedatives, De Wit et al.[4] reported that flurazepam, a benzodiazepine derivative readily self-administered by monkeys,[5] induced aversion instead of the expected euphoria in healthy volunteers. De Wit et al.[4] attribute the apparent contradiction to the fact that in animal studies the subjects are preselected for readiness to self-administer substances of abuse.

Actually, it is really the case. In these human studies[1,4] young, healthy volunteers were examined, i.e., subjects with serious drug abuse were not included in the examinations. On the other hand, it is well established[6,7] that reliable preference for benzodiazepines is only shown in subjects with a history of abuse of sedatives.

In fact, a similar bias applies to animal experiments as well. Namely, unless the acquisition of drug self-administration behavior specifically is examined, in the everyday practice, even the experimental animals are preselected and only subjects showing ''stable baseline activity'', i.e., strong individual propensity to drug addiction, are further examined.

Thus the question remains whether or not the wealth of experimental data accumulated during the last decades might discover something about the neurobiological basis of the individual differences in propensity to drug addiction.

Shortly after the discovery of endorphins, Goldstein[8] postulated that ''...in some people a genetically determined endorphin deficiency could predispose to narcotic addiction''. A further proposal of the same author[9] was that ''...By 'predispose' is meant that for a given availability of heroin and given conditions of social stress some individuals will be more vulnerable to addiction than others...''

But why should the eventual deficiency in endorphin production result in drug consumption? The answer was given still before their discovery. Martin et al.[10] observed that ''...It has been observed clinically that opiate addicts perceive themselves as being inept, ineffectual and unpopular... They also tend to be withdrawn, hypochondriacal, and critical of others... This constellation of feeling states has been called hypophoria... These feeling states are decreased by single doses of narcotic analgesics, amphetamines and barbiturates...''

Furthermore, practically predicting many findings reviewed in Chapter 9, Goldstein[9] proposed that ''continuous use of an exogenous opiate would cause a suppression of endorphin production, by analogy to the feedback regulation of other hormones. Then a part of the withdrawal syndrome could be related to persistent endorphin deficiency...''

In general, there is very little to add to these statements even today. The point is that the more recent observations, experimental, clinical, and epidemiological findings provide a rational framework for interpretation of these serendipitous predictions.

II. ANIMAL DATA ON INDIVIDUAL PROPENSITY TO DRUG ADDICTION

Giving them free choice over tap water or drug solution, nonpretreated naive animals do not select morphine-containing solution, presumably due to its bitter taste. However, rats voluntarily consume such solution if they are made previously dependent on morphine[11] and/or they receive this solution as the only source of fluid for various periods.[12–14] Nevertheless, even after this kind of ''forced feeding'' in choice tests, when the rats can choose between pure water and morphine solution, some individuals consume relatively high amounts of morphine, others relatively little.[12–14] The former ones obviously are more vulnerable to drug addiction.

The genetic origin of these individual differences is indicated by the finding that selective breeding of the animals showing high preference for morphine will generate subsequent generation(s) which consume even more morphine.[12,15] Analogously, mice can be selectively bred for high and low sensitivity to the analgesic effect of opiates.[16]

An interesting case of selective bidirectional breeding of rats is the development of two strains, one of which shows a very high (LC2-Hi) and the other one a rather low (LC2-Lo) rate of intracranial (hypothalamic) electrical self-stimulation (ICSS).[17] As reported by Dymshitz and Lieblich,[18] a much higher degree of conditioned place preference (CPP) to morphine and heroin could be established in LC2-Hi than in LC2-Lo rats. Thus there must be at least one common factor determining the individual sensitivity to opiates and rewarding experimental manipulations (ICSS, CPP). Therefore, it was logical to propose, as done by Goldstein,[8,9] that sensitive individuals are relatively deficient in endogenous opioids. Surprisingly, few such animal studies have been conducted in relation to ethanol. Blum et al.[19] reported that comparing an ethanol-preferring (C57BL/6J) and an ethanol-nonpreferring strain (C57BL/6N) of mice, the whole-brain enkephalin content was lower in the former one. Examining individual animals, the same authors[20] found negative correlation between the spontaneous ethanol intake and the brain enkephalin content. Unfortunately, these challenging findings in animal experiments have not been confirmed and/or extended ever since. Probably the relative insufficiency of endogenous opioids manifests in much subtler functional changes than its baseline level. Alterations in the turnover, transcription or translation, modification in reactivity to appropriate stimuli, altered opiate receptor density, or coupling the receptors to the appropriate effector systems, etc., all might result in functional deficiency of endogenous opioid system.

A further factor determining individual vulnerability to drug addiction might be the sensitivity to stress. Namely, as discussed in Chapter 8, stress activates the mesolimbic dopamine neurons, i.e., it enhances the dopamine turnover in them. For example, as both stressful manipulations and amphetamine facilitate the dopamine turnover, they are interchangeable in the process of sensitization.[21] Therefore, individuals more responsive to stress are expected to be more predisposed to drug addiction. The phenomenon was experimentally demonstrated by Piazza et al.,[22] who found that rats showing more intense locomotion in a new environment (novelty stress) also react with stronger locomotor response to amphetamine itself. Such animals have also been found to acquire amphetamine self-administration behavior more rapidly than the less sensitive, reactive individuals.[23]

Summarily, these data indicate that there must be some kind(s) of genetically determined neurobiological factor(s) explaining the individual sensitivity (and insensitivity) to drug addiction, even in animals. Nevertheless, the nature of these factors has not been clarified yet. The attractive hypothesis on deficiency of endogenous opioids has not been substantiated either.

III. NEUROHORMONAL PREDISPOSING FACTORS IN HUMANS

Following the same line of reasoning, some groups examined addicted humans for eventual changes in the functioning of endogenous opioids.

In a frequently quoted paper, Ho et al.[24] reported that in heroin addicts the plasma β-endorphin content was much lower than in healthy controls. Others[25] found the plasma β-lipotropin (β-LPH), i.e., its precursor, lowered in heroin addicts. These findings were considered as proof of endorphin deficiency in opiate and ethanol addiction. However, in the subsequent studies normal β-endorphin level was detected in the plasma of methadone-maintained opiate addicts.[26–28] Furthermore, in withdrawn[29] and naltrexone-treated addicts the plasma β-endorphin level was found just elevated.[27,30] The main problem with these data is that plasma β-endorphin and β-LPH are secreted by the anterior and neurointermediate lobes of the pituitary, cleaving a precursor(proopiomelanocortin) generating simultaneously ACTH, i.e., they are cosecreted with ACTH. Thus the elevated level of β-endorphin in withdrawn opiate antagonist-treated patients might simply reflect the stressful nature of these interventions and the findings might be unrelated to the central, mesolimbic opioid functions. Furthermore, the lower β-endorphin level in untreated addicts[24] might reflect the exhaustion of the pituitary-adrenal system and not the functional state of the endogenous opioid system. A bit more informative might be the endorphin concentration in the CSF. However, contrary to expectations it has been found just elevated in methadone maintainance or withdrawal.[28,31,32] Unfortunately, in the former studies[31,32] the ill-defined, so-called fractions I and II have been determined.

In alcholics, one group recorded significant lowering of CSF β-endorphin level.[33] In another study in subjects with a three-generation history of alcohol abuse the baseline β-endorphin level in the plasma was found lower and the ethanol-induced elevation of its level more intense than in controls.[34] These high-risk subjects also showed more conspicuous elevation of plasma cortisol level upon ethanol consumption than the controls.[34] The latter finding rather confirms another point, the enhanced stress sensitivity of addicts.

Thus in general the original claims[24,33] on deficiency of the endorphin system in the brain in opiate addicts and alcoholics have hardly been confirmed. Furthermore, there are virtually no data relating to the main physiologically rewarding opioids, enkephalins. On the other hand, the data reviewed above summarily indicate that in addicts the sensitivity to stress might be somehow enhanced. In view of the cross sensitization between drugs of abuse and stressful manipulations (see Chapter 8), the genetic factor underlying addiction, among others, might not be simply a deficiency of endogenous opioids, but rather an enhanced responsiveness to stress. On the other hand, besides a certain hyperresponsivity to stress (confirmed by psychometric studies as well; vide infra), for the time being the alleged neurochemical, neurohormonal factors underlying propensity to drug addiction cannot be specified yet.

Reviewing the alleged neurohormonal basis of addiction, a recent paper must not be overlooked. Blum et al.[35] claimed that in 77% of alcoholics the dopamine$_2$ receptor gene shows certain abnormalities. In view of the central role of dopamine in all kinds of rewarding mechanisms, the significance of this finding is tremendous. Nevertheless, confirmation is needed from independent laboratories.

IV. COMORBIDITY OF DRUG ABUSE WITH MENTAL DISORDERS

Examining the psychogenic factors making individuals liable to drug abuse before all the coprevalence of mental disorders must be taken into consideration. Obviously, if two disorders co-occur more frequently than the chance level, something common might be looked for in the etiology. Thus in the case of significant comorbidity our knowledge of the genesis of mental disease might somehow apply to the etiology of drug abuse as well. Reviewing the related literature, the reader has the very impression that drug abuse and mental disorders are not separate entities.

According to a recent epidemiological study[36] based on interviewing more than 20,000 people in the U.S.A., the lifetime prevalence rate of all kinds of mental disorders was 22.5%. Among the subjects with any kind of mental disorders, the odds ratio of having addictive disorder too was 2.7, with a lifetime prevalence of 29% (including an overlapping 22% of

TABLE 1
Lifetime Prevalence of Mental Disorders in Treatment-Seeking Cocaine Abusers

Diagnosis	Prevalence %
Major depression	30.5
Any affective disorder	60.7
Obsessive-compulsive disorder	0.3
Phobia	13.4
Panic disorders	1.7
Any anxiety disorder	20.8
Schizophrenia	0.3
Antisocial personality	7.7
Any mental disorder	73.5

Data from Rounsaville et al., *Arch. Gen. Psychiatry,* 48, 43, 1991.

alcoholism and 15% of other drug disorders).[36] Before going into detail, some pitfalls of dual diagnosis (mental disorder + drug abuse) must be taken into consideration. First is the ''self-medication theory'' that patients might resort to abused substances to overcome the symptoms of mental disorders; depressed patients might drink alcohol to overcome depression, anxiety, and dysphoria, i.e., for the main symptoms of affective disorders, a schizophrenic may take amphetamine to counteract either the side-effects of neuroleptics or the negative symptoms of the primary disease, etc.[37,38] Another source of faulty dual diagnosis might be the ability of abused substances to induce mental disorders by themselves, e.g., the induction of psychoses by psychotomimetics and stimulants,[38–40] the onset of rebound depression in cocaine abusers[41–43] and heroin users.[44,45] Thus the two processes are interactive whereupon substance abuse might aggravate mental disorders and mental disorders increase the risk of drug abuse.

Nevertheless, even if keeping in mind the limitations of the dual diganoses, the statistical data are frightening. As summarized in Table 1 (compiled from data Rounsaville et al.),[46] among the treatment-seeking cocaine abusers the lifetime prevalence of all kinds of psychiatric disorders was 73.5%, including primarily various kinds of affective disorders (60.7%); anxiety disorders (20.8%); personality disorders (mainly antisocial personality); and others. According to the same group (Table 2, compiled from data of Rounsaville et al.),[47] among opiate addicts the prevalence of major depression is (male/female): 48.9/69.2%; any mood disorder: 70.7/85.4%; anxiety disorders: 13.2/25.4%; antisocial personality 29.5/16.9%. According to many studies (for review see Cox et al.)[48] various kinds of panic disorders and phobias occur especially frequently among drug users. Another class of frequently diagnosed anxiety disorders is the obsessive-

TABLE 2
Lifetime Prevalence of Mental Disorders in Opiate Addicts

Diagnosis	Prevalence % in Males	Prevalence % in Females
Major depression	48.9	69.2
Any affective disorder	70.7	85.4
Schizophrenic disorders	0.7	0.8
Obsessive-compulsive disorders	1.7	2.3
Any anxiety disorder	13.2	25.4
Antisocial personality	29.5	16.9

Data from Rounsaville et al., *Arch. Gen. Psychiatry,* 39, 161, 1982.

compulsive disorders.[47,49] The diagnosis of obsessive-compulsive disorders is especially frequent among alcoholics, and it is probably the only psychiatric diagnosis which is associated with a single particular class of abused substances, namely, alcohol.[50–53]

Probably it is worth mentioning here that according to follow up studies of schoolchildren, those rated especially aggressive when first graders were the more likely to use drugs as teenagers.[54] According to a recent survey among the inpatients of mental hospitals, substance abusers tend to be more suicidal, homicidal, and destructive than nonusers.[55]

Of course, one of the main questions about the comorbidity, is whether or not abuse of specific classes of drugs is associated with certain selective diagnoses. Namely, beside a presumably common way of inducing euphoria and addiction, are the pharmacological mechanisms of action of various types of abused substances different, correlating, with specific mental disorders differing from each other in etiology and pathology? The related data are surprisingly scarce. As mentioned above, obsessive-compulsive disorders are especially frequent among alcoholics. Furthermore, Quitkin et al.[56] reported on the association between panic disorders and abuse of sedato-hypnotics. Recently, Mirin et al.[57] provided data on association of alcoholism and affective disorders, showing common genetic origin and sex-related manifestation whereupon drinking predominates in males with depression and females, but both are derived from the same family pedigree. However, the majority of abused substances cannot be associated with any class of mental disorders.

For the reviewer, two points are striking:

- The prevalence of affective, anxiety, and personality disorders among the diagnoses (schizophrenia is hardly mentioned)

- The lack of strong correlation between the type of drugs abused and the diagnoses made

In view of the diversity of psychiatric disorders found associated with drug abuse, there might hardly be any common underlying biological factor. Thus, rather, those proposals might indicate that the typical addict suffers from dysregulation of affect and is vulnerable to dysphoria; therefore, the subject resorts to any kind of abused substance, whatever is available in the particular social milleu.[37,47,58]

V. THE ADDICTIVE PERSONALITY

As discussed in Sections III and IV, individual propensity to drug abuse cannot be closely associated with any specific neurohormonal alteration, and not even with a well-specified psychiatric diagnosis. On the other hand, a wealth of clinical and psychological studies prove that some subjects are nevertheless vulnerable to drug addiction, i.e., there must be some kind of genetic factor. Namely, numerous studies indicate that descendants of alcoholics have greater risk of becoming alcoholics themselves than the children of abstinent parents (for recent reviews see Mirin et al.,[57] Levenson et al.,[59] and Hwu et al.).[60] An alternative hypothesis is that stressful, anxiety-provoking life situations might lead to drug addiction. However, contrary to such expectations, e.g., a study on identical twins shows that prior military service in the Vietnam conflict hardly modified the inherited genetic propensity to alcohol abuse.[61] On the other hand, a reduction of the amplitude of the so-called P300 wave of the event-related evoked potentials could be detected in subjects at risk for alcoholism for being the descendants of alcoholic parents[62–64] and in patients at risk for drug abuse because of juvenile delinquency.[65,66] Furthermore, creating stressful situations in laboratory conditions, the stress-reducing effect of alcohol was stronger in descendants of alcoholics than in controls.[59] Thus the genetic factors are very important in determining the liability to drug abuse. The identification of these factors is, however, for the time being rather a hope than reality.

As reviewed below, the individual propensity to drug abuse can be relatively closely associated with certain personality traits. For objective assessment of the specific personality traits of addicts, mostly various psychometric examinations were conducted. (The tests are generally checklists or questionnaires where various subgroups of questions or items relate to single personality traits, and cohorts of subgroups reflect ''tendencies'' or factors related to diagnoses or syndromes.) Probably in the most famous classical studies the Minnesota Multiphasic Personality Inventory (MMPI) was used for a factor analytic study of alcoholics, narcotic addicts, and criminals.[67,68] They differentiated four basic types of addictive personalities, including ''primary psychopaths'', characterized primarily with psychopathic deviance, depression, psychasthenia, and hypomania; ''neurotic psychopaths'' with elevation on the depression scale and hypochondriac and hysteric tendencies; ''schizoid psychopaths'' and ''unclassifiable'' subjects.[67] The main factor present in this kind of typology, psychopathic deviance (as a kind of diagnosis), can be broken down to such personality traits as low self-esteem, hypersensitivity, social maladaptation, poor impulse control (impulsivity), and emotional deprivation.[68] Subsequently, the same group[10,70] redefined the main factors comprising the psychopathic personality traits as: impulsivity, high basic needs, hypophoria (meaning negative affective states, including depression), sociopathy (social maladaptation), and a search for (drug-provided) highs. Subsequent studies using MMPI, but carried out by independent groups,[71–73] basically confirmed the findings of Hill and co-workers.[67,68] Loper et al.[71] conducted a followup study and found that college freshmen subsequently becoming alcoholics scored high in psychopathic deviance, hypomania, and they were more impulsive, gregarious, and nonconforming than their fellow students remaining abstinent. Craig,[72] reviewing the numerous studies conducted on addicts and using the MMPI lists, defined the main personality traits of addicts as: impulsivity, hostility, depression, anxiety, low frustration and pain tolerance, desire for sensation-seeking experience, poor socialization, difficulty in forming lasting relationships, hostility, irritability, irresponsibility, and feeling of insecurity and inadequacy. Administration of the abused substances might somewhat moderate the symptoms, i.e., they lower the scores on the MMPI scale, but during withdrawal the scores increase even higher than the original (abnormal) level.[74]

The use of other types of questionnaires and/or checklists gave more or less similar results. Applying the Hopkins Symptom Checklist, Harris and Linn[75] found the addicts anxious, obsessive-compulsive, sensitive in interpersonal relationships, depressed, and prone to somatization. Using the Beck Depression Inventory and Symptom Check List-90, addicts were characterized with somatization, obsessive compulsiveness, interpersonal sensitivity, depression, hostility, anxiety, phobic anxiety, paranoid ideation, and psychoticism.[76] An important point is that self-report and various laboratory measures of aggression are good predictors of substance abuse.[77]

An important predictor of substance abuse is the level of anxiety. Examining college students, high-anxiety students have been found to consume more alcohol than their less anxious counterparts.[78]

Examining the antecendents of initiation of drug use among adolescents the main precipitating factor has been found to be psychological distress. As emphasized by Kandel,[79] based on his own experience and review of the literature,''...The use of illicit (or licit) drugs as coping mechanisms to deal with feeling of psychosocial distress is suggested...''.

A further contributing factor might be ''sensation seeking''. The latter expression indicates a personality trait, a part of the extroverted personality, which prompts the subject to seek novel, varied, complex sensations, since the arousal level of the subject is below a level perceived as optimal.[80] Some studies, based on measurement of sensation seeking, claim that drug addicts consume drugs to increase the level of arousal.[81–83] Thus some subjects might consume to reduce the level of arousal (i.e., dysphoria, anxiety, repressed hostility), whereas others should do the same for the opposite reason.

An interesting finding is (implicitly involved in many other studies as well) that the abnormal personality traits are more or less similar in addicts abusing various kinds of drugs, e.g., heroin or cocaine.[84] Thus neither the psychiatric diagnosis (see Section IV) nor various subtypes of addictive personalities can be associated with the abuse of particular subclasses of addictive drugs.

Another approach to the addictive personality is provided by psychoanalytical studies with basically the same final conclusions. Freud wrote in 1893 (cit post. Salzmann)[85] "...Not everyone who has occasion to take morphine, cocaine, chloral hydrate and so on for a period acquires in this way an addiction to them. Closer inquiry usually shows that narcotics are meant to serve directly or indirectly as a substitute for lack of sexual satisfaction...". Some years later he said, "...I have gained the insight that masturbation is a primary addiction...". and drug addiction only serves as substitution for it (cit. post Salzman).[85] Thus, drug addiction is perceived by the psychoanalytics as a kind of surrogate gratification and conversion.[85,86] Others detect low frustration tolerance leading to depression and hostility in alcoholics, emphasize their inability to endure tension and anxiety, their low self-esteem, their feelings of inadequacy, and tendency to act out.[87] Ultimately, the drug is perceived as "...the chemical equivalent of mother...", as noted by Sternberg and Cohen.[87] Hatterer[86] emphasizes that "...the process is initially a coping mechanism which enables the person to adapt, but ultimately proves to be obsessive and compulsive in nature and practiced to such an excessive degree that it becomes maladaptive and syndrome producing". Thus a process initiated to gain immediate gratification for an immature person for whom drugs mean "...substitution of substance nutrient for love...",[87] winds up as obsessive-compulsive drug-seeking behavior. In a paper of Wurmser,[58] drugs are consumed "...as an attempt of self-treatment...as artificial and surrogate defense against overwhelming affects... as regressive narcissistic state of self-satisfaction".

Apparently, the outcome of psychometric and psychoanalytical studies mutually confirm each other.

Summarily, all these data confirm that

- Some subjects are genetically prone to drug addiction.
- Liability to drug addiction is associated with rather characteristic personality traits.

However, discounting enhanced sensitivity to stress, so far no data help to associate the clinical, psychiatric, and clinical psychological observations with the alleged neurohormonal basis of addiction.

REFERENCES

1. **Johanson, C. E. and Uhlenhuth, E. H.,** Drug preferences in humans, *Fed. Proc. Fed. Am. Soc. Exp. Biol.*, 41, 228, 1982.
2. **Erickson, P. G. and Alexander, B. K.,** Cocaine and addictive liability, *Soc. Pharmacol.*, 3, 249, 1989.
3. **Keegan, A.,** Tobacco may provide gateway to drug, alcohol abuse, *NIDA Notes*, 6, 23, 1991.
4. **De Wit, H., Uhlenhuth, E. H., and Johanson, C. E.,** Lack of preference for flurazepam in normal volunteers, *Pharmacol. Biochem. Behav.*, 21, 865, 1984.
5. **Johanson, C. E.,** Reinforcing properties of lorazepam and flurazepam in rhesus monkeys, *Fed. Proc. Fed. Am. Soc. Exp. Biol.*, 42, 345, 1983.
6. **Griffiths, R. R., Bigelow, G. E., Liebson, I., and Kaliszak, J. E.,** Drug preference in humans: double choice comparison of pentobarbital, diazepam and placebo, *J. Pharmacol. Exp. Ther.*, 215, 649, 1980.
7. **Haley, M. L. and Pickens, R. W.,** Diazepam dose-preference in humans, *Pharmacol. Biochem. Behav.*, 18, 449, 1983.
8. **Goldstein, A.,** Opioid peptides (endorphins) in pituitary and brain, *Science*, 193, 1081, 1976.
9. **Goldstein, A.,** Endorphins: physiology and clinical implications, *Ann. N.Y. Acad. Sci.*, 311, 49, 1978.
10. **Martin, W. R., Hewett, B. B., Baker, A. J., and Haertzen, C. A.,** Aspects of the psychopathology and pathophysiology of addiction, *Drug. Alcohol Depend.*, 2, 185, 1977.
11. **Dai, S., Hui, S.-C. G., and Ogle, C. W.,** Morphine preference in rats previously morphine dependent, *Pharmacol. Res. Commun.*, 16, 495, 1984.
12. **Nichols, J. R. and Hsiao, S.,** Addiction liability of albino rats: breeding for quantitative differences in morphine drinking, *Science*, 157, 561, 1967.
13. **Zeichner, J., Rosengren, L., Wronski, A., Rönnbäck, L.,** A new ingestion method for long-term morphine-intoxication in rat, *Pharmacol. Biochem. Behav.*, 17, 495, 1982.
14. **Rönnbäck, L.,** Morphine preference in individual rats after morphine ingestion, *Psychopharmacology*, 102, 257, 1990.
15. **Rönnbäck, L.,** Is there a genetic control of morphine preference in rat?, *Pharmacol. Biochem. Behav.*, 35, 15, 1990.
16. **Belknap, J. K., Danielson, P. W., Laursen, S. E., and Noordewier, B.,** Selective breeding for levorphanol-induced antinociception on the hot plate assay: commonalities in mechanisms of action with morphine, pentazocine, ethylketocyclazocine, U-50488H and clonidine in mice, *J. Pharmacol. Exp. Ther.*, 241, 477, 1987.

17. **Lieblich, I., Cohen, E., and Beiles, A.,** Selection for high and for low rates of self-stimulation in rats, *Physiol. Behav.*, 21, 843, 1978.
18. **Dymshitz, J. and Lieblich, I.,** Opiate reinforcement and naloxone aversion, as revealed by place preference paradigm, in two strains of rats, *Psychopharmacology*, 92, 473, 1987.
19. **Blum, K., Briggs, A. H., DeLallo, L., Elston, S. F. A., and Ochoa, R.,** Whole brain methionine-enkephalin of ethanol-avoiding and ethanol-preferring C57BL mice, *Experientia*, 38, 1469, 1982.
20. **Blum, K., Elston, S. F. A., DeLallo, L., Briggs, A. H., and Wallace, J. E.,** Ethanol acceptance as a function of genotype amounts of brain [Met]enkephalin, *Proc. Natl. Acad. Sci. U.S.A.*, 80, 6510, 1983.
21. **Antelman, S. M., Eichler, A. J., Black, C. A., and Kocan, D.,** Interchangeability of stress and amphetamine in sensitization, *Science*, 207, 329, 1980.
22. **Piazza, P. V., Deminière, J.-M., Le Moal, M., and Simon, H.,** Factors that predict individual vulnerability to amphetamine self-administration, *Science*, 245, 1512, 1989.
23. **Piazza, P. V., Deminière, J.-M., Le Moal, M., and Simon, H.,** Stress- and pharmacologically-induced behavioral sensitization increases vulnerability to acquisition of amphetamine self-administration, *Brain Res.*, 514, 22, 1990.
24. **Ho, W. K. K., Wen, H. L., and Ling, N.,** Beta-endorphin immunoactivity in the plasma of heroin addicts and normal subjects, *Neuropharmacology*, 19, 117, 1980.
25. **Facchinetti, F., Grasso, A., Petraglia, F., Parrini, D., Volpe, A., and Genazzani, A. R.,** Impaired circadian rhythmicity of β-lipotropin, β-endorphin and ACTH in heroin addicts, *Acta Endocrinol. (Copenhagen)*, 105, 149, 1984.
26. **Kreek, M. J., Wardlaw, S. L., Hartman, N., Raghunath, J., Friedman, J., Schneider, B., and Frantz, A. G.,** Circadian rhythms and levels of β-endorphin, ACTH, and cortisol during chronic methadone maintenance treatment in humans, *Life Sci.*, 33 (Suppl. I), 409, 1983.
27. **Cushman, P., Morris, D., Adams, M., and Dewey, W.,** Opiate addiction and plasma β-endorphin-like immunoreactivity: methadone maintained, recently detoxified and early naltrexone treated ex-addicts, *Alcohol Drug Res.*, 7, 533, 1987.
28. **Kosten, T. R., Kreek, M. J., Swift, C., Carney, M. K., and Ferdinands, L.,** Beta-endorphin levels in CSF during methadone maintenance, *Life Sci.*, 41, 1071, 1987.
29. **Emmrich, H. M., Nusslet, L., Gramsch, C., and John, S.,** Heroin addiction: beta-endorphin immunoreactivity in plasma increases during withdrawal, *Pharmakopsychiatria*, 16, 93, 1983.
30. **Kosten, T. R., Kreek, M. J., Ragunath, J., and Kleber, H. D.,** A preliminary study of beta endorphin during chronic naltrexone maintenance treatment in ex-opiate addicts, *Life Sci.*, 39, 55, 1986.
31. **Holmstrand, J., Gunne, L. M., Wahlström, A., and Terenius, L.,** CSF-endorphins in heroin addicts during methadone maintenance and during withdrawal, *Pharmakopsychiatria*, 14, 126, 1981.

32. **O'Brien, C. P., Terenius, L. Y., Nyberg, F., and McLellan, A. T.,** CSF endorphins in chronic opioid-dependent humans, *NIDA Res. Monogr.*, 67, 373, 1985.
33. **Genazzani, A. R., Nappi, G., Facchinetti, F., Mazzella, G. L., Parrini, D., Sinforiani, E., Petraglia, F., and Savoldi, F.,** Central deficiency of β-endorphin in alcohol addicts, *J. Clin. Endocrinol. Metab.*, 55, 583, 1982.
34. **Gianoulakis, C., Béliveau, D., Angelogianni, P., Meaney, M., Thavundayil, J., Tawar, V., and Dumas, M.,** Different pituitary β-endorphin and adrenal cortisol response to ethanol in individuals with high and low risk for future development of alcoholism, *Life Sci.*, 45, 1097, 1989.
35. **Blum, K., Noble, E. P., Sheridan, P. J., Montgomery, A., Ritchie, T., Jagdeeswaran, P., Nogami, H., Briggs, A. H., and Cohn, J. B.,** Allelic association of human dopamine D_2 receptor gene in alcoholism, *JAMA*, 263, 2055, 1990.
36. **Regier, D. A., Farmer, M. E., Rae, D. S., Locke, B. Z., Keith, S. J., Judd, L. L., and Goodwin, F. K.,** Comorbidity of mental disorders with alcohol and other drug abuse. Results from the epidemiologic catchment area (ECA) study, *JAMA*, 264, 2511, 1990.
37. **Khantzian, E. J.,** The self-medication hypothesis of addictive disorders: focus on heroin and cocaine dependence, *Am. J. Psychiatry*, 142, 1259, 1985.
38. **Lehman, A. F., Myers, C. P., and Corty, E.,** Assessment and classification of patients with psychiatric and substance abuse syndromes, *Hosp. Commun. Psychiatry*, 40, 1019, 1989.
39. **McLellan, A. T., Woody, G. E., and O'Brien, C. P.,** Development of psychiatric disorders in drug abusers, *N. Engl. J. Med.*, 301, 1310, 1979.
40. **Vardy, M. M. and Kay, S. R.,** LSD psychosis or LSD-induced schizophrenia? A multimethod inquiry, *Arch. Gen. Psychiatry*, 40, 877, 1983.
41. **Gawin, F. H. and Kleber, H. D.,** Abstinence symptomatology and psychiatric diagnosis in cocaine abusers: clinical observations, *Arch. Gen. Psychiatry*, 43, 107, 1986.
42. **Weiss, R. D., Mirin, S. M., Michael., J. L., and Sollogub, A. C.,** Psychopathology in chronic cocaine abusers, *Am. J. Drug Alcohol Abuse*, 12, 17, 1986.
43. **Kleinman, P. H., Miller, A. B., Millman, R. B., Woody, G. E., Todd, T., Kemp, J., and Lipton, D. S.,** Psychopathology among cocaine abusers entering treatment, *J. Nerv. Ment. Dis.*, 178, 442, 1990.
44. **Dorus, W. and Senay, E. C.,** Depression, demographic dimensions, and drug abuse, *Am. J. Psychiatry*, 137, 699, 1980.
45. **Rounsaville, B. and Weissman, M.,** Diagnosis and symptoms of depression in opiate addicts, *Arch. Gen. Psychiatry*, 39, 151, 1982.
46. **Rounsaville, B. J., Anton, S. F., Carroll, K., Budde, D., Prusoff, B. A., and Gawin, F.,** Psychiatric diagnoses of treatment seeking cocaine abusers, *Arch. Gen. Psychiatry*, 48, 43, 1991.
47. **Rounsaville, B. J., Weissman, M. M., Kleber, H., and Wilber, C.,** Heterogeneity of psychiatric diagnosis in treated opiate addicts, *Arch. Gen. Psychiatry*, 39, 161, 1982.

48. **Cox, B. J., Norton, G. R., Swinson, R. P., and Endler, N. S.,** Substance abuse and panic-related anxiety: a critical review, *Behav. Res. Ther.*, 28, 385, 1990.
49. **Nemiah, J. C. and Uhde, T. W.,** Obsessive-compulsive disorder, in *Comprehensive Textbook of Psychiatry,* Kaplan, H. I. and Sadock, B. J., Eds., Williams & Wilkins, Baltimore, 1989, 984.
50. **Goldhaber, S. Z., Buron, C., and Minkoff, K.,** An obsessive compulsive neurotic patient in crisis, *West. J. Med.*, 127, 120, 1977.
51. **Zivich, J. M.,** Alcoholic subtypes and treatment effectiveness, *J. Consulting Clin. Psychol.*, 49, 72, 1981.
52. **Zohar, J. and Insel, T. R.,** Obsessive-compulsive disorder and alcoholism, *J. Clin. Psychiatry*, 47, 153, 1986.
53. **Schaefer, M. R.,** Severity of alcohol dependence and its relationship to additional psychiatric symptoms in male alcoholic inpatients, *Am. J. Drug Alcohol Abuse*, 13, 435, 1987.
54. **Jaffe, J. H.,** Drug dependence: opioids, nonnarcotics, nicotine (tobacco), and caffeine, in *Comprehensive Textbook of Psychiatry,* Kaplan, H. I. and Sadock, B. J., Eds., Williams & Wilkins, Baltimore, 1989, 642.
55. **Kay, S. R., Kalathara, M., and Meinzer, A. E.,** Diagnostic behavioral characteristics of psychiatric patients who abuse substances, *Hosp. Commun. Psychiatry*, 40, 1062, 1989.
56. **Quitkin, F. M., Rifkin, A., Kaplan, J., and Klein, D. F.,** Phobic anxiety syndrome complicated by drug dependence, *Arch. Gen. Psychiatry*, 27, 159, 1972.
57. **Mirin, S. M., Weiss, R. D., Griffin, M. L., and Michael, J.,** Psychopathology in drug abusers and their families, *Compr. Psychiatry*, 32, 36, 1991.
58. **Wurmser, L.,** Psychoanalytic considerations of the etiology of compulsive drug use, *J. Am. Psychoanal. Assoc.*, 22, 820, 1974.
59. **Levenson, R. W., Oyama, O. N., and Meek, P. S.,** Greater reinforcement from alcohol for those at risk: parental risk, personality risk, and sex, *J. Abnorm. Psychol.*, 96, 242, 1987.
60. **Hwu, H.-G., Yeh, Y.-L., and Wang, J.-D.,** Risk factors of alcoholism among Taiwan aborigines: implications for etiological models and the nosology of alcoholism, *Acta Psychiatr. Scand.*, 83, 267, 1991.
61. **Goldberg, J., Eisen, S. A., True, W. R., and Henderson, W.,** A twin study of the effects of Vietnam conflict on alcohol drinking patterns, *Am. J. Public Health*, 80, 570, 1990.
62. **Begleiter, H., Porjesz, B., Rawlings, R., and Eckhardt, M.,** Auditory recovery function and P3 in boys at high risk for alcoholism, *Alcohol*, 4, 315, 1987.
63. **O'Connor, S., Hesselbrock, V., Tasman, A., and DePalma, N.,** P3 amplitude in two different tasks are decreased in young men with a history of paternal alcoholism, *Alcohol*, 4, 323, 1987.
64. **Patterson, B. W., Williams, H. L., McLean, G. A., Smith, L. T., and Schaefer, K. W.,** Alcoholism and family history of alcoholism: effects of visual and auditory event-related potentials, *Alcohol*, 4, 265, 1987.

65. **Herning, R. I., Hickey, J. E., Pickworth, W. B., and Jaffe, J. H.,** Auditory event related potentials in adolescents at risk for drug abuse, *Biol. Psychiatry*, 25, 598, 1989.
66. **Pickworth, W. B., Brown, B. S., Hickey, J. E., and Muntaner, C.,** Effects of self-reported drug use and antisocial behavior on evoked potentials in adolescents, *Drug Alcohol Depend.*, 25, 105, 1990.
67. **Hill, H. E., Haertzen, C. A., and Glaser, R.,** Personality characteristics of narcotic addicts as indicated by the MMPI, *J. Gen. Psychol.*, 62, 127, 1960.
68. **Hill, H. E., Haertzen, C. A., and Davis, H.,** An MMPI factor analytic study of alcoholics, narcotic addicts and criminals, *Q. J. Stud. Alcohol*, 23, 411, 1962.
69. **Astin, A. W.,** A factor study of the MMPI psychopathic deviate scale, *J. Consult. Psychol.*, 23, 550, 1959.
70. **Haertzen, C. A., Martin, W. R., Ross, F. E., and Neidert, G. L.,** Psychopathic state inventory (PSI): development of a short test for measuring psychopathic states, *Int. J. Addict.*, 15, 137, 1980.
71. **Loper, R. G., Kemmeier, S. M. L., and Hoffmann, H.,** MMPI characteristics of college freshman males who later became alcoholics, *J. Abnorm. Psychol.*, 82, 159, 1973.
72. **Craig, R. J.,** Personality characteristics of heroin addicts: A review of the empirical literature with critique — part II, *Int. J. Addict.*, 14, 607, 1979.
73. **Calsyn, D. A., Roszell, D. K., and Chaney, E. F.,** Validation of MMPI profile subtypes among opioid addicts who are beginning methadone maintenance treatment, *J. Clin. Psychol.*, 45, 991, 1989.
74. **Haertzen, C. A. and Hooks, N. T., Jr.,** Changes in personality and subjective experience associated with chronic administration and withdrawal of opiates, *J. Nerv. Mental Dis.*, 118, 606, 1969.
75. **Harris, R. and Linn, M. W.,** Differential response of heroin and nonheroin abusers to inpatient treatment, *Am. J. Drug Alcohol Abuse*, 5, 179, 1978.
76. **Hendriks, V. M., Steer, R. A., Platt, J. J., and Metzger, D. S.,** Psychopathology in Dutch and American heroin addicts, *Int. J. Addict.*, 25, 1051, 1990.
77. **Muntaner, C., Walter, D., Nagoshi, C., Fishbein, D., Haertzen, C. A., and Jaffe, J. H.,** Self-report vs. laboratory measures of aggression as predictors of substance abuse, *Drug Alcohol Depend.*, 25, 1, 1990.
78. **Kalodner, C. R., Delucia, J. L., and Ursprung, A. W.,** An examination of the tension reduction hypothesis: the relationship between anxiety and alcohol in college students, *Addictive Behav.*, 14, 649, 1989.
79. **Kandel, D. B.,** Epidemiological and psychosocial perspectives on adolescent drug use, *J. Am. Acad. Child Psychiatry*, 21, 328, 1982.
80. **Zuckerman, M., Bone, R. N., Neary, R., Mangelsdorff, D., and Brustman, B.,** What is the sensation seeker? Personality trait and experience correlates of the sensation-seeking scales, *J. Consult. Clin. Psychol.*, 39, 308, 1972.
81. **Khavari, K. A., Mabry, E., and Humes, M.,** Personality correlates of hallucinogen use, *J. Abnorm. Psychol.*, 86, 172, 1977.

82. **Skolnick, N. J. and Zuckerman, M.,** Personality change in drug abusers: a comparison of therapeutic community and prison groups, *J. Consult. Clin. Psychol.*, 47, 768, 1979.
83. **Zuckerman, M.,** Sensation seeking and the endogenous deficit theory of drug abuse, *NIDA Res. Monogr.*, 74, 59, 1986.
84. **Calsyn, D. A. and Saxon, A. J.,** Personality disorder subtypes among cocaine and opioid addicts using the Milton multiaxial inventory, *Int. J. Addict.*, 25, 1037, 1990.
85. **Freud, S.** cit. post **Salzman, L.,** The neurotic compromise of addiction, in *Addiction — Theory and Treatment*, Goldman, G. D. and Milman, D. S., Eds., Kendall/Hunt, Dubuque, IA, 1980, 3.
86. **Hatterer, L.,** The metamorphosis of addiction, in *Addiction — Theory and Treatment,* Goldman, G. D. and Milman, D. S., Eds., Kendall/Hunt, Dubuque, IA, 1980, 73.
87. **Sternberg, D. and Cohen, A.,** Drug dependency: the substitution of substance nutrients for love, in *Addiction — Theory and Treatment*, Goldman, G. D. and Milman, D. S., Eds., Kendall/Hunt, Dubuque, IA, 1980, 21.

12 New Strategies in the Treatment of Opiate and Cocaine Addiction

I. TRADITIONAL APPROACHES

Even a short overview of the traditionally used techniques in the treatment of opiate and cocaine addiction would exceed the scope of this book. Therefore, in this respect only some recent papers will be mentioned here, reflecting the present position of clinicians.

Undoubtedly, the *methadone replacement therapy* of heroin addiction seems to work in a significant proportion of the patients.[1] Its success is probably related to its pharmacokinetic properties. Namely, administering it once daily provides a relatively stable blood level, i.e., the lack of abrupt decline in its plasma concentration prevents any serious withdrawal symptoms.[1] Prevention of the withdrawal symptoms enables the patients to resume their normal everyday activities; thus it facilitates their social adaptation. However, it is important to emphasize that replacement therapy does not reduce the desire for opiates and does not eliminate craving, but only makes the opiate-taking behavior socially adaptable. Even in the case of successful methadone replacement therapy of heroin addiction, it must be administered continuously for years. If methadone administration is omitted or replaced by placebo, the patients stop attending the methadone clinic and resort again to heroin.[2] Thus its therapeutic application does not mean more than replacement of a very strong, but short-acting opiate agonist with another one which is less active, but suppresses withdrawal by providing a more stable blood level.

Another traditional approach is the use of *opiate antagonists*. Naltrexone treatment also means blockade of opiate receptors by a strong antagonist with a rather long half-life upon per os, therefore it seems to be ideal for drug therapy.[3,4] Thus, until naltrexone is eliminated from the body (at least a full day), the regularly abused opiate agonists do not induce euphoria. Hence the "conditional component" of drug abuse, where the drug-taking "ceremony" serves as conditional stimulus and the drug-induced euphoria is the reinforcement, is extinguished. However, by definition, opiate antagonists do not reduce craving, actually they enhance it, which may explain that the great majority of heroin addicts ''maintained'' on naltrexone relapse.[1]

A new procedure, the *clonidine* detoxification program initiated not very long ago by Gold et al.,[5,6] seemed to be very promising at the beginning. First, because of being a strong agonist of α_2-receptors, it suppresses the majority of autonomic symptoms of opiate withdrawal.[5,6] Second, some basic data indicate that in some areas of the brain, opiate and α_2-receptors function synergistically,[7] i.e., clonidine might be perceived as a replacement for opiates. However, as stated by the initiators themselves (Washton et al.)[8] of this new approach, ''...Clonidine may enable more addicts to reach an opiate-free state, but it does nothing to prevent relapse after an initial abstinence has been achieved...''.

Thus, though methadone (and some other kinds of replacement therapy with long-acting μ-opiate agonists), opiate antagonists (almost exclusively naltrexone), and clonidine are widely used in the pharmacotreatment of opiate addicts, clinicians keep actively looking for more efficient ways of treating the opiate/cocaine and alcohol addicts.

II. ANTIDEPRESSANTS IN THE TREATMENT OF ADDICTION

After an initial report of Tennant and Rawson,[9] Gawin and Kleber[10] reported on successful treatment of cocaine addiction with imipramine. In this open study[10] the patients decreased or completely stopped cocaine use during and after a 12-week treatment, their craving for cocaine was reduced, and they experienced an overall alleviation of chronic anhedonia which prompted them before treatment to resort to ''crack''. In view of the clinical significance of this finding, many groups tried to replicate these data using various classical tricyclic antidepressants, but the findings are rather divergent. Some of the subsequent studies (several of them blind) confirmed that desipramine really reduced the craving for cocaine.[11–15] Nevertheless, in the majority of independent works, desipramine (and imipramine) failed to prevent the relapse to cocaine abuse.[16,17]

The controversy might at least in part be attributed to differences in duration of treatment, dose regimen, and selection of patients. Namely, Gawin et al.[13] emphasize that the difference between the control and drug-treated groups only becomes evident after a couple of weeks, or rather months. A further explanation for the controversial findings was provided by Fischman et al.,[15] who reported that healthy male volunteers with a history of cocaine use did not reduce its i.v. self-administration upon pretreatment with desipramine, but they did report on reduced craving (under blind conditions). Furthermore, desipramine potentiated the deleterious cardiovascular effects of cocaine, its hypertensive and cardioaccelerant actions.[15] Probably it is worth mentioning here that controversial findings were obtained in Rhesus monkeys as well. As reported by Mello et al.,[18] chronic treatment with desipramine attenuated the self-administration of cocaine in some subjects, but facilitation of the behavior was seen in others.

Not less intriguing is the eventual mechanism of interaction of desipramine with cocaine. As expected from a dopamine-uptake inhibitor, cocaine increases the concentration of 3-methoxytyramine, i.e., the extracellular metabolite of dopamine without altering the level of its intracellular metabolite, homovanillic acid.[19] As is well established, tricyclic antidepressants also inhibit the reuptake of monoamines, including that of dopamine. Thus their mode of action is at least partially similar. Dackis and Gold[20] proposed that cocaine addiction is actually due to depletion of dopamine after extensive use. Actually, as discussed in Chapter 7, chronic cocaine treatment initially enhances the dopamine release in the nucleus accumbens, but subsequently just a decrease in its release can be recorded, and this phenomenon is attributed to exhaustion of dopamine stores. On the other hand, it is well established that tricyclic antidepressants downregulate the presynaptic adrenergic autoreceptors.[21] They supposedly downregulate the dopaminergic autoreceptors too.[22] Thus they might somehow facilitate the dopamine release from the ''exhausted'' nerve terminals and obviously this effect develops only after repeated treatments. On the other hand, Spyraki and Fibiger[23] provided evidence in behavioral experiments for supersensitivity of postsynaptic dopaminergic receptors in the mesolimbic dopaminergic system upon chronic administration of desipramine. Therefore, it is not surprising that after repeated treatments with antidepressants such as desipramine or citalopram, the amphetamine- and apomorphine-induced locomotor facilitation was found enhanced.[23,24] Downregulation of the autoreceptors on the dopamine-releasing terminals and upregulation of the postsynaptic ones in the mesolimbic system mean functional hyperactivity due to enhancement of dopamine liberation and augmented responsivity. Such changes are equivalent with the activation of the mesolimbic dopaminergic reward system. Actually, Fibiger and Phillips[25] reported on a significant decrease in the threshold of rewarding hypothalamic brain stimulation upon long-term desipramine treatment. Thus the beneficial effect of chronic desipramine treatment in the treatment of cocaine addiction might be due to upregulation of the mesolimbic dopamine system, leading to elimination of the anhedonia elicited by excessive cocaine use.

According to another line of clinical and experimental data, selective inhibitors of serotonin (5-HT) uptake blockers might also have a beneficial effect. Both fluoxetine[26] and zimelidine[27] are selective inhibitors of 5-HT (re)uptake. In humans there are positive data on the treatment of alcoholism by zimelidine[28] and fluoxetine,[28] as well. In the corresponding animal experiments in rats, fluoxetine reduced the self-administration of cocaine[30] and amphetamine,[31,32] the preference for ethanol,[33] and intracranial self-stimulation[34] as well. Zimelidine reduces the voluntary consumption of morphine.[35] In conjunction with these data it must be added that tryptophan, the precursor of 5-HT, has also been found to attenuate the self-administration of amphetamine.[31,36] The beneficial effect of 5-HT uptake blockers is not easy to interpret. The eventual role of 5-HT in the function of the mesolimbic rewarding system is not clarified yet. Since fluoxetine also suppresses the oral consumption of highly palatable glucose + saccharin solution, it was proposed[30] that it inhibits self-administration by producing mild illness or malaise, i.e., its addition to abused substances reduces their rewarding potency. The same might apply to carbamazepine, another atypical antidepressant agent, which was claimed to be helpful in the treatment of alcohol withdrawal in humans[37] and to suppress both the i.v. self-administration of cocaine and oral consumption of palatable glucose + saccharin solution in rats.[38]

Nevertheless, it is rather strange that different classes of antidepressants can be used in the treatment of addiction, but their modes of action are claimed to be different. Anyhow, the great interindividual differences in responsivity and the long latency of eventual beneficial effect of antidepressants make their use in the treatment of addiction impractical. On the other hand, the clinical experience accumulated with their use provides additional information on the pathology of substance abuse.

III. THE USE OF DIRECT DOPAMINE RECEPTOR STIMULANTS

Several clinical studies indicate that dopamine receptor agonists provide direct relief in the treatment of cocaine addiction.

Dackis and Gold[39] reported on decrease in craving for cocaine and its reduced consumption upon bromocriptine treatment. Subsequent studies confirmed that bromocriptine reduces cocaine craving and its consumption, and improves the mood of addicts deprived of it.[11,14,40,41] In the only related study, coadministration of desipramine and apomorphine gave better results than bromocriptine alone.[11] Being a direct stimulant of dopamine receptors, it is not surprising that bromocriptine reduced craving in acute examinations too,[39] i.e., its beneficial effect is not indirect as in the case of desipramine. Nevertheless, it is reasonable to suppose that long-term treatment with bromocriptine reverses the process of postsynaptic sensitization of dopamine receptors, i.e., the increase in B_{max} due to the exhaustion of presynaptic dopamine stores upon chronic abuse of cocaine.[11] In agreement with the clinical data, bromocriptine has been found to diminish the self-administration of cocaine in rats.[42]

Nevertheless, both the mode of action and the usefulness of bromocriptine as standard procedure to treat addiction require further investigation. As generally known of ergot alkaloids, bromocriptine induces a lot of unpleasant autonomic side-effects.[43] Being a direct dopamine agonist, bromocriptine has abuse potential by its own right: it induces conditioned place preference[44] and it is self-administered by monkeys.[45] As for its mode of action, some issues require further clarification. Namely, bromocriptine antagonizes the behavioral stimulant effects of cocaine in rats,[46] which is obviously not expected of a dopamine agonist. This antagonism might be due to interaction with the presynaptic dopamine (auto)receptors.[47] Another possibility is that it is only a partial agonist of the dopamine receptors, namely its stereotypy-inducing effect is rather weak.[46]

Some studies indicate that some other dopamine receptor agonists might also have beneficial effect. Cocaine craving could be reduced by both apomorphine[48] and a carbidopa + l-DOPA combination.[49] Anyhow, the feasibility of treatment of addiction with dopamine agonists requires further clinical and experimental examinations.

Finally, it is logical to mention here that in one open study[50] a long-acting neuroleptic agent, flupenthixol decanoate, was found beneficial in the treatment of cocaine abuse. The authors attribute their unexpected positive findings to inhibition of presynaptic dopamine autoreceptors (resulting in enhanced dopamine release) with simultaneous blockade of the postsynaptic one, i.e., abolition of the positive reinforcing effect of cocaine.

IV. THE USE OF BUPRENORPHINE, A κ-ANTAGONIST IN THE TREATMENT OF OPIATE AND COCAINE ADDICTION

In their pioneering work postulating the existence of κ-receptors, Martin et al.[51] qualified buprenorphine a partial μ-agonist, but an eventual interaction with κ-ones was not suggested. Nevertheless, two points were indicated. In chronic spinal dogs,

- Buprenorphine induced slight physical dependence if given alone.
- Though it elicited withdrawal symptoms in morphine-dependent dogs, buprenorphine alleviated the symptoms of morphine withdrawal.[51]

Subsequently, Jasinski et al.[52] described that in spite of its strong analgesic potency, buprenorphine blocked the effects of classical opiates and induced only a very low level of physical dependence. For this reason, buprenorphine was advised to be used for opiate detoxification instead of many other known partial μ-agonists-antagonists.[52]

While the clinical examination of buprenorphine was initiated, it also was further examined in animal experiments and, actually, until now the two lines of investigations (pharmacological and clinical) have not been correlated conclusively.

Buprenorphine substitutes for morphine in drug-discrimination studies[53,54] and maintains self-administration behavior[54,55] as μ-agonists do. It induces conditioned place preference and in subthreshold doses synergizes the similar effect of cocaine.[56,57] Furthermore, buprenorphine decreases the threshold for intracranial self-stimulation.[58] On the other hand, buprenorphine does not induce physical dependence in animal experiments and it precipitates withdrawal syndrome in opiate-dependent mice,[59] rats,[60] and monkeys[61] as expected of a μ-receptor antagonist. In line with these findings, buprenorphine in humans dose dependently suppresses the pupillary constriction and respiratory depression induced by hydromorphone, a pure μ-agonist.[62] Furthermore, buprenorphine, like other drugs of abuse, enhances the release of dopamine from the nucleus accumbens and potentiates the similar effect of cocaine.[56]

However, buprenorphine also binds to the κ-receptors.[63–65] Furthermore, as demonstrated in various operant conditioning paradigms in rats and pigeons, it has distinct κ-receptor antagonist properties. Notably, in rats it suppresses the stimulus discriminative effect of the κ-agonist, U-50488H.[66–68] Furthermore, it antagonizes the bremozocine-induced diuresis in rats.[66] In mice it antagonizes the antinociceptive effect of U-50488H in the writhing test.[69] Thus its κ-antagonist properties are well established in animal studies, but they have eventually been overlooked by clinicians.

As for its application in the treatment of drug abuse, Mello and coworkers[70,71] first described the usefulness of buprenorphine in detoxification of heroin addicts. Subsequently, the same group reported on significant suppression of heroin self-administration in Macaque monkeys upon chronic administration of buprenorphine, whereas methadone had no effect.[72] Since the first preliminary studies a series of clinical examinations confirmed the feasibility of buprenorphine in heroin detoxification.[73–79] In most of these studies the relative ease of withdrawing heroin is emphasized. In addition, eventually its moderate euphoric and antidysphoric effects[77] and the lack of withdrawal symptoms upon termination of buprenorphine is emphasized.[78]

As reported by Kosten et al.[80,81] in heroin addicts buprenorphine maintenance also reduces cocaine consumption. Naltrexone maintenance also reduced the abuse of cocaine by heroin addicts, but methadone did not help.[81] An ongoing study[82] indicates that buprenorphine suppresses the abuse of both heroin and cocaine in addicts. The eventual modulation by buprenorphine of cocaine abuse in patients not abusing opiates has not been examined yet. However, several studies have been carried out in animals. According to Mello's group,[82–84] buprenorphine does reduce the self-administration of cocaine in Rhesus monkeys. This effect is selective, since tolerance developed quickly to the initial suppression of responding for food, but the suppression of operant responses for cocaine remained greatly reduced in the animals till buprenorphine was regularly administered.[82,84]

The beneficial effect of buprenorphine cannot simply be due to its μ-antagonist effect, since, as discussed above, opiate antagonists such as naltrexone do not reduce craving. It cannot be explained by μ-agonist action either, since it does not induce physical dependence. This effect cannot be due to an eventual κ-agonists action either, since κ-agonists are distinctly dysphoric (see Chapter 4).

On the other hand, the suppression of both heroin and cocaine self-administration in animals and their abuse in humans can easily be explained by its κ-antagonist property. As discussed in Chapter 9, withdrawal-induced hypophoria is supposed to be due to a compensatory hyperactivity/hypertrophy of the endogenous κ-opioid system. Buprenorphine should abolish the consequences of enhanced release of dynorphin by occupying its receptors. This is, however, equivalent to the suppression of the subjective symptoms of opiate withdrawal. The same might apply to cocaine withdrawal, which results in dopamine release and in secondary disinhibition of endogenous κ-agonists. On the other hand, its own weak euphoric effect might be due to either

- its partial μ-agonist property and/or
- the relative functional preponderance of endogenous enkephalins/and β-endorphin, i.e., the ligands of μ-receptors and receptors inducing euphoria.

Thus, based on the latest data in the literature it is tempting to speculate that the drugs of the future in the treatment of addiction will be the κ-antagonists, since they suppress the mechanism of dysphoria elicited by interruption of dopamine release in the mesolimbic system (Table 1). Furthermore, this mechanism of action should be beneficial whether dopamine is released directly, as by cocaine, or indirectly as seen with opiates.

TABLE 1
The Main Recently Used Strategies in the Treatment of Opiate and Cocaine Addiction

The Procedure is relatively:	Basic concept of the therapeutic approach	Disadvantage of the procedure	Ref. (a recent review)
Old	Methadone maintenance, i.e., replacement of a strong, but short-acting opiate (heroin) with a weaker, but long-acting one	Methadone also is an opiate; there is no withdrawal only replacement, making the addict more sociable	1,2
Old	Administration of an opiate antagonist, mostly naltrexone	Elimination of euphoria without reducing craving; most addicts relapse	1,3, 4

Old	Clonidine, i.e., suppression of the withdrawal by the help of an α_2 receptor agonist	It does not prevent relapse, only suppresses certain, mostly autonomic symptoms of withdrawal	5,6, 8
Relatively new	Chronic administration of antidepressants, mostly desiparmine which apparently act by making more dopamine available for the dopamine receptors	The effect (reduction of dopamine receptor supersensitivity) develops slowly and the clinical data are rather equivocal	10–17, 20, 22–24
Relatively new	Use of dopamine receptor stimulants such as bromocriptine, apomorphine, or L-DOPA to counterbalance the hypoactivity of endogenous dopamine-releasing neurons	Clinical efficacy is questionable; many adverse effects	11,14, 39–41, 48,49
New	Buprenorphine treatment, which is reported to reduce craving for heroin and cocaine as well	Continuous (once a day or alternate day) treatment is needed; allegedly efficacious treatment without considerable side-effects	57,70, 71,73–78 80,81, 83,84

REFERENCES

1. **Jaffe, J. H.,** Psychoactive substance use disorders, in *Comprehensive Textbook of Psychiatry/V,* Vol. 1, Kaplan, H. I. and Sadock, B. J.,Eds., Williams & Wilkins, Baltimore, 1989, 642.
2. **Madden, C., Singer, G., Jagoda, J., Jethwa, J., and Croxford, R.,** Opiate seeking behavior under conditions of methadone detoxification and removal, *Int. J. Addict.*, 21, 947, 1986.
3. **Resnick, R. B., Schuyten-Resnick, E., and Washton, A. M.,** Narcotic antagonists in the treatment of opioid dependence: review and commentary, *Comp. Psychiatry*, 20, 116, 1979.
4. **Gold, M. S., Dackis, C. A., Potash, A. L. C., Sternbach, H. H., Annitto, W. J., Martin, D., and Paternak, M.,** Naltrexone, opiate addiction, and endorphins, *Med. Res. Rev.*, 2, 211, 1982.
5. **Gold, M. S., Pottash, A. L., Sweeney, D. R., and Kleber, H.,** Opiate withdrawal using clonidine; a safe, effective, and rapid nonopiate treatment, *JAMA*, 243, 343, 1980.
6. **Gold, M. S., Pottash, A. L., and Extein, I.,** Clinical utility of clonidine in opiate withdrawal, *NIDA Res. Monogr.*, 34, 95, 1981.
7. **Aghajanian, G. K.,** Tolerance of locus coeruleus neurones to morphine and suppression of withdrawal response with clonidine, *Nature*, 276, 186, 1978.
8. **Washton, A. M., Gold, M. S., and Pottash, A. C.,** Opiate and cocaine dependence. Techniques to help counter the rising tide, *Postgrad. Med.*, 77, 293, 1985.
9. **Tennant, F. S. and Rawson, R. A.,** Cocaine and amphetamine dependence treated with desipramine, *NIDA Res. Monogr.*, 43, 351, 1982.
10. **Gawin, F. H. and Kleber, H. D.,** Cocaine abuse treatment. Open pilot trial with desipramine and lithium carbonate, *Arch. Gen. Psychiatry*, 41, 903, 1984.
11. **Giannini, A. J. and Billett, W.,** Bromocriptine-desiparmine protocol in treatment of opiate addiction, *J. Clin. Pharmacol.*, 27, 549, 1987.
12. **Kosten, T. R., Schuman, B., Wright, D., and Gawin, F. H.,** A preliminary study of imipramine in the treatment of cocaine abuse in the methadone maintenance addicts, *J. Clin. Psychiatry*, 48, 442, 1987.
13. **Gawin, F. H., Kleber, H. D., Byck, R., Rounsaville, B. J., Kosten, T. R., Jatlow, P. I., and Morgan, C.,** Desipramine facilitation of initial cocaine abstinence, *Arch. Gen. Psychiatry*, 46, 117, 1989.
14. **Extein, I. L., Sharman, A., Gold, M. S., Morales, A., Jr., Goodnick, P. J., and Gross, D. A.,** Comparison of bromocriptine and desipramine in cocaine withdrawal, *Ann. Clin.*, 1, 193, 1989.
15. **Fischman, M. W., Foltin, R. W., Nestadt, G., and Pearlson, G. D.,** Effects of desipramine maintenance on cocaine self-administration by humans, *J. Pharmacol. Exp. Ther.*, 253, 760, 1990.
16. **Baxter, L. R.,** Desipramine in the treatment of hypersomnolence following abrupt cessation of cocaine use, *Am. J. Psychiatry*, 140, 1525, 1983.
17. **Weiss, R. D.,** Relapse to cocaine abuse after initiating cocaine treatment, *JAMA*, 260, 2545, 1988.

18. **Mello, N. K., Lukas, S. E., Bree, M. P., and Mendelson, J. H.,** Desipramine effects on cocaine-self-administration by rhesus monkeys, *Drug Alcohol Depend.*, 26, 103, 1990.
19. **Fekete, M. and Borsy, J.,** Chlorpromazine-cocaine antagonism. Its relation to changes of dopamine metabolism in the brain, *Eur. J. Pharmacol.*, 16, 171, 1971.
20. **Dackis, C. A. and Gold, M. S.,** New concepts in cocaine addiction: the dopamine depletion hypothesis, *Neurosci. Biobehav. Rev.*, 9, 469, 1985.
21. **Banerjee, S. P., Kung, L. S., Riggi, S. J., and Chanda, S. K.,** Development of beta adrenergic receptor subsensitivity by antidepressants, *Nature*, 268, 455, 1977.
22. **Chiodo, L. A. and Bunney, B. S.,** Typical and atypical neuroleptics: differential effects of chronic administration on the activity of A-9 and A-10 midbrain dopaminergic neurons, *J. Neurosci.*, 3, 1607, 1983.
23. **Spyraki, C. and Fibiger, H. C.,** Behavioral evidence for supersensitivity of postsynaptic dopamine receptors in the mesolimbic dopamine system after chronic administration of desipramine, *Eur. J. Pharmacol.*, 74, 195, 1981.
24. **Maj, Z., Rogoz, Z., Skuza, G., and Sowinska, H.,** Repeated treatment with antidepressant drugs increases the behavioral response to apomorphine, *J. Neural. Transm.*, 60, 273, 1984.
25. **Fibiger, H. C. and Phillips, A. G.,** Increased intracranial self-stimulation in rats after long-term administration of desipramine, *Science*, 214, 683, 1981.
26. **Fuller, R. W.,** Pharmacology of central serotonin neurons, *Annu. Rev. Pharmacol. Toxicol.*, 20, 111, 1980.
27. **Heel, R. C., Morley, P. A., Brogden, R. N., Carmine, A. A., Speight, T. M., and Avery, G. S.,** Zimelidine: a review of its pharmacologic properties and therapeutic efficacy in depressive illness, *Drugs*, 24, 169, 1982.
28. **Naranjo, C. A., Sellers, E. M., Roach, C. A., Woodley, D. V., Sanchez-Craig, M., and Sykora, K.,** Zimelidine-induced variations in alcohol intake by non-depressed heavy drinkers, *Clin. Pharm. Ther.*, 35, 374, 1984.
29. **Gorelick, D. A.,** Effect of fluoxetine on alcohol consumption, *Alcoholism*, 10, 113, 1986.
30. **Carroll, M. E., Lac, S. T., Asenico, M., and Kragh, R.,** Fluoxetine reduces cocaine self-administration in rats, *Pharmacol. Biochem. Behav.*, 35, 237, 1990.
31. **Leccese, A. P. and Lyness, W. H.,** The effects of putative 5-hydroxytryptamine receptor active agents on D-amphetamine self-administration in controls and rats with 5,7-dihydroxytryptamine median forebrain bundle lesions, *Brain Res.*, 303, 153, 1984.
32. **Yu, D. S. L., Smith, F. L., Smith, D. G., and Lyness, W. H.,** Fluoxetine-induced attenuation of amphetamine self-administration in rats, *Life Sci.*, 39, 1383, 1986.
33. **Zabik, J. E., Roache, J. D., Sidor, R., and Nash, J. F., Jr.,** The effects of fluoxetine on ethanol preference in the rat, *Pharmacologist*, 24, 204, 1982.
34. **Cazala, P.,** Effects of Lilly 110140 (fluoxetine) on self-stimulation behavior in the dorsal and ventral regions of the lateral hypothalamus in the mouse, *Psychopharmacology*, 71, 143, 1980.

35. **Rockman, G. E., Amit, Z., Bourque, C., Brown, Z. E., and Ogren, S.-O.,** Reduction in voluntary morphine consumption following treatment with zimelidine, *Arch. Int. Pharmacodyn.*, 244, 123, 1980.
36. **Smith, F. L., Yu, D. S., Smith, D. G., Lecesse, A. P., and Lyness, W. H.,** Dietary tryptophan supplements attenuate amphetamine self-administration in the rat, *Pharmacol. Biochem. Behav.*, 25, 849, 1986.
37. **Malcolm, R., Bellenger, J. C., Sturgis, E. T., and Anton, R.,** Double-blind controlled trial comparing carbamazepine to oxazepam treatment of alcohol withdrawal, *Am. J. Psychiatry*, 146, 617, 1989.
38. **Carroll, M. E., Lac, S. T., Asenico, M., Halikas, J. A., and Kragh, R.,** Effects of carbamazepine on self-administration of intravenously delivered cocaine in rats, *Pharmacol. Biochem. Behav.*, 37, 551, 1990.
39. **Dackis, C. A. and Gold, M. S.,** Bromocriptine as treatment of cocaine abuse, *Lancet*, 1, 1152, 1985.
40. **Dackis, C. A., Gold, M. S., Sweeney, D. R., Byron, J. P., Jr., and Climko, R.,** Single dose bromocriptine reverses cocaine craving, *Psychiatry Res.*, 20, 261, 1987.
41. **Tennant, F. S., Jr. and Shagherian, A. A.,** Double-blind comparison of amantadine and bromocriptine for ambulatory withdrawal from cocaine dependence, *Arch. Intern. Med.*, 147, 109, 1989.
42. **Hubner, C. B. and Koob, G. F.,** Bromocriptine produces decreases in cocaine self-administration in the rat, *Neuropsychopharmacology*, 3, 101, 1990.
43. **Reynolds, J. E. F., Ed.,** *Martindale The Extra Pharmacopoeia,* Pharmaceutical Press, London, 1011, 1989.
44. **Hoffman, D. C., Dickson, P. R., and Beninger, R. J.,** The dopamine D2 receptor agonists, quinpirole and bromocriptine produce conditioned place preferences, *Prog. Neuropsychopharmacol. Biol. Psychiatry*, 12, 315, 1988.
45. **Wolwerton, W. L., Goldberg, L. I., and Ginos, J. Z.,** Intravenous self-administration of dopamine receptor agonists by rhesus monkeys, *J. Pharmacol. Exp. Ther.*, 230, 678, 1984.
46. **Campbell, A., Phil, M., Baldessarini, R. J., Cremens, C., Teicher, M. H., Marsch, E., and Kula, N. S.,** Bromocriptine antagonizes behavioral effects of cocaine in the rat, *Neuropsychopharmacology*, 2, 209, 1989.
47. **Goldstein, M., Lew, J. Y., Nakamura, S., Battista, A. F., Lieberman, A., and Fuxe, K.,** Dopaminephilic properties of ergot alkaloids, *Fed. Proc. Fed. Am. Soc. Exp. Biol.*, 37, 2202, 1978.
48. **Hollander, E., Nunes, E., DeCaria, C. M., Quitkin, F. M., Cooper, T., Wager, S., and Klein, D. F.,** Dopaminergic sensitivity and cocaine abuse: response to apomorphine, *Psychiatry Res.*, 33, 161, 1990.
49. **Rosen, H., Flemenbaum, A., and Slater, V. L.,** Clinical trial of carbidopa-L-dopa combination for cocaine abuse, *Am. J. Psychiatry*, 143, 1493, 1986.
50. **Gawin, F. H., Allen, D., and Humblestone, B.,** Outpatient treatment of "crack" cocaine smoking with flupenthixol decanoate, *Arch. Gen. Psychiatry*, 46, 322, 1989.
51. **Martin, W. R., Eades, C. G., Thompson, J. A., Huppler, R. E., and Gilbert, P. E.,** The effects of morphine and nalorphine-like drugs in the non-dependent and morphine-dependent chronic spinal dog, *J. Pharmacol. Exp. Ther.*, 197, 517, 1976.

52. **Jasinski, D. R., Pevnick, J. S., and Griffith, J. D.,** Human pharmacology and abuse potential of analgesic buprenorphine, *Arch. Gen. Psychiatry*, 35, 501, 1978.
53. **Shannon, H. E., Cone, E. J., and Gorodetzky, C. W.,** Morphine-like discriminative stimulus effects of buprenorphine and demethoxybuprenorphine in rats: quantitative antagonism by naloxone, *J. Pharmacol. Exp. Ther.*, 229, 768, 1984.
54. **Young, A. M., Stephens, K. R., Hein, D. W., and Woods, J. H.,** Reinforcing and discriminative stimulus properties of mixed agonist-antagonist opioids, *J. Pharmacol. Exp. Ther.*, 229, 118, 1984.
55. **Mello, N. K., Bree, M. P., and Mendelson, J. H.,** Buprenorphine self-administration by rhesus monkey, *Pharmacol. Biochem. Behav.*, 15, 215, 1981.
56. **Brown, E. E., Finlay, J. M., Wong, J. T. F., Damsma, G., and Fibiger, H. C.,** Behavioral and neurochemical interactions between cocaine and buprenorphine: implications for the pharmacotherapy of cocaine abuse, *J. Pharmacol. Exp. Ther.*,256, 119, 1991.
57. **Kosten, T. A., Marby, D. W., and Nestler, E. J.,** Cocaine place preference is attenuated by chronic buprenorphine treatment, *Life Sci.*, 49, PL201, 1991.
58. **Hubner, C. B. and Kornetsky, C.,** The reinforcing properties of the mixed agonist-antagonist buprenorphine as assessed by brain-stimulation reward, *Pharmacol. Biochem. Behav.*, 30, 195, 1988.
59. **Cowan, A., Lewis, J. W., and Macfarlane, I. R.,** Agonist and antagonist properties of buprenorphine, a new antinociceptive agent, *Br. J. Pharmacol.*, 60, 537, 1977.
60. **Dum, J. E. and Herz, A.,** In vivo receptor binding of the opiate partial agonist buprenorphine is correlated with its agonist and antagonistic actions, *Br. J. Pharmacol.*, 74, 627, 1981.
61. **Aceto, M. D.,** Characterization of prototypical opioid antagonists, agonist-antagonists, and agonists in the morphine dependent rhesus monkey, *Neuropeptides*, 5, 15, 1984.
62. **Bickel, W. K., Stitzer, M. L., Bigelow, G. E., Liebson, I. A., Jasinski, D. R., and Johnson, R. E.,** Buprenorphine: dose-related blockade of opioid challenge effects in opioid dependent humans, *J. Pharmacol. Exp. Ther.*, 247, 47, 1988.
63. **Wood, P. L., Charleson, S. E., and Hudgin, R. L.,** Multiple opiate receptors: differential binding of μ, κ, and δ agonists, *Neuropharmacology*, 20, 1215, 1981.
64. **Lewis, J. W.,** Buprenorphine, *Drug Alcohol Depend.*, 14, 363, 1985.
65. **Su, T.-P.,** Further demonstration of kappa opioid binding sites in the brain: evidence for heterogeneity, *J. Pharmacol. Exp. Ther.*, 232, 144, 1985.
66. **Leander, J. D.,** Buprenorphine is a potent kappa opioid receptor antagonist, *Neuropharmacology*, 26, 1445, 1987.
67. **Negus, S. S. and Dykstra, L. A.,** κ Antagonist properties of burpenorphine, *Eur. J. Pharmacol.*, 156, 77, 1988.
68. **Negus, S. S., Picker, M. J., and Dykstra, L. A.,** Kappa antagonist properties of buprenorphine in non-tolerant and morphine-tolerant rats, *Psychopharmacology*, 98, 141, 1989.

69. **Leander, J. D.,** Buprenorphine is a potent κ-opioid receptor antagonist in pigeons and mice, *Eur. J. Pharmacol.*, 151, 457, 1988.
70. **Mello, N. K. and Mendelson, J. H.,** Buprenorphine suppresses heroin use by heroin addicts, *Science*, 207, 567, 1980.
71. **Mello, N. K., Mendelson, J. H., and Kuehnle, J. C.,** Buprenorphine effects on human self-administration: an operant analysis, *J. Pharmacol. Exp. Ther.*, 223, 30, 1982.
72. **Mello, N. K., Bree, M. P., and Mendelson, J. H.,** Comparison of buprenorphine and methadone effects on opiate self-administration in primates, *J. Pharmacol. Exp. Ther.*, 225, 378, 1983.
73. **Reisinger, M.,** Buprenorphine as new treatment for heroin dependence, *Drug Alcohol Depend.*, 16, 257, 1985.
74. **Seow, S. S. W., Quigley, A. J., and Ilett, K. F.,** Buprenorphine: a new maintenance opiate?, *Med. J. Austr.*, 144, 407, 1986.
75. **Kosten, T. R. and Kleber, H. D.,** Buprenorphine detoxification from opioid dependence: a pilot study, *Life Sci.*, 42, 635, 1988.
76. **Bickel, W. K., Stitzer, M. L., Bigelow, G. E., Liebson, I. A., Jasinski, D. R., and Johnson, R. E.,** A clinical trial of buprenorphine: comparison with methadone in detoxification of heroin addicts, *Clin. Pharmacol. Ther.*, 43, 72, 1988.
77. **Johnson, R. E., Cone, E. J., Henningfield, J. E., and Fudala, P. J.,** Use of buprenorphine in the treatment of opiate addiction. I. Physiologic and behavioral effects during a rapid dose induction, *Clin. Pharmacol. Ther.*, 45, 335, 1989.
78. **Fudala, P. J., Jaffe, J. H., Dax, E. M., and Johnson, R. E.,** Use of buprenorphine in the treatment of opioid addiction. II. Physiologic and behavioral effects of daily and alternate-day administration and abrupt withdrawal, *Clin. Pharmacol. Ther.*, 47, 525, 1990.
79. **Lange, W. R., Fudala, P. J., Dax, E. M., and Johnson, R. E.,** Safety and side-effects of buprenorphine in the clinical management of addiction, *Drug Alcohol Depend.*, 26, 19, 1990.
80. **Kosten, T. R., Kleber, H. D., and Morgan, C.,** Role of opioid antagonists in treating cocaine abuse, *Life Sci.*, 44, 887, 1989.
81. **Kosten, T. R., Kleber, H. D., and Morgan, C.,** Treatment of cocaine abuse with buprenorphine, *Biol. Psychiatry*, 26, 637, 1989.
82. **Mello, N. K., Lukas, S. E., Kamien, J. B., Mendelson, J. H., Drieze, J., and Cone, E. J.,** The effect of chronic buprenorphine treatment on cocaine and food-self-administration by rhesus monkeys, *J. Pharmacol. Exp. Ther.*, 260, 1185, 1992.
83. **Mello, N. K., Mendelson, J. H., Bree, M. P., and Lukas, S. E.,** Buprenorphine suppresses cocaine self-administration by rhesus monkeys, *Science*, 245, 859, 1989.
84. **Mello, N. K., Mendelson, J. H., Bree, M. P., and Lukas, S. E.,** Buprenorphine and naltrexone effects on cocaine self-administration by rhesus monkeys, *J. Pharmacol. Exp. Ther.*, 254, 926, 1990.

INDEX

A

D

E

F

G

H

I

K

L

M

P

R

S

T

U